322

Topics in Current Chemistry

Editorial Board:
K.N. Houk • C.A. Hunter • M.J. Krische • J.-M. Lehn
S.V. Ley • M. Olivucci • J. Thiem • M. Venturi • P. Vogel
C.-H. Wong • H. Wong • H. Yamamoto

Topics in Current Chemistry
Recently Published and Forthcoming Volumes

Constitutional Dynamic Chemistry
Volume Editor: Mihail Barboiu
Vol. 322, 2012

**EPR Spectroscopy: Applications
in Chemistry and Biology**
Volume Editors: Malte Drescher,
Gunnar Jeschke
Vol. 321, 2012

Radicals in Synthesis III
Volume Editors: Markus R. Heinrich,
Andreas Gansäuer
Vol. 320, 2012

Chemistry of Nanocontainers
Volume Editors: Markus Albrecht,
F. Ekkehardt Hahn
Vol. 319, 2012

**Liquid Crystals: Materials Design and
Self-Assembly**
Volume Editor: Carsten Tschierske
Vol. 318, 2012

**Fragment-Based Drug Discovery and X-Ray
Crystallography**
Volume Editors: Thomas G. Davies,
Marko Hyvönen
Vol. 317, 2012

**Novel Sampling Approaches in Higher
Dimensional NMR**
Volume Editors: Martin Billeter,
Vladislav Orekhov
Vol. 316, 2012

Advanced X-Ray Crystallography
Volume Editor: Kari Rissanen
Vol. 315, 2012

**Pyrethroids: From Chrysanthemum to Modern
Industrial Insecticide**
Volume Editors: Noritada Matsuo, Tatsuya Mori
Vol. 314, 2012

**Unimolecular and Supramolecular
Electronics II**
Volume Editor: Robert M. Metzger
Vol. 313, 2012

**Unimolecular and Supramolecular
Electronics I**
Volume Editor: Robert M. Metzger
Vol. 312, 2012

Bismuth-Mediated Organic Reactions
Volume Editor: Thierry Ollevier
Vol. 311, 2012

Peptide-Based Materials
Volume Editor: Timothy Deming
Vol. 310, 2012

Alkaloid Synthesis
Volume Editor: Hans-Joachim Knölker
Vol. 309, 2012

Fluorous Chemistry
Volume Editor: István T. Horváth
Vol. 308, 2012

**Multiscale Molecular Methods in Applied
Chemistry**
Volume Editors: Barbara Kirchner,
Jadran Vrabec
Vol. 307, 2012

Solid State NMR
Volume Editor: Jerry C. C. Chan
Vol. 306, 2012

Prion Proteins
Volume Editor: Jörg Tatzelt
Vol. 305, 2011

**Microfluidics: Technologies and
Applications**
Volume Editor: Bingcheng Lin
Vol. 304, 2011

Constitutional Dynamic Chemistry

Volume Editor: Mihail Barboiu

With Contributions by

T. Aastrup · M. Barboiu · R. Custelcean · A. Deratani ·
N. Giuseppone · S. Ladame · S. Lecommandoux · J.-M. Lehn ·
E. Mahon · K. Meguellati · B.L. Miller · E. Moulin · G.D. Pantoş ·
N. Ponnuswamy · D. Quémener · J. Ramírez · O. Ramström ·
M. Sakulsombat · J.K.M. Sanders · A.-M. Stadler ·
A.R. Stefankiewicz · Y. Zhang

Editor
Dr. Mihail Barboiu
Institut Européen des Membranes –
ENSCM-UMII-CNRS 5635
Place Eugene Bataillon CC047
34095, Montpellier, Cedex 5
France

ISSN 0340-1022   e-ISSN 1436-5049
ISBN 978-3-642-28343-7   e-ISBN 978-3-642-28344-4
DOI 10.1007/978-3-642-28344-4
Springer Heidelberg Dordrecht London New York

Library of Congress Control Number: 2012932056

© Springer-Verlag Berlin Heidelberg 2012
This work is subject to copyright. All rights are reserved, whether the whole or part of the material is concerned, specifically the rights of translation, reprinting, reuse of illustrations, recitation, broadcasting, reproduction on microfilm or in any other way, and storage in data banks. Duplication of this publication or parts thereof is permitted only under the provisions of the German Copyright Law of September 9, 1965, in its current version, and permission for use must always be obtained from Springer. Violations are liable to prosecution under the German Copyright Law.
The use of general descriptive names, registered names, trademarks, etc. in this publication does not imply, even in the absence of a specific statement, that such names are exempt from the relevant protective laws and regulations and therefore free for general use.

Printed on acid-free paper

Springer is part of Springer Science+Business Media (www.springer.com)

Volume Editor

Dr. Mihail Barboiu

Institut Européen des Membranes –
ENSCM-UMII-CNRS 5635
Place Eugene Bataillon CC047
34095, Montpellier, Cedex 5
France

Editorial Board

Prof. Dr. Kendall N. Houk

University of California
Department of Chemistry and Biochemistry
405 Hilgard Avenue
Los Angeles, CA 90024-1589, USA
houk@chem.ucla.edu

Prof. Dr. Christopher A. Hunter

Department of Chemistry
University of Sheffield
Sheffield S3 7HF, United Kingdom
c.hunter@sheffield.ac.uk

Prof. Michael J. Krische

University of Texas at Austin
Chemistry & Biochemistry Department
1 University Station A5300
Austin TX, 78712-0165, USA
mkrische@mail.utexas.edu

Prof. Dr. Jean-Marie Lehn

ISIS
8, allée Gaspard Monge
BP 70028
67083 Strasbourg Cedex, France
lehn@isis.u-strasbg.fr

Prof. Dr. Steven V. Ley

University Chemical Laboratory
Lensfield Road
Cambridge CB2 1EW
Great Britain
Svl1000@cus.cam.ac.uk

Prof. Dr. Massimo Olivucci

Università di Siena
Dipartimento di Chimica
Via A De Gasperi 2
53100 Siena, Italy
olivucci@unisi.it

Prof. Dr. Joachim Thiem

Institut für Organische Chemie
Universität Hamburg
Martin-Luther-King-Platz 6
20146 Hamburg, Germany
thiem@chemie.uni-hamburg.de

Prof. Dr. Margherita Venturi

Dipartimento di Chimica
Università di Bologna
via Selmi 2
40126 Bologna, Italy
margherita.venturi@unibo.it

Prof. Dr. Pierre Vogel

Laboratory of Glycochemistry
and Asymmetric Synthesis
EPFL – Ecole polytechnique féderale
de Lausanne
EPFL SB ISIC LGSA
BCH 5307 (Bat.BCH)
1015 Lausanne, Switzerland
pierre.vogel@epfl.ch

Prof. Dr. Chi-Huey Wong

Professor of Chemistry, Scripps Research
Institute
President of Academia Sinica
Academia Sinica
128 Academia Road
Section 2, Nankang
Taipei 115
Taiwan
chwong@gate.sinica.edu.tw

Prof. Dr. Henry Wong

The Chinese University of Hong Kong
University Science Centre
Department of Chemistry
Shatin, New Territories
hncwong@cuhk.edu.hk

Prof. Dr. Hisashi Yamamoto

Arthur Holly Compton Distinguished
Professor
Department of Chemistry
The University of Chicago
5735 South Ellis Avenue
Chicago, IL 60637
773-702-5059
USA
yamamoto@uchicago.edu

Topics in Current Chemistry
Also Available Electronically

Topics in Current Chemistry is included in Springer's eBook package *Chemistry and Materials Science*. If a library does not opt for the whole package the book series may be bought on a subscription basis. Also, all back volumes are available electronically.

For all customers with a print standing order we offer free access to the electronic volumes of the series published in the current year.

If you do not have access, you can still view the table of contents of each volume and the abstract of each article by going to the SpringerLink homepage, clicking on "Chemistry and Materials Science," under Subject Collection, then "Book Series," under Content Type and finally by selecting *Topics in Current Chemistry*.

You will find information about the

– Editorial Board
– Aims and Scope
– Instructions for Authors
– Sample Contribution

at springer.com using the search function by typing in *Topics in Current Chemistry*.

Color figures are published in full color in the electronic version on SpringerLink.

Aims and Scope

The series *Topics in Current Chemistry* presents critical reviews of the present and future trends in modern chemical research. The scope includes all areas of chemical science, including the interfaces with related disciplines such as biology, medicine, and materials science.

The objective of each thematic volume is to give the non-specialist reader, whether at the university or in industry, a comprehensive overview of an area where new insights of interest to a larger scientific audience are emerging.

vii

Thus each review within the volume critically surveys one aspect of that topic and places it within the context of the volume as a whole. The most significant developments of the last 5–10 years are presented, using selected examples to illustrate the principles discussed. A description of the laboratory procedures involved is often useful to the reader. The coverage is not exhaustive in data, but rather conceptual, concentrating on the methodological thinking that will allow the non-specialist reader to understand the information presented.

Discussion of possible future research directions in the area is welcome.

Review articles for the individual volumes are invited by the volume editors.

In references *Topics in Current Chemistry* is abbreviated *Top Curr Chem* and is cited as a journal.

Impact Factor 2010: 2.067; Section "Chemistry, Multidisciplinary": Rank 44 of 144

Preface

Constitutional Dynamic Chemistry (CDC) and its application Dynamic Combinatorial Chemistry (DCC) are new evolutional approaches to produce chemical diversity. In contrast to the classical methods, they allow for the simple generation of functional systems amplified from a mixture of inter-exchanging architectures which result from sets of building blocks interacting reversibly. Kinetic or thermodynamic resolution, self-assembly followed by covalent modification, and phase-change processes all shed light on useful strategies to control and create convergence between self-organization and constitutional functions. Such dynamic libraries have special relevance for a very diverse range of applications such as drug-, catalyst-, and material discovery. CDC implements a dynamic molecular/supramolecular reversible interface between interacting constituents, mediating the structural self-correlation of different domains of generated systems by virtue of their basic constitutional behaviors. The self-assembly of the components controlled by mastering molecular/supramolecular interactions may allow the flow of structural information from molecular level toward nanoscale dimensions. This volume of *Topics in Current Chemistry* focuses on constitutional methods for understanding and controlling such upscale propagation of structural information. These methods show potential to impose further precise order at the mesoscale and to discover new routes to obtain highly ordered ultradense arrays over macroscopic distances. During the last decade, CDC has become increasingly interesting for Dynamic Interactive Systems (DIS). Networks of continuously exchanging and reversibly reorganizing objects form the core of DIS. For example, molecules, supermolecules, polymers, biomolecules, biopolymers, nanotubes, surfaces, nanoparticles, liposomes, materials, and cells are all operating under the natural selection to allow spatial/temporal and structural/functional adaptability in response to constitutional internal or stimulant external factors. The contributions to this volume open new horizons, shortening the essential steps from molecular to functional nano-objects. These steps are sometimes too long and the research strategies should expand the fundamental understanding of complex dynamic structures and properties as it relates to creating products and manufacturing processes. Combined dynamic strategies to produce constitutional/combinatorial systems can be effectively shared as soon merged marketable technology to benefit most research laboratories and industrial producers. The 11 chapters are structured in three groups: 1. *Evolutional Approaches to Produce Chemical Diversity and the Development of the Constitutional Dynamic Systems* (Lehn, Miller, Ramström et al., and Ladame), 2. *Constitutional Self-Assembly Toward Complex Architectures* (Pantos and Sanders, Custelcean, and Quemener et al.), and 3. *Constitutional Dynamic Chemistry Toward Dynamic Interactive Systems* (Barboiu, Mahon et al., Giussepone et al., and Stadler et al.).

I would like to thank all the authors, as well as all those who have facilitated this volume, and I hope that readers will find answers to key questions concerning basic principles and related evolutional approaches that have been used in Constitutional Dynamic Chemistry (CDC). The most revolutionary consequences may reflect the fascinating possibilities offered by selection, evolution, amplification, molecular recognition, and replication processes. This volume is not a comprehensive treatise, but is a timely objective snapshot of the CDC field from which the reader can get a broader insight into this and hopefully a future source of inspiration.

Montpellier   Mihail Barboiu

Contents

Constitutional Dynamic Chemistry: Bridge from Supramolecular Chemistry to Adaptive Chemistry 1
Jean-Marie Lehn

Multistate and Phase Change Selection in Constitutional Multivalent Systems 33
Mihail Barboiu

Dynamic Systemic Resolution 55
Morakot Sakulsombat, Yan Zhang, and Olof Ramström

Dynamic Combinatorial Self-Replicating Systems 87
Emilie Moulin and Nicolas Giuseppone

DCC in the Development of Nucleic Acid Targeted and Nucleic Acid Inspired Structures 107
Benjamin L. Miller

Dynamic Nanoplatforms in Biosensor and Membrane Constitutional Systems 139
Eugene Mahon, Teodor Aastrup, and Mihail Barboiu

Dynamic Assembly of Block-Copolymers 165
D. Quémener, A. Deratani, and S. Lecommandoux

Dynamic Chemistry of Anion Recognition 193
Radu Custelcean

Supramolecular Naphthalenediimide Nanotubes 217
Nandhini Ponnuswamy, Artur R. Stefankiewicz,
Jeremy K. M. Sanders, and G. Dan Pantoş

xi

**Synthetic Molecular Machines and Polymer/Monomer Size Switches
that Operate Through Dynamic and Non-Dynamic Covalent Changes** ... 261
Adrian-Mihail Stadler and Juan Ramírez

**Reversible Covalent Chemistries Compatible with the Principles
of Constitutional Dynamic Chemistry: New Reactions to Create
More Diversity** 291
Kamel Meguellati and Sylvain Ladame

Index 315

Constitutional Dynamic Chemistry: Bridge from Supramolecular Chemistry to Adaptive Chemistry

Jean-Marie Lehn

Abstract Supramolecular chemistry aims at implementing highly complex chemical systems from molecular components held together by non-covalent intermolecular forces and effecting molecular recognition, catalysis and transport processes. A further step consists in the investigation of chemical systems undergoing *self-organization*, i.e. systems capable of spontaneously generating well-defined functional supramolecular architectures by self-assembly from their components, thus behaving as programmed chemical systems. Supramolecular chemistry is intrinsically a *dynamic chemistry* in view of the lability of the interactions connecting the molecular components of a supramolecular entity and the resulting ability of supramolecular species to exchange their constituents. The same holds for molecular chemistry when the molecular entity contains covalent bonds that may form and break reversibility, so as to allow a continuous change in constitution by reorganization and exchange of building blocks. These features define a *Constitutional Dynamic Chemistry* (CDC) on both the molecular and supramolecular levels.

CDC introduces a paradigm shift with respect to constitutionally static chemistry. The latter relies on design for the generation of a target entity, whereas CDC takes advantage of dynamic diversity to allow variation and selection. The implementation of selection in chemistry introduces a fundamental change in outlook. Whereas *self-organization by design* strives to achieve full control over the output molecular or supramolecular entity by explicit programming, *self-organization with selection* operates on dynamic constitutional diversity in response to either internal or external factors to achieve *adaptation*.

The merging of the features: -information and programmability, -dynamics and reversibility, -constitution and structural diversity, points to the emergence of *adaptive* and *evolutive chemistry*, towards a *chemistry of complex matter*.

J.-M. Lehn (✉)
Institut de Science et d'Ingénierie Supramoléculaires, Université de Strasbourg, 8 allée Gaspard Monge, 67000 Strasbourg, France
e-mail: lehn@unistra.fr

Keywords Adaptive chemistry · Dynamic networks · Dynamic polymers · Molecular recognition · Multiple dynamics · Self-organization · Supramolecular chemistry

Contents

1 From Molecular Recognition to Self-Organization 2
2 Self-Organization: From Design to Selection 4
3 From Supramolecular Chemistry to Constitutional Dynamic Chemistry 5
4 Implementation of Constitutional Dynamic Chemistry 8
    4.1 Dynamic Generation of Receptors 9
    4.2 Dynamic Search for Biologically Active Substances 11
    4.3 Constitutional Dynamic Materials – Dynamic Polymers/Dynamers 11
    4.4 Constitutional Dynamic Biopolymers: Biodynamers 13
5 Adaptation in Constitutional Dynamic Systems: Molecular and Supramolecular 16
    5.1 Adaptation to a Phase Change 16
    5.2 Adaptation to a Change in Medium/Environment 16
    5.3 Adaptation to Molecular Structure Formation: Folding and Effector Binding 17
    5.4 Adaptation to Morphological Change: Shape Switching 19
6 Multiple Dynamics and Dynamic Networks 20
    6.1 Multiple Constitutional Dynamic Processes 21
    6.2 Multiple Dynamics for Information Processing Devices 21
    6.3 Constitutional Dynamic Networks 23
7 Conclusion and Outlook 25
References 27

Abbreviations

CDC   Constitutional dynamic chemistry
CDL   Constitutional dynamic library
CDN   Constitutional dynamic networks
DCC   Dynamic covalent chemistry or dynamic combinatorial chemistry
DNCC   Dynamic non-covalent chemistry
VCL   Virtual combinatorial library

1 From Molecular Recognition to Self-Organization

Supramolecular chemistry has experienced an extraordinary development in the last forty years or so, at the triple meeting point of chemistry with biology and physics. It has given rise to numerous review articles, special issues of journals and books [1–6]. The intention here is to provide a view emphasizing its development towards constitutional dynamic chemistry, together with some conceptual considerations and an outlook, along and beyond the lines earlier horizons, towards

Fig. 1 Self-organization by design involves programming through molecular information storage and supramolecular processing. Self-organization with selection takes advantage of constitutional dynamics to generate diversity and implement constitutional variation to allow for adaptation

the emergence of adaptive chemistry, as seen mainly from our own perspective (Fig. 1) [3, 7–9].

Supramolecular chemistry has first relied on the development of preorganized molecular receptors for effecting *molecular recognition*, catalysis, and transport processes on the basis of the *molecular information* stored in the covalent framework of the components and processed at the supramolecular level through specific interactional algorithms.

Through the appropriate manipulation of the intermolecular non-covalent interactions to achieve molecular recognition, supramolecular chemistry has paved the way towards regarding chemistry not only as the science of the structure and transformation of matter, but also as an *information science,* involving the storage of information at the molecular level, in the structural features, and its retrieval, transfer, and processing at the supramolecular level via specific spatial relationships and interaction patterns (hydrogen bonding arrays, sequences of donor and acceptor groups, metal ion coordination units, etc.).

Supramolecular chemistry has thus emphasized the perception of chemistry as the *science of informed matter*, with the aim of gaining progressive control over the organization of matter, over its spatial (structural) and temporal (dynamical) features. It has led to the ever clearer perception, deeper analysis, and more deliberate application of information features in the elaboration and transformation of matter, tracing the path from merely condensed matter to more and more highly organized matter, towards systems of increasing complexity.

Achieving optimal molecular recognition rests on the derivation of receptor-substrate pairs presenting complementarity in geometry and interactions, through correct construction of one (or both) of the interacting species. Beyond mastering such preorganization and taking advantage of it, supramolecular chemistry has been actively exploring the design of systems undergoing *self-organization*, i.e., systems capable of spontaneously generating well-defined, organized supramolecular architectures by self-assembly from their components [3, 7–17].

2 Self-Organization: From Design to Selection

Self-organization is the fundamental cosmic process that has led to the generation of complex matter, from particles to the thinking organism, in the course of the evolution of the universe. Unraveling the mechanisms of the self-organization of matter offers a most challenging task to chemistry [7]. Along the way, as progress is being made, implementation in non-natural, abiotic chemical systems may be performed. The spontaneous but controlled generation of complex supramolecular entities by means of suitable components and interactions amounts to performing self-organization *by design* (Fig. 1).

Self-organization processes may be directed via the molecular information stored in the covalent framework of the components and read out at the supramolecular level through specific interaction/recognition patterns that define processing algorithms. They thus represent the operation of *programmed chemical systems* [3, 10, 11], and are of major interest for supramolecular science and engineering. They give access to advanced *functional supramolecular materials*, such as supramolecular polymers [18–21], liquid crystals, and lipid vesicles [22, 23], as well as solid-state assemblies [24, 25].

A great variety of (functional) supramolecular architectures has been generated by complex self-assembly procedures, of purely "organic" as well as of "inorganic" natures [3–5, 12–17, 26–30]. The latter have led to a range of metallosupramolecular entities presenting original structural, physical, and chemical properties.

A particularly demonstrative case of programming is represented by the generation of two very different metallosupramolecular architectures from the same ligand molecule constructed from two types of subunits (coding, respectively, for helicate [3, 27, 31] and for grid formation [30]), depending on the set of coordinating metal cations used. It amounts to the generation of two *different* supramolecular entities by processing the *same* molecular information through two *different* algorithms (the coordination features of the metal cations). It represents an intriguing case of multiple expression of molecular information resulting in the generation of two different outputs (products) from a single program composed of two subroutines (Fig. 2) [32].

As the self-organization of supramolecular entities takes place following a multistep pathway towards the progressive build-up of the final entities, it has to explore the available structure/energy combinations. Consequently, self-organization processes are in principle able to select the correct molecular components for the generation of a given supramolecular entity from a collection of building blocks. Thus, beyond self-organization by design, relying on programming, self-organization may take place *with selection*, by virtue of a basic feature inherent in supramolecular chemistry, its *dynamic character* (Fig. 1).

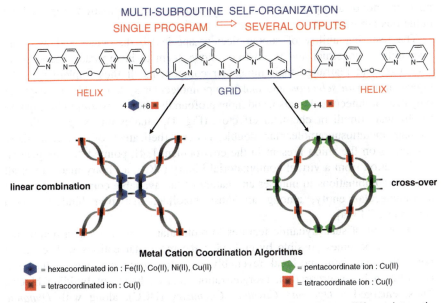

Fig. 2 The processing of the same molecular information through two different interaction algorithms induces the self-organization of two different metallo-supramolecular architectures, amounting to double expression of molecular information. A single ligand molecule composed of subunits of two different types (coding respectively for double helicate and for [2×2] grid formation) yields two different metallosupramolecular architectures depending on the coordination features of the two different sets of metal cations used. In other words, a single program composed of two different subroutines leads to two different outputs depending on the processing algorithms

3 From Supramolecular Chemistry to Constitutional Dynamic Chemistry

Supramolecular chemistry has, from the start, been defined in its structural (matter) and bonding (energy) features as "chemistry beyond the molecule," its entities being constituted of molecular components held together by non-covalent interactions [1, 3, 33]. A third basic feature resides in its dynamic nature, that was always implicit and operating in all processes investigated over the years, but has been explicitly taken advantage of and implemented only in more recent years. Thus, novel perspectives are opened when considering that supramolecular chemistry is intrinsically a *dynamic chemistry* in view of the lability of the non-covalent interactions connecting the molecular components of a supramolecular entity. This dynamic character is essential for the self-assembly of the supramolecular entities from their molecular components through more or less rapid exploration of the structure/energy hypersurface. It is at the basis of the generation of the highly complex architectures held together by electrostatic, hydrogen bonding, donor-acceptor interactions or metal ion coordination, reported by numerous laboratories.

The resulting ability of supramolecular species to dissociate and associate reversibly, deconstruct and reconstruct allows them to incorporate, decorporate, exchange, and rearrange their molecular components. It thus implements self-organization *with selection* and makes supramolecular systems able to adapt not only their architecture, but also, and more profoundly, their constitution in response to physical stimuli or chemical effectors (Fig. 1). Studies on the generation of circular metallosupramolecular double helices (helicates) of different sizes, depending on the anion present in the environment [34], pointed "to the generation/selection from a virtual combinatorial library (VCL, a library made up of all possible combinations in number and nature of the available components) of that supramolecular entity, among all those possible, that best binds a given substrate..." [34a].

Extension of such dynamic features to molecular chemistry was a major step forward. It becomes possible by endowing the molecular entities with covalent bonds that may form and break reversibly, so as to allow a continuous change in constitution by incorporation, reorganization, and exchange of building blocks. Thus, emerged a *Dynamic Covalent Chemistry* (DCC), along with *Dynamic Non-Covalent Chemistry* (DNCC) [7, 9, 34–41]. Both areas were brought together under the general concept of *Constitutional Dynamic Chemistry* [7, 9] (CDC) that covers *both* the molecular and supramolecular levels.

CDC introduces a paradigm shift with respect to constitutionally static chemistry. The latter relies on design for the generation of a target entity, whereas CDC takes advantage of its dynamic character for the generation of constitutional molecular and supramolecular dynamic diversity to allow for variation and selection. The implementation of selection in chemistry introduces a fundamental change in outlook. CDC operates selection on dynamic constitutional diversity in response to the pressure of either chemical or physical, internal or external factors to achieve *adaptation*, thus enabling adaptive chemistry [3, 7–9] (Fig. 1).

Thus, in addition to the classical dynamic chemical features, reactional dynamics and motional dynamics, a third novel type of dynamic processes is to be considered, that opens wide perspectives: *constitutional dynamics*, whereby a chemical entity, be it molecular or supramolecular, may undergo continuous change in its constitution through dissociation into various components and reconstitution into the same or different entities.

The emergence of this third type of dynamic chemistry stems from the recognition of the dynamic features intrinsic to supramolecular entities and their explicit introduction into molecules. Importing such dynamic features into molecular chemistry requires shifting from "static" to "dynamic" covalent bonds, so as also to endow molecular species with the ability to undergo similar dynamic exchange and reorganization processes by virtue of the reversible formation and breaking of covalent connections. It implies looking at molecules as labile entities, in contrast to the usual search for stability, and opens novel perspectives to covalent chemistry.

It requires searching for reversible reactions and catalysts that allow the making and breaking of covalent bonds, preferentially under mild conditions. Whereas supramolecular entities are dynamic by nature, molecular entities are dynamic by intent.

This radical change in outlook builds on the richness of constitutional diversity and the benefits of variability. It stresses the virtues of instructed mixtures [3, 31], such as was revealed in the self-selection processes occurring in the side-by-side self-assembly of double helical metal complexes (helicates), whereby only the correctly paired double helicates were produced from a mixture of ligands and metal ions in dynamic coordination equilibrium [31, 37c]. It is this work that first led us in the early 1990s to envisage a dynamic chemistry bringing into play the constitution of chemical species.

The formation and dynamic character of molecular and supramolecular constitutional dynamic entities result respectively from reversible condensation of components through *complementary functional groups* (molecular, covalent, chemical, *functional recognition*) and from recognition-directed reversible association of components through *complementary interactional groups* (supramolecular, non-covalent, physical, *interactional recognition*). They may thus be considered either as *chemically dynamic*, involving chemical reactions or as *physically dynamic*, based on physical non-covalent interactions (Fig. 3).

As at both levels the processes involve component recombination, they define a *dynamic combinatorial chemistry (DCC)* of supramolecular as well as of molecular nature [35–38]. This denomination stresses their combinatorial character, whereas CDC highlights the fact that they concern the basic feature of chemical entities, their very constitution. Whereas the usual "static" combinatorial chemistry is based on extensive libraries of prefabricated molecules, DCC implements the reversible connection of sets of basic components to give access to virtual combinatorial libraries (VCLs) [35, 42], whose constituents comprise all possible combinations that may potentially be generated (but do not need to be present initially). It represents a powerful means for generating dynamic, effector-responsive diversity. The constituent(s) actually expressed/selected among all those accessible is (are) expected, under thermodynamic control, to be that (those) presenting the strongest interaction with the target, that is, the highest receptor/substrate molecular

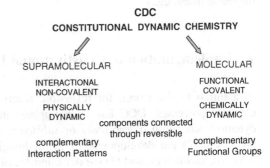

Fig. 3 Constitutional dynamic chemistry at both the supramolecular (*left*) and the molecular (*right*) levels

recognition. The overall process is thus instructed (target-driven), combinatorial, and dynamic. It bypasses the need actually to synthesize the constituents of a combinatorial library by letting the target perform the assembly of the optimal partner from a virtual set of constituents.

Along these lines and as a proof of principle, reversible imine formation was implemented in 1997 for the generation of enzyme (carbonic anhydrase) inhibitors from a dynamic covalent library [43] and reversible covalent selection approaches to catalytic systems were presented [44].

Constitutional dynamics imply changes in constitution concerning the nature, number, and arrangement of the components of molecular or supramolecular entities, thus generating *molecular and supramolecular diversity* through continuous recomposition, recombination, reorganization, construction, and deconstruction by either external (incorporation, decorporation, exchange of components) or internal (rearrangement, reshuffling of components) processes, under the pressure of internal factors or external environmental stimuli. The system may respond to such effects by expressing the constituent(s) presenting best adaptation to a given situation, through selection of the most suitable components among those available.

A set of interconverting supramolecular or molecular entities represents a real or virtual [35, 42] *constitutional dynamic library* (CDL). The full ensemble of all constituents, i.e., of all possible combinations of the components, may be termed the *combinome* of the system under consideration. The set of components may be considered as the *genotype* of the dynamic system and the sets of constituents generated from these components in given conditions its *phenotypes*.

A CDL may modify its composition, i.e., the relative amounts of its constituents, and be characterized by three main parameters: *conversion, composition*, and *expression* [45].

The simultaneous modulation of these three parameters of a CDL results in the expression of different constituents, through component selection driven by chemical effectors or physical stimuli. Such is, for instance, the case for sets of imine constituents under changes in protonation and/or temperature [45].

Variations in expression of the different constituents as a factor of external parameters represent an adaptation of the system to environmental conditions, such as medium (solvent), presence of interacting species (protons, metal ions, substrate molecules, etc.), or physical factors (temperature, pressure, electric or magnetic fields, etc.).

4 Implementation of Constitutional Dynamic Chemistry

In addition to the search for reversible reactions of potential use in dynamic covalent processes, DCC has been implemented in three main areas: (1) the dynamic generation of *receptors* or *substrates* driven by molecular recognition processes, (2) the development of methodologies of interest in the search for *bioactive substances*, and (3) the development of *dynamic materials*.

4.1 Dynamic Generation of Receptors

Numerous studies have been performed on the dynamic generation from DCLs of either molecular receptors for substrates or substrates for receptors [35–39, 46]. Work in our group has been directed in particular towards the influence of physical and chemical stimuli on the behavior of DCLs, as for instance in the case of the constitutional dynamic reorganization exerted by temperature and protonation on imine libraries [45], the induction of liquid crystal properties by an electric field acting on a dynamic library [47], or component selection induced by cation binding in a folded dynamer chain [48] (see also below).

A particularly illustrative case is that of induced fit expression/amplification of a receptor by selection among an interconverting set of entities, enforced by substrate binding, in a triple dynamic library (constitutional, configurational, and conformational) (Fig. 4) [49].

On the other hand, in the formation of guanine quartet-based hydrogels, it was shown that the system selected those components that generated the most stable gel [50] (see Fig. 5 in [50]). Such a *self-optimization* behavior may be of much deeper general significance, namely for *prebiotic chemical evolution,* whereby selection is driven by phase cohesion, the entity selected being that giving the most stable

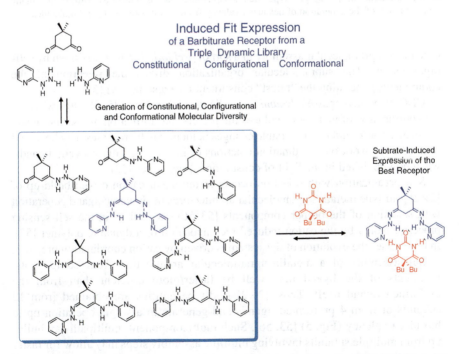

Fig. 4 Expression/amplification of the optimal constituent from a triple dynamic library of potential receptor molecules through induced fit interaction with a substrate molecule (a barbiturate). In addition, there is a non-covalent/supramolecular dynamic process involving substrate binding and exchange

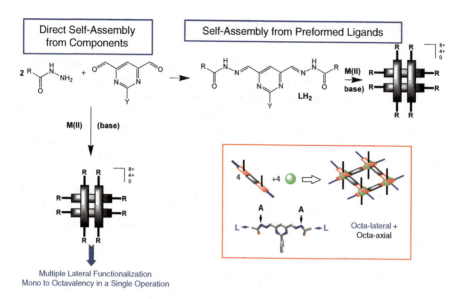

Fig. 5 Generation of multifunctionality/multivalency by self-assembly. Octa-valency from either hydrazide and dialdehyde components/subunits (*left*), or preformed ligands (*top right*) in the case of the formation of [2×2] grid-type metallosupramolecular architectures. *Insert*: Schematic representation of the generation of hexadeca-valency from tetrasubstituted ligand molecules

organized supramolecular assembly in a sort of *prebiotic Darwinism* driven by self-organization. The supramolecular organization drives the selection of the components generating the "fittest" constituent (see also below) [51].

CDC also encompasses *dynamic coordination chemistry* [35, 38, 40], whereby the coordination of metal ions induces the preferential formation of specific ligand molecules and/or induces reversible changes in them. Such processes may be traced back to early work on coordination reactions of imine-based macrocyclic ligands, when now revisited in the light of constitutional dynamics [52].

Self-organization with selection occurs in the metal cation driven build-up of [2×2] grid-type metallosupramolecular architectures by dynamic ligand generation with selection of the proper components [53–56], as well as in the self-sensing process enabled by the cation-induced recomposition of a dynamic polymer [57]. In these cases the evolution of the system is driven by cation coordination pressure.

Self-assembly of a metallosupramolecular architecture from components/fragments of the ligand may well be faster/more efficient than from the preformed ligand itself. Thus, [2×2] grid-type entities are obtained from 12 subunits or from 4 preformed ligands and generate in a single operation up to hexadeca-valency (Fig. 5) [53, 56]. Such multicomponent, multipathway build-up from multiple subunits involving multiple assembly steps may allow for faster exploration of the structure/energy hypersurface for more efficient expression of the final entity.

4.2 Dynamic Search for Biologically Active Substances

The discovery of bioactive substances amounts to searching for a molecular key to fit a biological lock. Apart from the random screening of natural or synthetic compounds, two "classical" approaches may be distinguished: (1) rational design, which resides in looking for and designing a single key, the correct one, and rests on the implementation of molecular recognition and (2) combinatorial chemistry, which relies on the generation of vast collections of candidate keys that may all be assayed by high throughput screening with fast robotics. A third approach results from the implementation of covalent CDC [35–37]: *dynamic combinatorial/covalent chemistry* (DCC), that relies on the dynamic generation of a collection of interconverting keys representing all the possible combinations of fragments of keys, with the goal that this virtual set of potential keys may contain one (or more) that fits the lock, under either *thermodynamic selection,* expressing the constituent/key that presents the strongest interaction with the target/lock, or *kinetic selection*, giving the key that forms fastest within the lock. In both cases, it is the supramolecular lock/key recognition interactions that direct the process (see Fig. 9 in [9]) [58–60]. DCC combines the diversity provided by combinatorics with the selection driven by recognition and the exploration enabled by constitutional dynamics. Proof of principle for such an approach has been obtained in the dynamic generation/amplification of enzyme inhibitors and of substrates for biological receptors [43, 58–60].

4.3 Constitutional Dynamic Materials – Dynamic Polymers/ Dynamers

One may define molecular and supramolecular *dynamic materials* as materials whose components are linked through reversible covalent or non-covalent connections and undergo spontaneous and continuous change in constitution by assembly/disassembly processes in a given set of conditions. Such materials may in principle respond to multiple physical or chemical stimuli by constitutional variation and thus act as *adaptive materials.*

Via recognition-directed association and self-organization processes, supramolecular chemistry has opened new perspectives in materials science towards the design and engineering of *supramolecular materials*. These, again, are dynamic by nature, whereas molecular materials must be rendered dynamic by introduction of reversible covalent connections between building blocks. Because of their intrinsic ability to exchange their components, they may in principle select them in response to external stimuli or environmental factors and therefore behave as *adaptive materials* of either molecular or supramolecular nature [7–9].

Applying such considerations to polymer chemistry leads to the definition of *constitutionally dynamic polymers, dynamers,* of both molecular and supramolecular types [18–21] that have a constitutional/combinatorial diversity determined by the number of different monomers. The components effectively incorporated into the polymers depend on the nature of the connections (recognition or functional groups) and core groups, as well as on the interactions with the environment, so that dynamers possess the possibility of adaptation by association/growth/dissociation sequences. The dynamic and combinatorial features of dynamic polymers give access to higher levels of behavior such as self-healing, adaptability response to external stimulants (heat, light, chemical additives, etc.).

A *supramolecular polymer chemistry* has developed, concerning polymers of supramolecular nature generated by the self-assembly of monomers interconnected through complementary interaction/recognition groups [18–21].

Covalent dynamers may also present a range of unusual properties such as crossover component recombination between neat films in dynamic polymer blends [61], soft-to-hard transformation of polymer mechanical properties through component incorporation [62], and dynamic modification of optical properties (Fig. 6) [63].

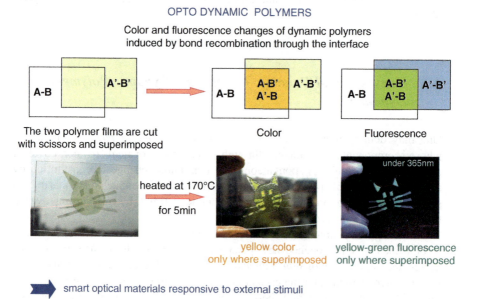

Fig. 6 Optical effects in dynamers. Color change and fluorescence generation induced by component recombination at the interface between two dynameric films. (*Top*) Schematic representation of the process yielding two new combinations at the interface of two superimposed films of dynamers. (*Bottom*) Photographs of an actual experiment, before (*left*) and after (*center* and *right*) recombination induced by heating

Self-Healing of a Dynamer Film

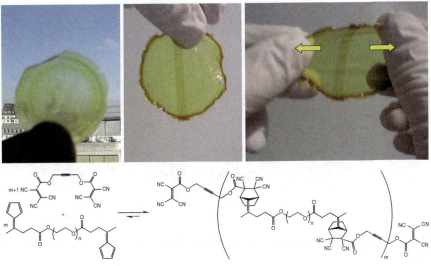

Fig. 7 Self-healing of a dynamer film (*top*) based on a room temperature-reversible Diels-Alder reaction (*bottom*)

Implementation of a self-contained Diels–Alder reaction reversible at room temperature [64a] allowed for self-healing of dynamer films. Indeed, when such a film was cut and the two resulting pieces superimposed and pressed gently to ensure a microscopic contact, the two pieces resealed and could no longer be separated by pulling them apart after only 10 s (Fig. 7) [64b]. The fact that the DA adducts linking the monomers can revert and reform constantly at room temperature, even in the condensed phase, accounts for the ability of the network to undergo self-healing. It also allows adaptation in response to *mechanical stress*, a further illustration of the ability of dynamic materials to respond to various physical effectors (temperature [45], electric field [47]). The dynamer constantly creates new chain ends and forms new connections thus building a material that can in principle self-repair across all its volume. These results extend to the molecular/covalent domain the properties displayed by supramolecular elastomers, for which regeneration of linkages between components is an intrinsic property due to the non-covalent nature of the interactions building up the dynamer [65].

4.4 *Constitutional Dynamic Biopolymers: Biodynamers*

Dynamers based on components of biological nature may generate dynamic analogs of biopolymers, *biodynamers*, that offer the possibility to combine the functional properties (recognition, catalysis) of naturally occurring polymers with

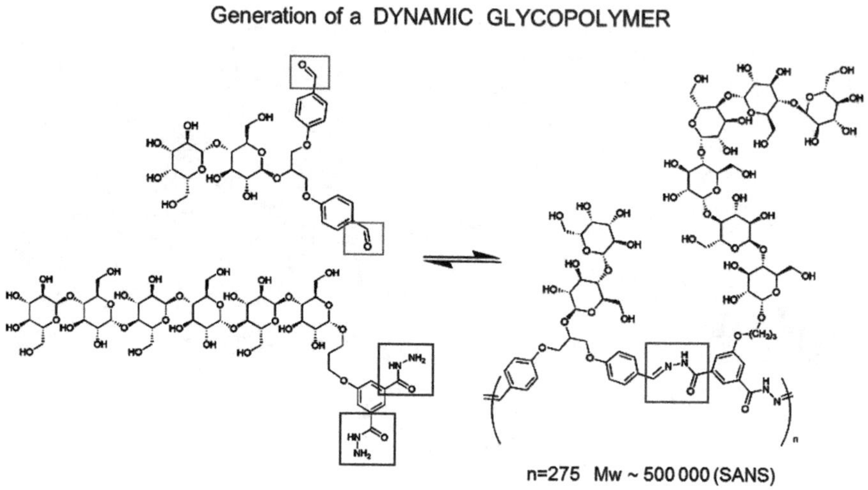

Generation of a DYNAMIC GLYCOPOLYMER

n=275 Mw ~ 500 000 (SANS)

Fig. 8 Glycodynamers: generation of a main-chain dynamic polymer bearing glycosidic side-chains

the adaptive behavior of constitutional dynamic systems. Therefore, in view of the prospects offered, the incorporation of biologically relevant moieties into dynamic polymers deserves close scrutiny. Thus, dynamic analogs of proteins, nucleic acids, and polysaccharides may be envisaged. They could provide, in addition to potential biorecognition properties, an adaptive character that would enable them to reorganize their sequence or constitution so as to select the preferential/optimal sequence, nature, and proportion of the bioactive monomeric subunits in response to external physical or chemical stimuli (temperature, pH, etc.) or to the presence of a chemical template or a biological target.

Glycodynamers, dynamic glycopolymers, have been generated by polycondensation of components bearing lateral bioactive oligosaccharide chains (Fig. 8). They present remarkable fluorescence properties with emission wavelengths that depend on the constitution of the polymer and are tunable by constitutional modification through exchange/incorporation of components, thus also demonstrating their dynamic character [66, 67]. Constitution dependent binding of these glycodynamers to a lectin, peanut agglutinin, has been demonstrated [67].

On the other hand, main-chain dynamic analogs of polysaccharides have been obtained and their reversibility was demonstrated by component exchange reaction, allowing tuning of their size and composition [68].

DyNAs, dynamic analogs of nucleic acids, were obtained as cationic main-chain dynamers bearing nucleobase groups and were shown to interact with polyanionic species displaying a marked increase in size with the negative charge of the substrate, thus demonstrating their adaptive character (Fig. 9) [69].

Finally, *dynamic peptoids*, dynamic analogs of peptides and proteins, have been generated by polycondensation of complementary difunctional monomers derived

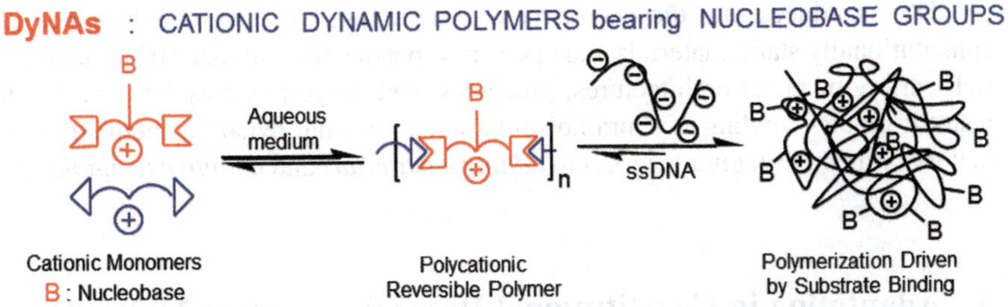

DyNAs : CATIONIC DYNAMIC POLYMERS bearing NUCLEOBASE GROUPS

Fig. 9 DyNAs: dynamic analogs of nucleic acids built on a main-chain cationic reversible polymer chain bearing nucleobase groups

DYNAMIC PEPTOIDS

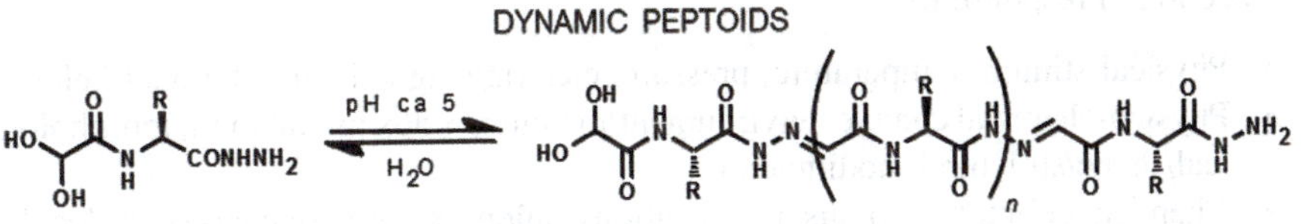

Fig. 10 Formation of dynamic analogs of peptides from hetero-difunctional amino acid derivatives

from amino acids. This process may be implemented for the induction of polymerization driven by self-organization and might operate with monomer selection, a most intriguing feature (Fig. 10) [70].

Responsive nanomaterials may be obtained by dynamic assembly of components derived from biomolecules [71]. Dynamic biomaterials are of interest for biodegradability, as well as for time-delayed, dynamic formulation and controlled release strategies, for instance of antimicrobial agents [72] or of fragrances [73].

By allowing for constitutional variation through rearrangement, incorporation, and exchange of components via reversible disconnection and reconnection, constitutional dynamics confer a range of properties to dynamers and to dynamic materials more generally such as:

- "Dial-in" properties: solubility, hydrophobicity, hydrophilicity, viscosity, castability, thermal behavior, mechanical, optical, electrical characteristics, etc.
- Self-mending, self-healing, self-repairing
- Dynamic blending of materials
- Expression/fine tuning of a virtual (latent) property for dynamic films or coatings
- Dynamic formulation for controlled release of bioactive ingredients (drugs, agrochemicals, flavors, fragrances, etc.)
- Dynamic biomaterials
- "Green" properties, environmental, and biological degradability

CDC introduces into the chemistry of materials a basic shift with respect to constitutionally static materials and opens new perspectives in materials science. A rich variety of novel architectures, processes, and properties may be expected to result from the blending of supramolecular and molecular dynamic chemistry with materials chemistry, giving access to *adaptive materials* and *adaptive technologies*.

5 Adaptation in Constitutional Dynamic Systems: Molecular and Supramolecular

CDC allows for adaptation through dynamic constitutional variation by component selection in response to:

- Physical stimuli: temperature, pressure, electric/magnetic/gravitational fields
- Physico-chemical change: environment/medium, redox modulation, morphological/shape/structural modification
- Chemical effectors: protons, metal cations, anions, neutral, or charged molecules

The processes may be described by a constitutional dynamic network, consisting of a set of linked interconverting constituents that may undergo either agonistic or antagonistic variations (see below).

5.1 Adaptation to a Phase Change

Clearcut adaptation to phase change may occur on passage from solution to solid state. As an illustrative early case, revisited in the present context, a bis-bipyridine ligand reacted with Cu(I) ions in solution to generate an equilibrating set of constituents by oligomeric self-assembly: a double helical, a triangular, and a square metallosupramolecular species. Upon crystallization, the full set transformed entirely into the double helicate as a single constituent in the solid state (Fig. 11) [74].

A related case is given by the component selection occurring in the generation of a hydrogel phase based on dynamically connected guanosine-quartet derivatives, driven by formation of the constituent that yields the most stable gel in a sort of "phase selection" process (see also Fig. 5 in [50]).

5.2 Adaptation to a Change in Medium/Environment

Depending on the interaction of the constituents of a dynamic library with the molecules in the medium, such as the molecules of the solvent or of an additive, a

Constitutional Dynamic Chemistry: Bridge from Supramolecular Chemistry 17

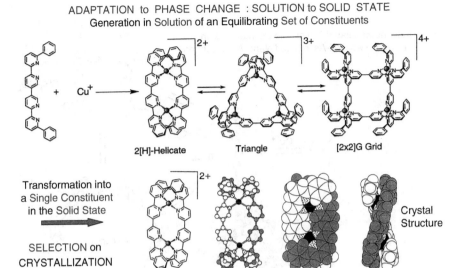

Fig. 11 Adaptation to phase change from solution to the solid state, with transformation of an equilibrating mixture in solution into a single crystalline entity

given member of the set may be preferentially generated or stabilized. Thus, a locked bipyridine ligand reacted with Cu(II) ions to form a [2×2] grid-type architecture in the weakly coordinating nitromethane solvent, but was converted in a reversible way into a hexagonal entity in acetonitrile, due to the binding of solvent molecules to the metal ion centers inside the cavity (Fig. 12) [75].

A striking example of medium driven component selection is given by the polycondensation of amphiphilic monomeric components in aqueous solution that yields a dynamer chain undergoing supramolecular self-organization with chain-folding into a helical rigid rod type superstructure, presenting a hydrophobic core and hydrophilic side-chains. In the presence of two different monomers, component selection, driven by medium induced hydrophobic effects, leads to incorporation of the component which presents the largest hydrophobic molecular area, thus displaying dynamic constitutional adaptation to the medium. Selectivity is lost on progressive addition of a miscible organic solvent to the aqueous solution (Fig. 13) [76].

5.3 *Adaptation to Molecular Structure Formation: Folding and Effector Binding*

Component selection may be driven by internal rather than external factors. Thus, access to a stable folded structure may drive the selection, within a diverse set of

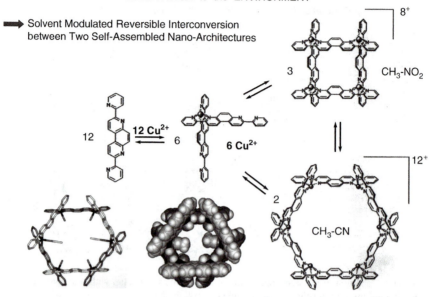

Fig. 12 Structural adaptation to the medium, from weakly to strongly coordinating solvent molecules. The binding of acetonitrile molecules to the Cu(II) centers inside the cavity, as revealed by the solid state structure (*bottom left*) of the hexagonal structure, drives the formation of a hexagonal superstructure

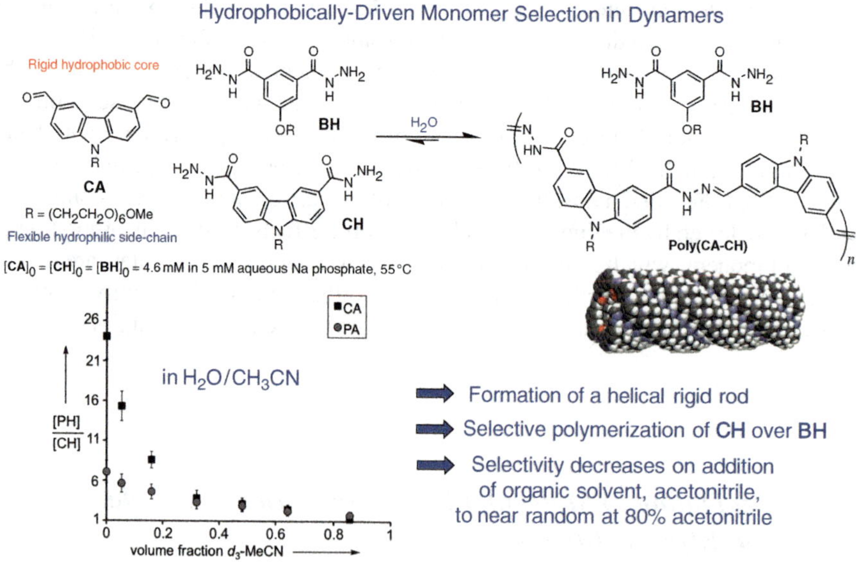

Fig. 13 Constitutional adaptation to the environment in the hydrophobically-driven component selection on formation of a rigid rod dynamer from amphiphilic monomers. Polycondensation occurs in aqueous solution with preferential incorporation of the monomer presenting the largest hydrophobic area. Selectivity is progressively lost in solutions containing increasing amounts of organic solvent, acetonitrile

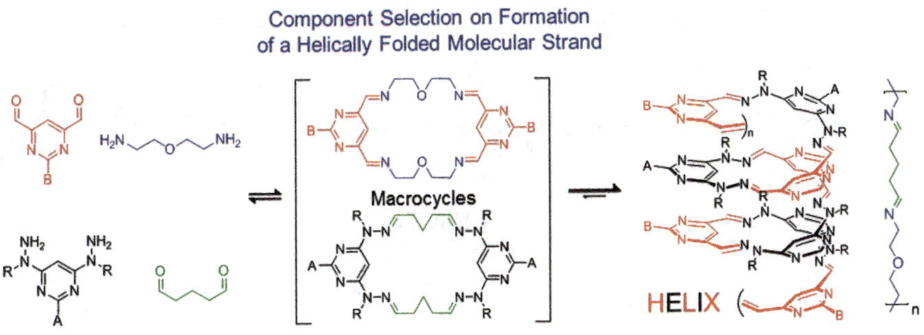

Constitutional Dynamic Chemistry: Bridge from Supramolecular Chemistry 19

Fig. 14 Component recombination under selection induced by formation of a helically folded molecular strand

components, of those which will lead to its generation. Such is the case in the formation of a helically folded strand by component exchange (Fig. 14) [77].

An intriguing case would be the selection of specific amino acid components to yield a given folded chain of a dynamic peptoid (see above). One may also point out the relationship between this case of selection driven by molecular self-organization to that of component selection driven by supramolecular self-organization in the formation of a gel of higher stability (see above) [50]. In more general terms, the latter case represents adaptation driven by the formation of supramolecular assemblies of higher organization.

Evolution of a constitutional dynamic library of dynamer strands may be driven by metal ion binding towards the generation of a specific metallo-supramolecular architecture, such as a [2×2] grid [54], or of a dynamer presenting specific interactions and properties (such as optical self-sensing) [57] (see also Figs. 7 and 8 in [9]).

Constitutional adaptation to effector binding has been demonstrated by component reorganization in a set of dynamers, leading to the generation of the dynamer chain that adopts a folded superstructure capable of binding metal cations in its folds (Fig. 15) [48].

5.4 Adaptation to Morphological Change: Shape Switching

A change in shape of a component may induce a recomposition of a CDL with generation of different constituents under component selection. Thus, the shape modification induced by cation binding to a ligand component leads to adaptation of the system to the two morphologies, displaying reversible switching between macrocyclic and polymeric states [78a, b], with selection of different components specific to each state (Fig. 16) [78c]. With suitable components, the system in addition is able to generate an optical signal [79] (see also below).

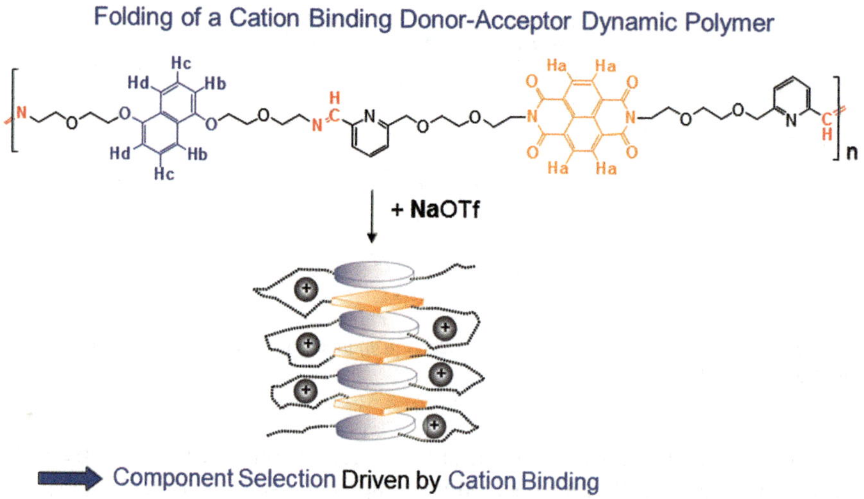

Fig. 15 A dynamer chain presenting a folded superstructure capable of binding sodium ions is formed from a set of four different dynamers (see [48]), under recomposition of monomeric components driven by cation coordination

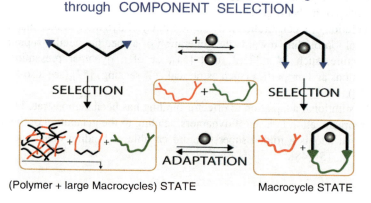

Fig. 16 Adaptation to shape change by component selection. Switching between W and U shapes of a component, under metal cation binding and removal, induces interconversion between polymer and macrocycle states with selection of the preferred component

6 Multiple Dynamics and Dynamic Networks

Multiple dynamic features are present in systems combining dynamic processes of different nature, either molecular or supramolecular, of motional, constitutional, and reactional types. These dynamic processes are orthogonal, under thermodynamic or kinetic control, and allow in principle for separate triggering by means of appropriate stimuli or effectors. Their combination opens paths in particular towards selective reactions and information storage devices.

6.1 Multiple Constitutional Dynamic Processes

CDC allows for "multiple dynamic" processes that combine and take advantage of both non-covalent and covalent dynamics, as in the assembling of metallo-architectures bearing functional groups [53, 80], Multiple constitutional dynamics can be envisaged by combining different, orthogonal, reversible covalent reactions [81] (imine and disulfide exchange has for instance been implemented in the design of a molecular walker [81]) together with different non-covalent processes (e.g., H-bonding, metal ion coordination). It also provides means for performing *constitutional dynamic synthesis*. Thus, supramolecular dynamics enable the assembly of functional components with suitable selection and structural control, whereas molecular covalent dynamics operate in post-assembly connection between the assembled components, resulting in molecular architectures of high complexity, as described in the formation of interlocked structures from metal-coordination [82a] or donor-acceptor interaction directed assembly [82b] combined with imine formation [82c].

6.2 Multiple Dynamics for Information Processing Devices

Imines and related compounds containing a carbon–nitrogen double bond $C=N$ (hydrazones, acylhydrazones, oximes) present the very attractive feature of being *double dynamic* entities capable of undergoing both:

- *Configurational dynamics*, by photochemical and thermal *cis–trans* isomerizations, as well as
- *Constitutional dynamics*, by exchange of the amine or carbonyl component [83]

The configurational isomerization occurs with conservation of the constitutional integrity of the substance and is comparatively fast (or may be made so), whereas component exchange generates a new molecule at a rate that may be chemically controlled and is usually much slower. Thus, imine-type entities allow for both short-lived and long-lived information storage, respectively, in the configuration and in the constitution of a molecule. *Short term storage* is structurally/physically borne by a molecular shape, and *long term storage* is inscribed in the formation of a novel molecule. Importantly, both processes are orthogonal and may be separately controlled/operated.

A simple bis-pyridyl hydrazone has been shown to lie at the core of a *triple dynamic* chemical system whose different states can be transformed one into another by a specific physical stimulus or chemical effector: (1) the *configurational states* E and Z may undergo photochemical and thermal interconversion, (2) the *constitutional states* are interconvertible by component exchange, and (3) *the coordination states* involve locking or unlocking by metal ion binding or removal. It therefore allows the storage and processing of both long term and short term (memory) information in a controlled fashion (Fig. 17) [84].

Acylhydrazones derived from the reaction of hydrazides with 2-pyridine carboxaldehyde, possess similar features and form a specially attractive class of compounds. Indeed, hydrazides are easily accessible from carboxylic acid groups and simple condensation with pyridine-2-carboxaldehyde or related moieties will allow for the introduction of triple dynamics into a great variety of entities of interest in materials science, biophysical chemistry, as well as information storage devices. For instance, the conversion of a carboxylic acid group in a biomolecule into a pyridyl-acyl hydrazone site will confer upon it triple dynamic features. On the other hand, replacing pyridyl-2-carboxaldehyde by molecules presenting diverse structural groups may be expected to enrich greatly the functional features.

Such entities display three levels of control and generate different physical properties (e.g., spectroscopic, absorption, fluorescence) in a switchable fashion. They thus represent components for the design of systems displaying a high level of functional complexity.

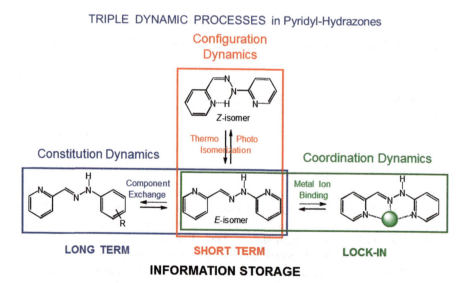

Fig. 17 Triple dynamic processes operating in pyridyl-hydrazones constitution dynamics by component exchange (*left*); configuration dynamics by photo and thermoinduced Z,E interconversion (*center*); coordination dynamics by metal cation bonding and release (*right*). The three processes allow in principle for long term, short term and locked information storage processes

Systems implementing multiple dynamics deserve active investigation, as they provide access to a rich set of properties and may extend to the modulation of the features of supramolecular assemblies [85].

6.3 Constitutional Dynamic Networks

In the progression towards systems presenting higher levels of complexity, CDC gives access to the generation of networks of dynamically interconverting constituents connected structurally (molecular and supramolecular arrays) and eventually also reactionally (sets of connected reactions). They define *constitutional dynamic networks (CDNs)* that may in particular couple to either reversible or irreversible thermodynamic processes and present a specific stability/robustness with respect to external perturbations. Connectivities between the constituents of a dynamic library define *agonistic* and *antagonistic* relationships depending on whether the increased expression of a given constituent respectively decreases or increases one or more of the others. Thus, feedback between two (or more) species (e.g., a substrate and its receptor) may lead to simultaneous optimization of both (some), e.g., the generation of a potential receptor favors the expression of the corresponding substrate and conversely.

Such dynamic sets of interconnected compounds may be represented by weighted graphs, where vertices, edges, and diagonals describe the connections between the members of a set, their agonistic or antagonistic relationships, as well as their relative weights. Figure 18 illustrates the simplest case, that of four components **A**, **B**, **C**, and **D** generating four constituents **AC**, **AD**, **BC**, and **BD** by reversible connection of **A,B** with **C,D**. Subjecting such a system to interaction with an effector **E** drives the upregulation of **AC** (and therefore of its agonist **BD** as well) and the down-regulation of the antagonists **AD** and **BC** (Fig. 19).

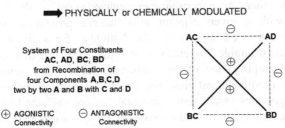

Fig. 18 Graphical representation of a constitutional dynamic network of four interconnected and interconverting constituents **AC**, **AD**, **BC**, and **BD**

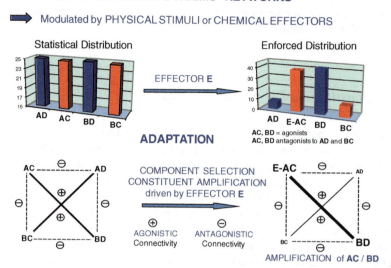

Fig. 19 Evolution of a constitutional dynamic system under the pressure of an effector **E**, leading to adaptation through generation of an enforced distribution (*top*). Graphical representation of the evolution of the corresponding dynamic network as a weighted square graph (*bottom*)

The action of two different effectors may switch a CDN from a given distribution to another one. Figure 20 represents such a constitutional dynamic switching in the simplest case of four constituents, responding to two different effectors. In addition, each state may be characterized by the generation of different properties, such as optical effects and/or substrate binding induced by the constitutional changes [79].

CDNs may in principle also perform connected evolution, whereby feedback between two (or more) species (e.g., a substrate and its receptor) leads to simultaneous optimization of both (some), a sort of *coevolution* process, where the generation of a potential receptor favors the expression of the corresponding substrate and conversely.

An especially intriguing consideration is that agonistically related constituents amplify each other. As a consequence, enhancement of the "fittest" through interactions with a given effector also induces promotion, survival of the "unfittest"! It may well happen that the "unfittest" for a given effector **E** may present specific desirable properties, so that the effector **E** may be used to drive indirectly the amplification of the "unfittest" and thus the generation of these properties. In a more general view, one could consider that evolution towards the "fittest" also provides a niche (environmental, "ecological", etc.) for the "unfittest". It represents an amplification of the constituents displaying the least competition for the same resources and occupying different ecological/medium/environmental

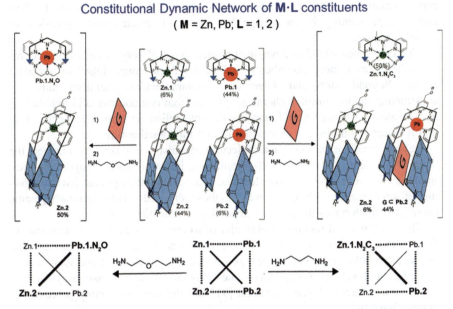

Fig. 20 Adaptation of a constitutional dynamic network in response to two different effectors. (*Top*): M-L system involving two ligand components L, two different metal cations M (Zn^{II} and Pb^{II}) and two effectors (diamine components); in addition, a guest G binds to one of the constituents. (*Bottom*): Weighted graph representation with agonistic (*solid line*) and antagonistic (*dotted line*) relationships between the four constituents. Addition of a diamine effector induces amplification/up-regulation of agonist constituents (*in bold*) and repression/down-regulation of the other two, antagonist, constituents, connected respectively by *a heavy and a light diagonal axis*. The opposite operation of the two diamine effectors represents a constitutional switching of the CDN. See [79] for details

niches. Along the same lines, antagonistic relationships lead to the extinction of partial competitors that thrive for some of the same resources (components) [86].

Highly interconnected networks (reactionally as well as constitutionally) relate to systems chemistry [87]. Networks may be considered to lie beyond patterns, which may be taken to represent different states of the dynamic network [88]. Analysis of weighted networks defines the states of complex systems at multiple scales [89, 90].

7 Conclusion and Outlook

In the context of the "big" problems challenging science, where physics addresses the origin and laws of the universe, and biology the rules of life, chemistry may claim to provide the means both for unraveling the progressive evolution towards complex matter by uncovering the processes that underlie self-organization [3, 7, 9] and for implementing the knowledge thus acquired to create novel expressions of complex matter.

Molecular chemistry has developed a wide range of very powerful procedures for building ever more complicated molecules from atoms linked by covalent bonds. Beyond molecular chemistry, supramolecular chemistry aims at constructing highly complex chemical systems from components held together by non-covalent intermolecular forces. Both have relied on design for generating highly organized molecular or supramolecular functional entities.

Beyond preorganization, supramolecular chemistry has actively pursued the design of systems undergoing self-organization into well-defined superstructures. Implementing reversibility of covalent and non-covalent linkages has led to exploring self-organization with selection and to the emergence of constitutional dynamic chemistry which traces the path towards adaptive chemistry (Fig. 21).

The structural and functional plasticity of its entities as well as of the networks that interconnect them bears a more or less close conceptual relation to a number of processes belonging to other areas of science. To consider just one such case, at the highest level of complexity, one may mention the assembly-forming connections scheme of brain function [91] and the adaptive coding processes in neuroscience [92].

Along another line of thought, one might also consider that the ability of a constitutional dynamic library to generate potentially all virtual constituents represents a sort of superposition of all states of the system corresponding to a given combination of the initial components. The conditions determine which state/combination/constituent becomes/is observable.

Further developments involve sequential, hierarchical self-organization on an increasing scale, with emergence of novel features/properties at each level [3, 7], self-organization in space as well as in time [93] and passage beyond reversibility, towards evolutive chemistry involving self-organization and constitutional dynamics in non-equilibrium systems.

Chemical evolution rests on selection operating on structural and functional diversity generated by the action of intra- and intermolecular electromagnetic

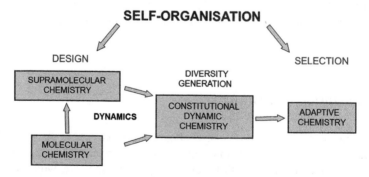

Fig. 21 Chemistry taking steps along the progressive evolution towards complex matter

forces on the components of matter. It is clear that, before there was Darwinian evolution of living organisms, there must have been a purely chemical evolution that progressively led to the threshold of life. Chemistry is taking the steps towards unraveling the complexification of matter from the atom to the thinking organism!

Acknowledgment This work was financed as part of the ANR 2010 BLAN 717 2.

References

1. Lehn JM (1988) Supramolecular chemistry – scope and perspectives molecules, supermolecules, and molecular devices. Angew Chem Int Ed Engl 27:89–112
2. (1994) Perspectives in supramolecular chemistry. Wiley, Chichester
3. Lehn J-M (1995) Supramolecular chemistry: concepts and perspectives. VCH, Weinheim
4. Atwood JL et al (1996) Comprehensive supramolecular chemistry. Pergamon, Oxford
5. Steed J, Atwood JL (2009) Supramolecular chemistry, 2nd edn. Wiley, Chichester
6. (a) Ariga K, Kunitake T (2006) Supramolecular chemistry – fundamentals and applications. Springer, Heidelberg. (b) Uhlenheuer DA et al (2010) Combining supramolecular chemistry with biology. Chem Soc Rev 39:2817–2826
7. (a) Lehn JM (2002) Toward complex matter: supramolecular chemistry and self-organization. Proc Natl Acad Sci U S A 99:4763–4768. (b) Lehn JM (2002) Toward self-organization and complex matter. Science 295:2400–2403
8. Lehn J-M (1999) Supramolecular science: where it is and where it is going. Kluwer, Dordrecht
9. Lehn JM (2007) From supramolecular chemistry towards constitutional dynamic chemistry and adaptive chemistry. Chem Soc Rev 36:151–160
10. Lehn JM (1990) Perspectives in supramolecular chemistry – from molecular recognition towards molecular information-processing and self-organization. Angew Chem Int Ed Engl 29:1304–1319
11. Lehn JM (2000) Programmed chemical systems: multiple subprograms and multiple processing/expression of molecular information. Chem Eur J 6:2097–2102
12. Whitesides GM et al (1991) Molecular self-assembly and nanochemistry – a chemical strategy for the synthesis of nanostructures. Science 254:1312–1319
13. Lawrence DS et al (1995) Self-assembling supramolecular complexes. Chem Rev 95:2229–2260
14. Philp D, Stoddart JF (1996) Self-assembly in natural and unnatural systems. Angew Chem Int Ed Engl 35:1155–1196
15. (a) Leininger S et al (2000) Self-assembly of discrete cyclic nanostructures mediated by transition metals. Chem Rev 100:853–907. (b) Seidel SR, Stang PJ (2002) High-symmetry coordination cages via self-assembhly. Acc Chem Res 35:972–983
16. Swiegers GF, Malefetse TJ (2000) New self-assembled structural motifs in coordination chemistry. Chem Rev 100:3483–3537
17. Lindoy LF, Atkinson IM (2000) Self-assembly in supramolecular systems. Royal Society of Chemistry, Cambridge
18. (a) Ciferri A (2005) Supramolecular polymers, 2nd edn. Dekker, New York. (b) J.-M. Lehn Ch. 1 in 18a
19. (a) Lehn JM (2002) Supramolecular polymer chemistry- scope and perspectives. Polym Int 51:825–839. (b)Lehn JM (1993) Supramolecular chemistry – molecular information and the design of supramolecular materials. Makromol Chem Macromol Symp 69:1–17
20. (a) Brunsveld L et al (2001) Supramolecular polymers. Chem Rev 101:4071–4097. (b) De Greef TFA et al (2009) Supramolecular polymerization. Chem Rev 109:5687–5754

21. (a) Lehn JM (2005) Dynamers: dynamic molecular and supramolecular polymers. Prog Polym Sci 30:814–831. (b)Lehn JM (2010) Dynamers: dynamic molecular and supramolecular polymers. Aust J Chem 63:611–623 (expanded version of [21a])

22. (a) Menger FM, Gabrielson KD (1995) Cytomimetic organic-chemistry – early developments. Angew Chem Int Ed 34:2091–2106. (b) Paleos CM et al (2001) Molecular recognition of complementary liposomes in modeling cell-cell recognition. Chembiochem 2:305–310

23. (a) Marchi-Artzner V et al (2001) Selective adhesion, lipid exchange and membrane-fusion processes between vesicles of various sizes bearing complementary molecular recognition groups. Chemphyschem 2:367–376. (b) Richard A et al (2004) Fusogenic supramolecular vesicle systems induced by metal ion binding to amphiphilic ligands. Proc Natl Acad Sci USA 101:15279–15284. (c) Gong Y et al (2008) Functional determinants of a synthetic vesicle fusion system. J Am Chem Soc 130:6196–6205. (d) Voskuhl J, Ravoo BJ (2009) Molecular recognition of bilayer vesicles. Chem Soc Rev 38:495–505. (e) Marsden HR et al (2011) Model systems for membrane fusion. Chem Soc Rev 40:1572–1585

24. Desiraju GR (1995) The crystal as a supramolecular entity, perspectives in supramolecular chemistry. Wiley, Chichester

25. Hosseini MW (2005) Molecular tectonics: from simple tectons to complex molecular networks. Acc Chem Res 38:313–323

26. (a) Reinhoudt DN, Crego-Calama M (2002) Synthesis beyond the molecule. Science 295:2403–2407. (b) Prins LJ et al (2001) Noncovalent synthesis using hydrogen bonding. Angew Chem Int Ed 40:2382–2426

27. (a) Constable EC (1992) Oligopyridines as helicating ligands. Tetrahedron 48:10013–10059. (b) Piguet C et al (1997) Helicates as versatile supramolecular complexes. Chem Rev 97:2005–2062. (c) Albrecht M (2001) "Let's twist again" – Double-stranded, triple-stranded, and circular helicates. Chem Rev 101:3457–3497. (d) Albrecht M, Frölich R (2007) Symmetry driven self-assembly of metallo-supramolecular architectures. Bull Chem Soc Jpn 80:797–808

28. (a) Fujita M (1998) Metal-directed self-assembly of two- and three-dimensional synthetic receptors. Chem Soc Rev 27:417–425. (b) Sun WY et al (2002) Multicomponent metal-ligand self-assembly. Curr Opin Chem Biol 6:757–764. (c) Ward MD (2002) Supramolecular coordination chemistry. Annu Rep Prog Chem A 98:285. (d) Fujita M et al. (2005) Coordination assemblies from a Pd(II)-cornered square complex. Acc Chem Res 38:369–378

29. (a) Caulder DL, Raymond KN (1999) Supermolecules by design. Acc Chem Res 32:975–982. (b) Saalfrank RW, Demleitner B (1999) Transition metals in supramolecular chemistry. Wiley, New York. (c) Swiegers GF, Malefetse TJ, (2000) New self-assembled structural motifs in coordination chemistry. Chem Rev 100:3483. (d) Leininger S, Olenyuk B, Stang PJ (2000) Self-assembly of discrete cyclic nanostructures mediated by transition metals. Chem Rev 100:853. (e) Albrecht M (2000) From molecular diversity to template-directed self-assembly – new trends in metallo-supramolecular chemistry. J Incl Phenom Macrocyl Chem 36:127–151. (f) Seidal SR, Stang PJ (2002) High symmetry coordination cages via self-assembly. Acc Chem Res 35:972. (g) Gianneschi NC et al (2005) Development of a coordination chemistry-based approach for functional supramolecular structures. Acc Chem Res 38:825-837. (h) Steel PJ (2005) Ligand design in multimetallic architectures: six lessons learned. Acc Chem Res 38:243–250. (i) Saalfrank RW et al (2008) Supramolecular coordination chemistry: the synergistic effect of serendipity and rational design. Angew Chem Int Ed 47:8794–8824

30. Ruben M et al (2004) Grid-type metal ion architectures: functional metallosupramolecular arrays. Angew Chem Int Ed Engl 43:3644–3662

31. Krämer R et al (1993) Self-recognition in helicate self-assembly – spontaneous formation of helical metal-complexes from mixtures of ligands and metal-ions. Proc Natl Acad Sci USA 90:5394–5398

32. Funeriu DP et al (2000) Multiple expression of molecular information: enforced generation of different supramolecular inorganic architectures by processing of the same ligand information through specific coordination algorithms. Chem Eur J 6:2103–2111

Constitutional Dynamic Chemistry: Bridge from Supramolecular Chemistry

33. (a) Lehn JM (1978) Cryptates – inclusion complexes of macropolycyclic receptor molecules. Pure Appl Chem 50:871–892. (b) Lehn JM (1978) Cryptates – chemistry of macropolycyclic inclusion complexes. Acc Chem Res 11:49–57
34. (a) Hasenknopf B et al (1996) Self-assembly of a circular double helicate. Angew Chem Int Ed Engl 35:1838–1840. (b) Hasenknopf B et al (1997) Self-assembly of tetra- and hexanuclear circular helicates. J Am Chem Soc 119:10956–10962. This case represents selection from a dynamic non-covalent (coordination) library
35. Lehn JM (1999) Dynamic combinatorial chemistry and virtual combinatorial libraries. Chem Eur J 5:2455–2463
36. (a) Rowan SJ et al (2002) Dynamic covalent chemistry. Angew Chem Int Ed 41:898–952. (b) Cheeseman JD et al (2005) Receptor-assisted combinatorial chemistry: Thermodynamics and kinetics in drug discovery. Chem Eur J 11:1708–1716
37. (a) Miller BL (2010) Dynamic combinatorial chemistry. Wiley, Chichester. (b) Reek JNH, Otto S (2010) Dynamic combinatorial chemistry. Wiley-VCH, Weinheim. (c) For self-sorting systems, see: Ghosh S, Isaacs L, Chap. 4, in [37a])
38. (a) Corbett PT et al (2006) Dynamic combinatorial chemistry. Chem Rev 106:3652–3711. (b) Ladame S (2008) Dynamic combinatorial chemistry: on the road to fulfilling the promise. Org Biomol Chem 6:219–226
39. Ramström O, Lehn JM (2002) Drug discovery by dynamic combinatorial libraries. Nature Rev Drug Discovery 1:26–36. (b) Hochgürtel M, Lehn J-M (2006) Fragment-based approaches in drug discovery. Wiley-VCH, Weinheim. (c) Ramström O, Lehn J-M (2007) Comprehensive medicinal chemistry II. Elsevier, Oxford. (d) Ramström O et al (2004) Dynamic combinatorial carbohydrate libraries: probing the binding site of the concanavalin A lectin. Chem Eur J 10:1711–1715. (e) Hotchkiss T et al (2005) Ligand amplification in a dynamic combinatorial glycopeptide library. Chem Commun:4264–4266
40. For reversible coordination processes, see for instance: (a) [34]; (b) Kruppa M, König B (2006) Reversible coordinative bonds in molecular recognition. Chem Rev 106:3520–3560. (c) De S et al (2010) Metal-coordination-driven dynamic heteroleptic architectures. Chem Soc Rev 39:1555–1575. (d) For the interconversion of a quadruple helicate and grid species see: Baxter PNW et al (2000) Self-assembly and structure of interconverting multinuclear inorganic arrays: a [4 x 5]-Ag-20(I) grid and an Ag-10(I) quadruple helicate. Chem Eur J 6:4510–4517. (e) see also Figures 11 and 12 hereafter
41. For dynamic processes involving hydrogen-bonded entities, see: (a) Calama MC et al (1998) Libraries of non-covalent hydrogen-bonded assemblies; combinatorial synthesis of supramolecular systems. Chem Commun:1021–1022. (b) Timmerman P et al. (1997) Noncovalent assembly of functional groups on Calix[4]arene molecular boxes. Chem Eur J 3:1823–1832. (c) Cai MM et al. (2002) Cation-directed self-assembly of lipophilic nucleosides: the cation's central role in the structure and dynamics of a hydrogen-bonded assembly. Tetrahedron 58:661–671
42. Severin K (2004) The advantage of being virtual-target-induced adaptation and selection in dynamic combinatorial libraries. Chem Eur J 10:2565–2580
43. Huc I, Lehn JM (1997) Virtual combinatorial libraries: Dynamic generation of molecular and supramolecular diversity by self-assembly. Proc Natl Acad Sci USA 94:2106–2110. This case represents selection from a dynamic covalent library.
44. Brady PA, Sanders JKM (1997) Selection approaches to catalytic systems. Chem Soc Rev 26:327–336
45. Giuseppone N, Lehn JM (2006) Protonic and temperature modulation of constituent expression by component selection in a dynamic combinatorial library of imines. Chem Eur J 12:1715–1722
46. (a) Otto S, Severin K (2007) Dynamic combinatorial libraries for the development of synthetic receptors and sensors. Top Curr Chem 277:267–288. (b) Lam RTS et al (2005) Amplification of acetylcholine-binding catenanes from dynamic combinatorial libraries. Science 308:667–669. (c) Au-Yeung HY et al (2009) Templated amplification of a naphthalenediimide-based receptor

from a donor-acceptor dynamic combinatorial library in water. Chem Commun:419–421. (d) Klein JM et al (2011) A remarkably flexible and selective receptor for Ba2+ amplified from a hydrazone dynamic combinatorial library. Chem Commun 47:3371–3373

47. Giuseppone N, Lehn JM (2006) Electric-field modulation of component exchange in constitutional dynamic liquid crystals. Angew Chem Int Ed Engl 45:4619–4624
48. Fujii S, Lehn JM (2009) Structural and functional evolution of a library of constitutional dynamic polymers driven by alkali metal ion recognition. Angew Chem Int Ed Engl 48:7635–7638
49. Berl V et al (1999) Induced fit selection of a barbiturate receptor from a dynamic structural and conformational/configurational library. Eur J Org Chem:3089–3094
50. Sreenivasachary N, Lehn JM (2005) Gelation-driven component selection in the generation of constitutional dynamic hydrogels based on guanine-quartet formation. Proc Natl Acad Sci U S A 102:5938–5943
51. For systems presenting a type of supramolecular Darwinism, see: Müller A et al (2001) Generation of cluster capsules (I-h) from decomposition products of a smaller cluster (Keggin-T-d) while surviving ones get encapsulated: species with core-shell topology formed by a fundamental symmetry-driven reaction. Chem Commun:657–658
52. (a) Nelson SM (1982) Binuclear complexes of macrocyclic schiff-base ligands as hosts for small substrate molecules. Inorg Chim Acta 62:39–50. (b) Drew MGB et al (1978) Template synthesis of a bimetallic complex of a 30-membered decadentate macrocyclic ligand – crystal and molecular-structure of a lead(Ii) complex. J Chem Soc Chem Commun:415–416. For recent examples, see for instance: (c) Storm O, Lüning U (2002) How to synthesize macrocycles efficiently by using virtual combinatorial libraries. Chem Eur J 8:793–798. (d) Lüning U (2008) Macrocycles in supramolecular chemistry: from dynamic combinatorial chemistry to catalysis. Pol J Chem 82:1161–1174
53. Nitschke JR, Lehn JM (2003) Self-organization by selection: Generation of a metallosupramolecular grid architecture by selection of components in a dynamic library of ligands. Proc Natl Acad Sci USA 100:11970–11974
54. Giuseppone N et al (2004) Generation of dynamic constitutional diversity and driven evolution in helical molecular strands under Lewis acid catalyzed component exchange. Angew Chem Int Ed Engl 43:4902–4906
55. Nitschke JR (2007) Construction, substitution, and sorting of metallo-organic structures via subcomponent self-assembly. Acc Chem Res 40:103–112
56. Cao XY et al (2007) Generation of [2X2] grid metallosupramolecular architectures from preformed ditopic bis(acylhydrazone) ligands and through component self-assembly. Eur J Inorg Chem:2944–2965.
57. Giuseppone N, Lehn JM (2004) Constitutional dynamic self-sensing in a zinc(II)/polyiminofluorenes system. J Am Chem Soc 126:11448–11449
58. Bunyapaiboonsri T et al (2001) Dynamic deconvolution of a pre-equilibrated dynamic combinatorial library of acetylcholinesterase inhibitors. Chembiochem 2:438–444
59. (a) Hochgürtel M et al (2002) Target-induced formation of neuraminidase inhibitors from in vitro virtual combinatorial libraries. Proc Natl Acad Sci USA 99:3382–3387. (b) Caraballo R et al (2010) Towards dynamic drug design: identification and optimization of beta-galactosidase inhibitors from a dynamic hemithioacetal system. Chembiochem 11:1600–1606
60. Valade A et al (2006) Target-assisted selection of galactosyltransferase binders from dynamic combinatorial libraries. An unexpected solution with restricted amounts of the enzyme. Chembiochem 7:1023–1027
61. Ono T et al (2005) Dynamic polymer blends – component recombination between neat dynamic covalent polymers at room temperature. Chem Commun:1522–1524
62. Ono T et al (2007) Soft-to-hard transformation of the mechanical properties of dynamic covalent polymers through component incorporation. Chem Commun:46–48
63. Ono T et al (2007) Optodynamers: expression of color and fluorescence at the interface between two films of different dynamic polymers. Chem Commun:4360–4362.

64. (a) Boul PJ et al (2005) Reversible diels-alder reactions for the generation of dynamic combinatorial libraries. Org Lett 7:15–18. (b) Reutenauer P et al (2009) Room temperature dynamic polymers based on diels-alder chemistry. Chem Eur J 15:1893–1900
65. Cordier P et al (2008) Self-healing and thermoreversible rubber from supramolecular assembly. Nature 451:977–980
66. Ruff Y, Lehn JM (2008) Glycodynamers: dynamic analogs of arabinofuranoside oligosaccharides. Biopolymers 89:486–496
67. Ruff Y, Lehn JM (2008) Glycodynamers: fluorescent dynamic analogues of polysaccharides. Angew Chem Int Ed Engl 47:3556–3559
68. Ruff Y et al (2010) Glycodynamers: dynamic polymers bearing oligosaccharides residues – generation, structure, physicochemical, component exchange, and lectin binding properties. J Am Chem Soc 132:2573–2584
69. Sreenivasachary N et al (2006) DyNAs: constitutional dynamic nucleic acid analogues. Chem Eur J 12:8581–8588
70. A. Hirsch, J.-M. Lehn unpublished results
71. Lu Y, Liu JW (2007) Smart wanomaterials inspired by biology: dynamic assembly of error-free manomaterials in response to multiple chemical and biological stimuli. Acc Chem Res 40:315–323
72. Hetrick EM, Schoenfisch MH (2006) Reducing implant-related infections: active release strategies. Chem Soc Rev 35:780–789
73. (a) Levrand B et al (2006) Controlled release of volatile aldehydes and ketones by reversible hydrazone formation – "classical" profragrances are getting dynamic. Chem Commun:2965–2967. (b) Levrand B et al (2007) Controlled release of volatile aldehydes and ketones from dynamic mixtures generated by reversible hydrazone formation. Helv Chim Acta 90:2281–2314. (c) Godin G et al (2010) Reversible formation of aminals: a new strategy to control the release of bioactive volatiles from dynamic mixtures. Chem Commun 46:3125–3127
74. (a) Baxter PNW et al (1997) Generation of an equilibrating collection of circular inorganic copper(I) architectures and solid-state stabilisation of the dicopper helicate component. Chem Commun:1323–1324. (b) For a recently described case of crystallization-controlled dynamic self-assembly, see: Takahagi H, Iwasawa N (2010) Crystallization-controlled dynamic self-assembly and an on/off switch for equilibration using boronic ester formation. Chem Eur J 16:13680–13688
75. Baxter PNW et al (2000) Adaptive self-assembly: environment-induced formation and reversible switching of polynuclear metallocyclophanes. Chem Eur J 6:4140–4148
76. Folmer-Andersen JF, Lehn JM (2009) Constitutional adaptation of dynamic polymers: hydrophobically driven sequence selection in dynamic covalent polyacylhydrazones. Angew Chem Int Ed Engl 48:7664–7667
77. (a) Lao LL et al (2010) Evolution of a constitutional dynamic library driven by self-organisation of a helically folded molecular strand. Chem Eur J 16:4903–4910. For folding-driven processes, see also: (b) Oh K et al (2001) Folding-driven synthesis of oligomers. Nature 414:889–893. (c) Hill DJ et al (2001) A field guide to foldamers. Chem Rev 101:3893–4011
78. (a) Ulrich S et al (2009) Reversible constitutional switching between macrocycles and polymers induced by shape change in a dynamic covalent system. New J Chem 33:271–292. (b) Ulrich S, Lehn JM (2009) Adaptation to shape switching by component selection in a constitutional dynamic system. J Am Chem Soc 131:5546–5559
79. Ulrich S, Lehn JM (2009) Adaptation and optical signal generation in a constitutional dynamic network. Chem Eur J 15:5640–5645
80. Goral V et al (2001) Double-level "orthogonal" dynamic combinatorial libraries on transition metal template. Proc Natl Acad Sci USA 98:1347–1352
81. (a) Orrillo AG et al (2008) Covalent double level dynamic combinatorial libraries: selectively addressable exchange processes. Chem Commun:5298–5300. (b) Rodriguez-Docampo Z, Otto S (2008) Orthogonal or simultaneous use of disulfide and hydrazone exchange in dynamic covalent chemistry in aqueous solution. Chem Commun:5301–5303. (c) Sarma RJ et al (2007)

Disulfides, imines, and metal coordination within a single system: Interplay between three dynamic equilibria. Chem Eur J 13:9542–9546. (d) Imine and disulfide exchange has for instance been implemented in the design of a molecular walker: Barrell MJ et al (2011) Light-driven transport of a molecular walker in either direction along a molecular track. Angew Chem Int Ed 50:285–290

82. (a) Chichak KS et al (2004) Molecular Borromean rings. Science 304:1308–1312. (b) Northrop BH et al (2006) Template-directed synthesis of mechanically interlocked molecular bundles using dynamic covalent chemistry. Org Lett 8:3899–3902. (c) For a recent case, see for instance the formation of a cryptophane cage via dynamic imine formation: Givelet C et al (2011) Templated dynamic cryptophane formation in water. Chem Commun 47:4511–4513

83. Lehn JM (2006) Conjecture: imines as unidirectional photodriven molecular motors-motional and constitutional dynamic devices. Chem Eur J 12:5910–5915

84. Chaur MN et al (2011) Configurational and constitutional information storage: multiple dynamics in systems based on pyridyl and acyl hydrazones. Chem Eur J 17:248–258

85. Yagai S, Kitamura A (2008) Recent advances in photoresponsive supramolecular self-assemblies. Chem Soc Rev 37:1520–1529

86. (a) A related case is given by the generation of a hydrogel based on dynamically connected guanosine-quartet derivatives [50] (see also above). Amplification of the constituent that yields the most stable gel leads to the agonistic amplification of the "image" constituent, as free entity, not part of the network of the gel but present in the sol phase. In this case, self-organisation-driven CDC has two remarkable effects: it selects for the best gelator(s) and generates free constituent(s) that may undergo further chemistry in a sort of "phase selection" process; (b) for complex interactions and behaviour in a biological system, see for instance: Chuang JS et al (2010) Cooperation and Hamilton's rule in a simple synthetic microbial system. Mol Syst Biol 6

87. (a) Stankiewicz J, Eckardt LH (2006) Chembiogenesis 2005 and systems chemistry workshop. Angew Chem Int Ed 45:342–344. (b) Kindermann M et al (2005) Systems chemistry: kinetic and computational analysis of a nearly exponential organic replicator. Angew Chem Int Ed 44:6750–6755. (c) Hunta RAR, Otto S (2011) Dynamic combinatorial libraries: new opportunities in systems chemistry. Chem Commun 47:847–858. (d) Corbett PT et al (2007) Systems chemistry: pattern formation in random dynamic combinatorial libraries. Angew Chem Int Ed 46:8858–8861. (e) For the integration of replication based strategies in dynamic covalent systems, see: del Amo V, Philp D (2010) Integrating replication-based selection strategies in dynamic covalent systems. Chem Eur J 16:13304–13318. (f) For the behaviour of biological networks, see for instance: Barkai N, Leibler S (1997) Robustness in simple biochemical networks. Nature 387:913–917. (g) Elowitz MB, Leibler S (2000) A synthetic oscillatory network of transcriptional regulators. Nature 403:335–338

88. Serrano MA et al (2009) Extracting the multiscale backbone of complex weighted networks. Proc Natl Acad Sci USA 106:6483–6488

89. Whitesides GM, Grzybowski B (2002) Self-assembly at all scales. Science 295:2418–2421

90. Pattern formation in DCLs of metal-dye complexes provides a basis for sensing processes: Severin K (2010) Pattern-based sensing with simple metal-dye complexes. Curr Opin Chem Biol 14:737–742

91. Singer W (1995) Development and plasticity of cortical processing architectures. Science 270:758–764

92. Ridderinkhof KR, van den Wildenberg WPM (2005) Neuroscience – adaptive coding. Science 307:1059–1060

93. Self-organization in time may be considered to involve the generation of oriented (motor) motion by motional selection from random Brownian motion, see: (a) Kay ER et al (2007) Synthetic molecular motors and mechanical machines. Angew Chem Int Ed 46:72–191 and references therein; (b) see [83] and references therein

Top Curr Chem (2012) 322: 33–54
DOI: 10.1007/128_2011_196
© Springer-Verlag Berlin Heidelberg 2011
Published online: 28 June 2011

Multistate and Phase Change Selection in Constitutional Multivalent Systems

Mihail Barboiu

Abstract Molecular architectures and materials can be *constitutionally self-sorted* in the presence of different biomolecular targets or external physical stimuli or chemical effectors, thus responding to an external selection pressure. The high selectivity and specificity of different bioreceptors or self-correlated internal interactions may be used to describe the complex constitutional behaviors through multistate component selection from a dynamic library. The self-selection may result in the dynamic amplification of self-optimized architectures during the phase change process. The *sol–gel resolution of dynamic molecular/ supramolecular libraries* leads to higher self-organized *constitutional hybrid materials*, in which organic (supramolecular)/inorganic domains are reversibily connected.

Keywords Carbonic anhydrase · Dynamic constitutional chemistry · Dynamic interactive systems · Hybrid materials

Contents

1 Introduction 34
2 Multiple Expressions of Target-Encoded Dynamic Constitutional Libraries 35
  2.1 Enzyme-Encoded DCL: Towards the Discovery of Isozyme-Specific Inhibitors 35
  2.2 External Stimuli Recomposition/Selection of DCL: Towards the "Dynamic
     Interactive Systems" 39
3 Sol–Gel-Driven DCL Constitutional Amplification-Toward
  Constitutional Hybrid Materials 42
4 Conclusions 49
References 49

M. Barboiu
Institut Européen des Membranes – ENSCM-UMII-CNRS 5635, Place Eugène Bataillon, CC 047,
34095 Montpellier Cedex 5, France
e-mail: mihai.barboiu@iemm.univ-montp2.fr

Abbreviations

CA   Carbonic anhydrases
CDC   Constitutional dynamic chemistry
DCC   Dynamic combinatorial chemistry
DCL   Dynamic combinatorial library(ies)

1 Introduction

Constitutional dynamic chemistry (CDC) [1–3] and its application dynamic combinatorial chemistry (DCC) [4–10] are new evolutional approaches to produce *chemical diversity*. In contrast to the stepwise methodology of classic combinatorial techniques, DCC allows for the generation of large molecular libraries from small sets of building blocks based on reversible interconversion between the library species. With this DCC approach, the building elements are spontaneously assembled to virtually form all possible combinations using covalent or non-covalent interactions between the species. Compound libraries generated by DCC show special applications on biology and biomedicine. By virtue of the reversible interchanges, a DCL can adapt to the system constraints, for example allowing selection events driven by molecular recognition [4–6]. In this case, the target entity itself is used to select the active ligand, directly from a library pool, resulting in a screening process that is more efficient and greatly simplified [6–10]. As an added advantage, the screening signal is amplified, due to the adaptation process, facilitating detection and characterization.

The design and construction of supramolecular architectures has attracted intense interest during the last 40 years not only for their potential applications as new functional materials but also for their fascinating constitutional diversity. It is based on the structural organization and functional integration within a molecular/ supramolecular architecture of components presenting features such as functional-activity. Self-organization of supramolecular entities may be directed by *design* or by *constitutional selection* of *dynamic combinatorial systems* [2].

CDC has also been identified as an especially promising means to explore spatial/temporal supramolecular evolution and this concept can also been used on a range of applications. A specific advantage with dynamically generated libraries gives the possibility for the compounds to self-adjust to a chosen target species at a given time in a certain environment. If libraries are produced in the presence of a bioreceptor, new ligands can be selected that resemble the naturally occurring ligands and new, potentially useful affinity molecules can be generated.

We therefore considered addressing in the first part of this review some of most representative examples in which specific molecular architectures and materials are *constitutionally self-sorted* in the presence of different biomolecular targets or external physical stimuli/chemical effectors. They respond to an external selection pressure. The high selectivity and specificity of different bioreceptors may be used to describe a complex constitutional behavior through component selection from the DCLs, driven by the selective binding to the active sites. These multistate systems also point to the possibility of modulating the drug discovery methods by constitutional recomposition induced by the specific bioreceptor targets.

On the other hand, the *self-organization by design* is based on the implementation of compounds containing specific molecular information stored in the arrangement of suitable binding sites and of external components reading out the structural information through the algorithm defined by their interactional preferences. Thus, this might allow the generation of dynamic molecular or supramolecular libraries presenting features such colloidal [10–12], gel [13, 14], or solid-state selection [15–25] of a constituent of an equilibrating collection of components reversibly switching between different arrays.

In this context the supramolecular crystalline [15–22] or hybrid materials [23–35] can be prepared and *constitutionally self-sorted* by using an irreversible kinetic process like crystallization or sol–gel polymerization. The self-selection is based on constitutional internal interactions of library components, resulting in the dynamic amplification of self-optimized architectures, during the phase change process. With all this in mind, the second part will be devoted to *sol–gel resolution of dynamic molecular/supramolecular libraries,* emphasizing recent developments, especially as pursued in our laboratory.

2 Multiple Expressions of Target-Encoded Dynamic Constitutional Libraries

2.1 Enzyme-Encoded DCL: Towards the Discovery of Isozyme-Specific Inhibitors

DCC [1–10] has been extensively implemented during the last decade as a powerful approach in drug discovery [5] that gives access to rapid and attractive identification of ligands and inhibitors for biological receptors and enzymes. The dynamic combinatorial approach is based on a shift of chemical equilibrium of a library of reversibly connected molecular components encompassing all possible combinations, driven by a biomolecular (molecular) target that favors the amplification of the fittest constituent forming the most stable non-covalent supramolecular entities with the target [36]. DCC has successfully implemented in a variety of biological systems non-exhaustively including lectins [37–40], acetylcholinesterase [41, 42], neuraminidase [43, 44], galactosyltransferase [45], glycosidase [46], DNA [47, 48] etc.

Carbonic anhydrases (CA) have been one of the early addressed biological targets for which the DCC [49–53] may offer a complementary route to high-throughput combinatorial methods [54].

The first example in this field has been pioneered by Lehn et al. who reported a library of 12 constituents containing different Zn^{2+} complexing groups and various aromatic moieties connected by the reversible imino-bond, generating thus a hydrophobic sulphonamide inhibitor possessing high affinity toward the bovine carbonic anhydrase (bCA II, EC 4.2.1.1) [49]. Then the feasibility of this concept has been extended by Nguyen et al. (Fig. 1) [50] and Poulsen et al. [51–53] including a kinetic and a thermodynamic approach based on cross-metathesis reversible reaction, all of which address the same challenge: the discovery of small molecule inhibitors of bCA II, an easily accessible and inexpensive enzyme, but not very useful for discovering human CA inhibitors [55].

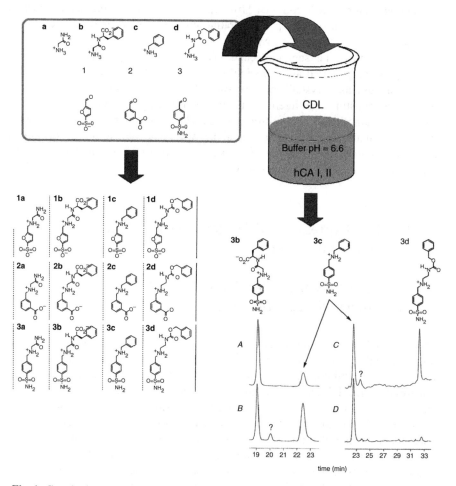

Fig. 1 Constitutional dynamic chemistry applied to bovine carbonic anhydrase bCA II isozyme and elaboration of constitutional dynamic library (CDL). Precursor amines **a–d** and aldehydes **1–3** and resulting components of the combinatorial library **1a–c–3a–c**. HPLC traces of the final reaction mixtures showing amplification of **3c** and **3d** (adapted from [49])

CA represent an important class of ubiquitously expressed zinc metalloenzymes catalyzing the reversible hydration of carbon dioxide to bicarbonate and a proton. Much progress has been achieved in the past decade in identifying selective CA inhibitors (CAIs) or activators by means of rational drug design [56–64]. The emergence of numerous families of selective CA inhibitors against several pharmacologically relevant isozymes are based on specific strategies including X-ray crystal structures for some enzyme-inhibitor complexes [58]. Among the 13 catalytically active α-CA isozymes currently known and studied as the drug targets, human carbonic anhydrases hCA I and hCA II are considered the most selective isoforms. Their inhibition has already offered important biomedical options in the development of antiglaucoma, antiepileptic, antiobesity, or anticancer drugs.

A recent study showed that a finer analysis can be performed to identify enzyme inhibitors and to evaluate their relative affinities toward the human hCA II, considered as one of the most active isoforms and studied as a drug target [65] (Fig. 2).

A DCL of 20 components has been generated under thermodynamic control by imine formation and exchange, combined with non-covalent bonding within the enzyme active site [65]. This method enabled the identification of a series of sulfonamide inhibitors **1D**, **1C**, and **2D** presenting a good inhibition and potent formation in the presence of hCA II isozyme (Fig. 2). Moreover, these data were beneficial to identify rapidly from a DCL of competitive components compound **4E**, which might represent a better compromise between entropic/enthalpic factors as a result of combined hydrophobic/H bonding binding effects of the component **4** present in a hydrophobic pocket. Finally, once the fittest structural features has been found, more precisely defined components can be developed in the next studies, allowing for the identification of enzyme inhibitors showing selectivity. Although the CA inhibitor field is a small one, these findings may be relevant to general drug design research, especially when enzyme families with a multitude of members and with similar active site features are targeted.

Indeed, the family of the CA, with a large number of representatives (13 catalytically active isoforms in mammals) playing fundamental physiological and pathological functions, can be used as a paradigm in non-conventional drug design studies aimed at obtaining compounds with selectivity for some isoforms, and thus drug candidates with reduced side effects. The observed high selectivity and specificity of hCA I and hCA II isozymes may be used to describe the complex behavior displayed by the constitutional recomposition of a dynamic library under the distinct and specific templating effect of the two enzymes.

A dynamic combinatorial library of six components can be generated under thermodynamic control by imine formation and exchange combined with non-covalent bonding within the enzyme binding site and DCL was evaluated for their relative affinities toward the physiologically relevant human carbonic anhydrase hCA I and hCA II isozymes [66].

In this context the constitutional dynamic library (CDL) is susceptible to change its composition (output expression) through component selection driven by the

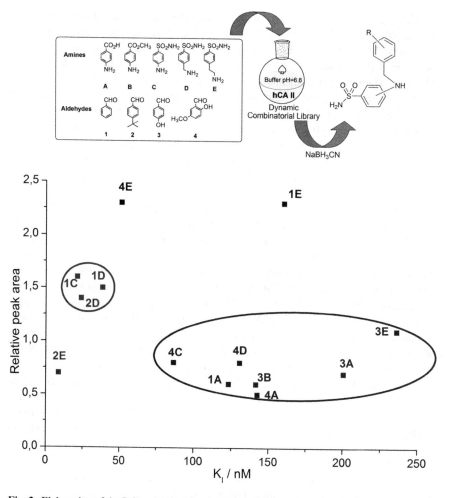

Fig. 2 Elaboration of the DCL of inhibitors and their relative peak area expressing the amplification relative to the free-enzyme DCL, function of inhibitory power against hCA II (adapted from [65])

selective binding to human hCAI and hCA II isozymes (Fig. 3). Among all possible imines formed, active compounds of appropriate geometry can be easily identified in competitive reactional conditions.

Similar studies by Beau et al. demonstrate the potential of a UDP-galactose library to search for selective binders to two galactosyltransferases enzymes using the same substrate. Despite the simplicity of the DCL composition, this adaptive DCL system is able to differentiate the two enzymes and identify very simple binders that may serve as starting points for the elaboration of selective inhibitors (Fig. 4) [45].

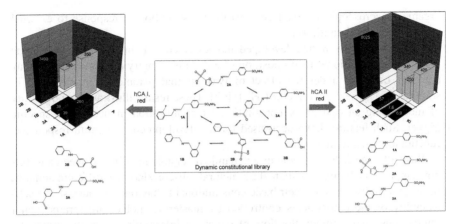

Fig. 3 Elaboration of the DCL of inhibitors' inhibition constants K_I and the amplification of the constitutional dynamic library (CDL) against catalytically active human hCA I and hCA II cytosolic isozymes (adapted from [66])

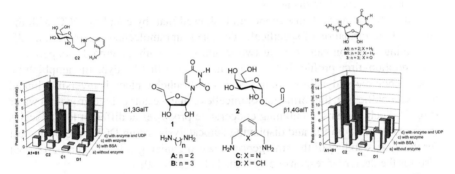

Fig. 4 Structure of the building blocks for a DCL designed to generate possible UDP-galactose mimics. Amplification of the constitutional dynamic library (CDL) against catalytically active α1,3GalT and β1,4GalT enzymes (adapted from [45])

2.2 External Stimuli Recomposition/Selection of DCL: Towards the "Dynamic Interactive Systems"

The reversibility of interactions between components of a system is a crucial factor and, accordingly, the dynamic interfaces might render the emergence system states self-adaptive, which mutually (synergistically) may adapt their spatial/temporal distribution based on their own structural constitution during the simultaneous formation of self-organized domains. The extension of the constitutional chemistry approach to nanoplatforms would be able to compete at multiple length scales within nanoscopic networks and to display variations in their sizes and functionality. Furthermore we can relate this behavior to purely synthetic compositions such as the "dynamic interactive systems" [19] characterized by their aptitude to organize macroscopically (self-control) their distribution in response to external stimuli in coupled equilibria.

These concepts were first developed and described by Lehn [67] and Giuseppone [11]. The constitutional recomposition of a dynamic library of imines can display complex behavior under the effect of two external parameters: a physical (T) stimulus and a chemical effector ($[H^+]$) [67]. These results illustrate the possibility of modulating an optical by constitutional recomposition induced by a specific trigger. Such features have been used for the development of stimuli-responsive, functional dynamic materials.

Basically, the CDC implements a dynamic reversible interface between interacting components. It might mediate the structural self-correlation of different domains of the system by virtue of their basic constitutional behaviors. In contrast, the self-assembly of the components controlled by mastering molecular/supramolecular interactions, may embody the flow of structural information from the molecular level to nanoscale dimensions. Understanding and controlling such up-scale propagation of structural information might offer the potential to impose further precise order at the mesoscale and create new routes to obtain highly ordered ultradense arrays over macroscopic distances.

Within this context, Giuseppone et al. showed that, by coupling DCC with the autocatalytic formation of specifically designed supramolecular assemblies, a self-replicating selection can occur at two length scales with a sigmoid (cooperative) concentration–time profile. Indeed, they have found that by dynamic amphiphilic block copolymers (dynablocks), in which a hydrophobic block is reversibly linked to a hydrophilic one, the formation of micelles can have autopoietic growth in water (Fig. 5). Such systems, combining cooperative processes at different length scales in networks of equilibria and displaying autocatalysis within DCLs, are of interest for the understanding of the emergence of self-organizing collective properties but also for the design of responsive systems [11, 12, 68, 69].

The selection of one or more components occurs as function of either internal (the nature and the geometry of the binding subunits, the stoichiometry, etc.) or external factors (nature of the solvent, the presence of specific molecules or ions, etc.). In view of the lability of the reversible molecular and supramolecular interactions (H-bonding, van der Waals, coordinative bonds, etc.) the self-assembly processes may present a number of novel features such as cooperativity, diversity, selection, or adaptation.

Within this context, the dynamic constitutional (i.e., covalent or supramolecular) systems can undergo constitutional recomposition under the effect of different parameters, marking changes in global properties and in the functional behaviors of the new evolving systems. Lehn and Giuseppone illustrate the selective response of this specific dynamic system to chemical effectors (Zn^{2+}) resulting in the constitutional recomposition of the system in response to a specific effector. In addition to inducing selection, Zn^{2+} ions also lead to a fluorescence shift/enhancement (Fig. 6).

On a conceptual level, both features brought together express a synergistic adaptive behavior: the addition of an external effector drives a constitutional

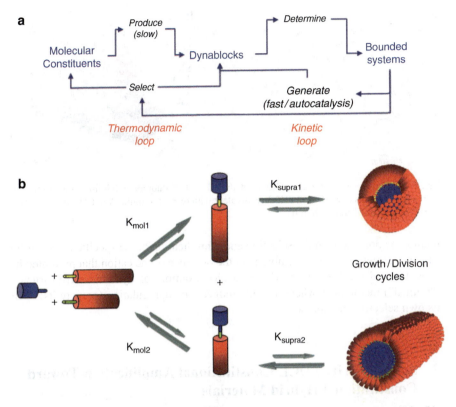

Fig. 5 (**a**) Synergistic constitutional relationships observed at two length scales within (**b**) a model minimal self-replicating DCL. For clarity, the growth/division cycles of micellar structures are not represented (adapted from [11])

evolution of the dynamic mixture towards the selection and amplification of the species that in return allows the generation of a signal indicating the presence of the very effector that promoted its generation in the first place [68, 69].

Such constitutional reorganization can be emphasized at supramolecular/nanometric level by designing columnar ion-channel architectures confined within scaffolding hydrophobic silica mesopores [70]. Evidence has been presented that such a membrane adapts and evolves its internal structure so as to improve its ion-transport properties: the dynamic non-covalent bonded macrocyclic ion-channel-type architectures can be morphologically tuned by alkali salts templating during the transport experiments or the conditioning steps. The dynamic character allied to reversible interactions between the continually interchanging components makes them respond to external ionic stimuli and adjust to form the most efficient transporting superstructure in the presence of the fittest cation, selected from a set of diverse less-selective possible architectures which can form by their self-assembly. From the conceptual point of view these membranes express a synergistic adaptive behavior: the addition of the fittest alkali ion drives a constitutional evolution of the

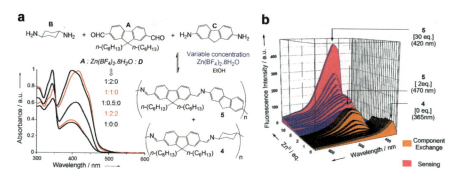

Fig. 6 Fluorescence spectra at in CHCl3 of the CDL II of fluorene polyimines, on addition of increasing amounts of Zn(BF4)2·8H2O (equivalent with respect to initial A in CDL II); excitation at 320 nm (adapted from [68, 69])

membrane pores toward the selection and amplification of the specific transporting superstructures within the membrane in the presence of the cation that promoted its generation in the first place. This is a nice example of dynamic self-instructed ("trained") membranes where a solute induces the upregulation (prepare itself) of its own selective membrane.

3 Sol–Gel-Driven DCL Constitutional Amplification-Toward Constitutional Hybrid Materials

Hybrid organic–inorganic materials produced by sol–gel processes are the subject of various investigations, offering the opportunity to achieve nanostructured materials first from robust organogel systems or second from self-organized supramolecular silsesquioxane systems [71]. Of special interest is the structure-directed function of biomimetic and bioinspired hybrid materials and control of their build-up from suitable units by self-organization. The main interest focuses on functional biomimetic materials in which the recognition-driven properties could be ensured by a well-defined incorporation of receptors of specific *molecular recognition and self-organization* functions, incorporated in hybrid solid dense or mesoporous materials [72–77]. Moreover, the different interconverting outputs resulting from such supramolecular systems may form by self-organization a dynamic polyfunctional diversity from which we may "extract selectively" a constitutional preferred hybrid architecture by sol–gel polymerization in the solid state, under the intrinsic stability of the system.

Considerable challenges lie ahead and the more significant one is the "dynamic marriage" between *supramolecular self-assembly* and the *sol–gel process*, which kinetically and sterochemically might communicate in order to converge toward self-organized functional hybrid materials. The weak interactions (H-bonds, coordination or van der Waals interactions, etc.) positioning of the molecular components to give the supramolecular architectures are typically less robust than the cross-linked covalent bonds formed in a specific polymerization process. Accordingly, the sole solution to overcome these difficulties is to improve the binding (association) efficiency of molecular components generating supramolecular assemblies. At least in theory, an increased number of interactions between molecular components and the right selection of the solvent might improve the stability of the templating supramolecular systems, communicating with the inorganic siloxane network.

Nucleobases oligomerization can be an advantageous choice to reinforce the controlled communication between interconnected "supramolecular" and "siloxane" systems. Moreover, the different interconverting outputs that nucleobases may form by oligomerization define a dynamic polyfunctional diversity which may be "extracted selectively" by sol–gel polymerization in solid state, under the intrinsic stability of the system. In this context, alkoxysilane nucleobases form in solution different types of hydrogen bonded aggregates which can be expressed in the solid state as discrete higher oligomers. Three heteroditopic nucleobase ureido-silsesquioxanes A_{Si}, U_{Si}, G_{Si} receptors have been recently reported by the Barboiu group [25–27] (Fig. 7). They generate self-organized continual superstructures in solution and in the solid state based on three encoded features: (1) the molecular recognition, (2) the supramolecular H-bond directing interactions, and (3) the covalently bonded triethoxysilyl groups.

The inorganic precursor moiety allows us, by sol–gel processes, to transcribe the solution self-organized dynamic superstructures in the solid heteropolysiloxane materials. The A_{Si} and U_{Si} compounds were designed as rigid H-bonding modules. For instance, by introducing bulky blocking alkoxysilanepropylcarboxamide groups in N9 (A) and N1 (U) positions we limit only the Watson–Crick and the Hoogsteen interactions as preferential H-bonding motifs. The A_{Si} and U_{Si} precursors generate self-organized superstructures based on two encoded features: (1) they contain a nucleobase moiety which can form ribbon-like oligomers via the combination of H-bond pairings and (2) the nucleobase moiety is covalently bonded to siloxane-terminated hydrophobic groups packing in alternative layers, allowing them, by sol–gel process, to transcribe their self-organization in the hybrids.

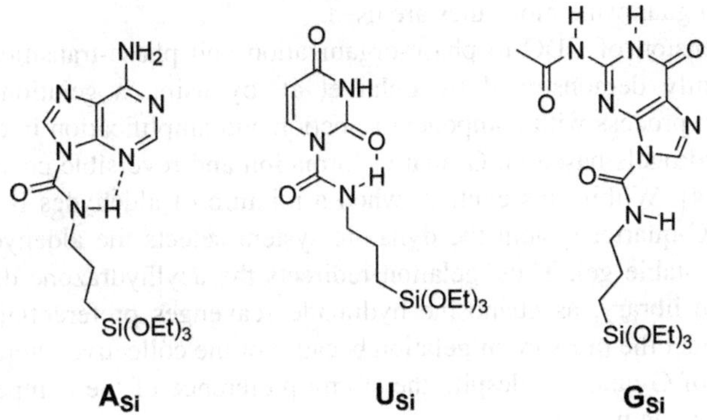

Fig. 7 Molecular structures of nucleobase ureido-silsesquioxanes A_{Si}, U_{Si}, G_{Si}

The dynamic self-assembly processes of such supramolecular systems undergoing continous reversible exchange between different self-organized entities in solution may in principle be connected to kinetically controled sol–gel process in order to extract and select an amplified supramolecular device under a specific set of experimental conditions. Such "dynamic marriage" between supramolecular self-assembly and in sol–gel polymerization processes which synergistically might communicate leads to "constitutionnal hybrid materials."

The generation of hybrid materials M_A, M_U, and M_{A-U} can be achieved using mild sol–gel conditions. X-ray powder diffraction experiments show that well-defined long-range order is present in the precursors A_{Si} and U_{Si}, but also in the hybrid materials M_A, M_U, and M_{A-U} after the sol–gel polymerization step. As a general rule, as proved by the differences between the values of interplanar Bragg diffraction distances, d_{Si-Si} the condensation process between the ethoxysilane groups during the sol–gel process results in the formation (extraction) of the *most compact hybrid materials* M_A, M_U and M_{A-U} compared with the unpolymerized A, U, and AU_{mix} powders (Fig. 8). After the sol–gel process, the constitutional preference for compact geometries in hybrid materials is most likely dictated by hydrophobic interactions and Hoogsteen H-bonding self-assembly. These examples unlock the door to the *self-organized constitutional hybrid materials*. This shows that the primary supramolecular dynamic systems generated under thermodynamic control can successfully be coupled with a secondary synthetic sol–gel resolution under kinetic resolution. The sol–gel dynamic resolution can also be related to synthetic innovative strategies for which a reduced need for purification of final materials is advantageous.

Another interesting nucleoside motif is the *G-quartet*, formed by the hydrogen-bonding self-assembly of four guanosine molecules and stabilized by alkali cations, which play an important role in biology in particular in nucleic acid telomers of potential interest to cancer therapy. [78, 79] The role of cation templating is to stabilize by coordination to the eight carbonyl oxygens of two sandwiched *G-quartets*, the *G-quadruplex*, the columnar device formed by the vertical stacking of four *G-quartets*. The *G-quadruplex* with a chiral twisted supramolecular architecture represents a nice example of a dynamic supramolecular system when guanine and guanosine molecules are used.

The extension of CDC to phase-organization and phase-transition events has been elegantly demonstrated by Lehn et al. by using a gelation-driven self-organization process with component selection and amplification in constitutional dynamic hydrogels based on G-quartet formation and reversible covalent connections [13, 14]. Within this context, when a mixture of aldehydes is employed to decorate a G-quartet system the dynamic system selects the aldehyde that leads to the most stable gel. Thus, gelation redirects the acylhydrazone distribution in the dynamic library, as guanosine hydrazide scavenges preferentially a specific aldehyde under the pressure of gelation because of the collective interactions in the assemblies of G-quartets, despite the strong preference of the competing components in the system.

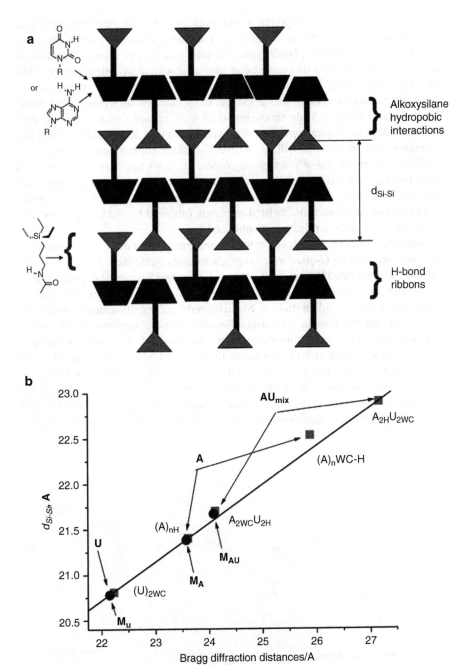

Fig. 8 Toward a constitutional transcription of base-pairing codes in hybrid materials. (**a**) Postulated model of self-organization of parallel H-bonded nucleobase aggregates and hydrophobic propyltriethoxysilane layers. (**b**) Guide to the eye interplanar d_{Si-Si} distances calculated from the geometry of minimized structures vs experimental interplanar Bragg diffraction distances. The *squares* correspond to the unpolymerized powders of precursors **A**, **U**, and their 1:1 mixture **AU**$_{mix}$, while *circles* correspond to hybrid materials **M**$_A$, **M**$_U$, and **M**$_{A-U}$

Barboiu et al. recently reported a new way to transcribe the supramolecular chirality and functionality of *G-quadruplex* at the nanometric and micrometric scale [26, 27]. *Molecular chirality* may be used as a tool to assemble molecules and macromolecules into supramolecular structures with dissymmetric shapes. The *supramolecular chirality*, which results from both the properties and the way in which the molecular components associate, is by constitution dynamic and therefore examples of large scale transcription of such *virtual chirality* remain rare. The generation of *G-quadruplex hybrid materials* can be achieved by mixing **G$_{Si}$** derivative with potassium triflate, where G-quartet superstructures have been amplified. Then the *sol–gel selection process* (Fig. 9) has been followed by a second *inorganic transcription* into inorganic silica replica materials by calcination (Fig. 10). Long-range amplification of the *G-quadruplex* supramolecular chirality into hybrid organic-inorganic twisted nanorods followed by the transcription into inorganic silica microsprings can be obtained.

Amazingly, these materials are, at the nanometric or micrometric scale, topologically analogous to its *G-quadruplex* supramolecular counterpart. After the sol–gel process, the preformed helical silica network has embedded probably enough chiral information to be irreversibly amplified (reinforced) during the calcination process when almost total condensation of Si–OH bonds occurs. By calcinations of the hybrid material, the templating twisted G-quadruplex architectures are eliminated and inorganic silica anisotropic microsprings are obtained. They present the same helical topology, without inversion inside the helix. These objects have a different helical pitch, which strongly depends on the self-correlation between hexagonal twisted mesophase domains at the nanometric level. Moreover, we obtain *chiral materials* by using a starting *achiral* guaninesiloxane **G$_{Si}$** as precursor of *achiral G-quartet* and of *chiral supramolecular G-quadruplex*. Figure 10 represents the first

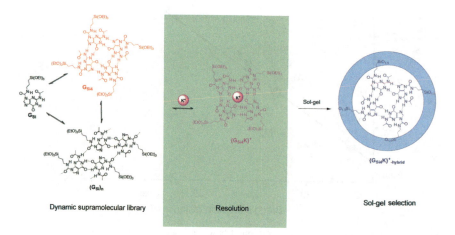

Fig. 9 Cation-template resolution of a dynamic supramolecular guanine system in which G-quartet is reversibly exchanging with linear ribbons followed by a secondary irreversible sol–gel selection of G-quadruplex hybrid materials

Multistate and Phase Change Selection in Constitutional Multivalent Systems 47

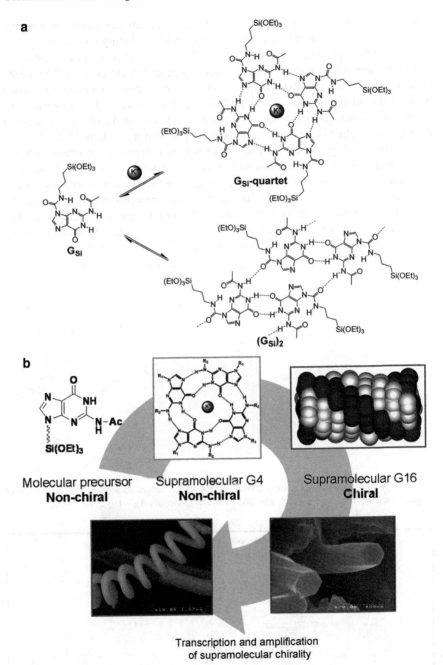

Fig. 10 (a) The cation-templated hierarchic self-assembly of guanine alkoxysilane gives the *G-quartet* in equilibrium with *G-ribbons*, (b) the chiral *G-quadruplex* transcribed in solid hybrid materials by sol–gel in the presence of templating K^+ cation

picture of the *dynamic G-quadruplex* constitutionally transcribed at the nanometric level; it unlocks the door to the new materials world paralleling that of biology.

Biomimetic-type hybrids can be generated by using another strategy to transcribe and to fix the self-assembly of the G-quadruplex architectures in self-organized nanohybrids which is based on a *double reversible covalent* iminoboronate connection between the guanosine moiety and the hybrid [32] or dynameric [27] matrix. This contributes to the high level of adaptability and correlativity of the self-organization of the supramolecular G-quadruplex and the inorganic siloxane systems (Fig. 11).

The same strategy to transcribe the supramolecular dynamic self-organization of the G-quadruplex and ureidocrown-ether ion-channel-type columnar architectures in constitutional hybrids has been applied by using a "dynamic reversible hydrophobic interface" which can render the emerging hybrid mesophases self-adaptive. The reversible hydrophobic interactions allow both supramolecular and inorganic silica components to adapt mutually (synergistically) their spatial constitution during simultaneous (collective) formation of micrometric self-organized hybrid domains (Fig. 12) [30]. Such "dynamic marriage" between supramolecular self-assembly and

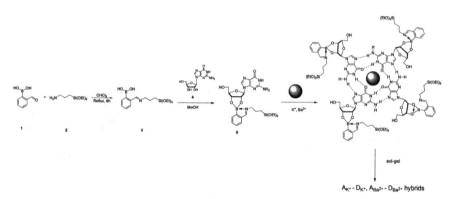

Fig. 11 Synthesis of iminoboronateguanosine precursor **5** followed by ion-template resolution of G-quartet architectures and sol–gel selection of hybrid materials $A_K^+\text{–}D_K^+$, $A_{Ba}^{2+}\text{–}D_{Ba}^{2+}$

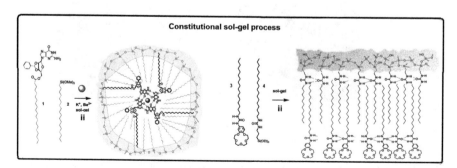

Fig. 12 Constitutional hybrid materials based on G-quadruplex and ureidocrown-ether architectures applied by using a *"dynamic reversible hydrophobic interface"* between the organic and inorganic phases

inorganic sol–gel polymerization process, which synergistically communicate, leads to higher self-organized hybrid materials with increased micrometric scales.

4 Conclusions

Complex dynamic and positive feedback between molecular/supramolecular partners in dynamic combinatorial libraries (DCLs) gives rise to emergent functional systems with a collective behavior. From the conceptual point of view, these systems express a synergistic constitutional self-reorganization (self-adaptation) of their configuration, producing an adaptive response in the presence of internal or external structural factors.

All the examples presented in this review shed light on the most major advantage with reversible DCLs over their irreversible systems [54], which is their potential adaptability to express the sorting constituent in response to an external selection pressure, based on constitutional dynamics within a confined enzymatic pocket, under the pressure of internal constitutional organization or by phase-change amplification.

Dynamic self-assembly of supramolecular systems prepared under thermodynamic control may in principle be connected to a kinetically controlled sol–gel process in order to extract and select the interpenetrated hybrid networks. Such "dynamic convergence" between supramolecular self-assembly and inorganic sol––gel processes, which synergistically communicate, leads to higher self-organized hybrid materials with increased micrometric scales.

Sol–gel constitutional resolution of constitutional hybrid architectures from DCLs toward *Dynamic Interactive Materials – systems materials* should expand the fundamental understanding of nanoscale structures and properties as it relates to creating products and manufacturing processes. More generally, applying such consideration to materials leads to the definition of *constitutional hybrid materials*, in which organic (supramolecular)/inorganic domains are reversibily connected. Considering the simplicity of this strategy, possible applications on the synthesis of more complex architectures might to be very effective, reaching close to novel expressions of complex matter.

Acknowledgments This work was financed as part of the Marie Curie Research Training Network – "DYNAMIC" (MRTN-CT-2005-019561), a EURYI scheme award. (www.esf.org/euryi) and ANR 2010 BLAN 717 2.

References

1. Lehn J-M (2007) From supramolecular chemistry towards constitutional dynamic chemistry and adaptative chemistry. Chem Soc Rev 36:151–160
2. Lehn J-M (2002) Toward complex matter: supramolecular chemistry and self-organization. Proc Natl Acad Sci USA 99:4763–4768

3. Barboiu M, Lehn JM (2002) Dynamic chemical devices: modulation of contraction/extension molecular motion by coupled-ion binding/pH change-induced structural switching. Proc Natl Acad Sci USA 99:5201–5206
4. Lehn J-M (1999) Dynamic combinatorial chemistry and virtual combinatorial libraries. Chem Eur J 5:2455–2463
5. Ramstrom O, Lehn J-M (2002) Drug discovery by dynamic combinatorial libraries. Nat Rev Drug Discov 1:26–36
6. Corbett PT, Leclaire J, Vial L, West KR, Wietor J-L, Sanders JKM, Otto S (2006) Dynamic combinatorial chemistry. Chem Rev 106:3652–3711
7. Coussins GRL, Poulsen S-A, Sanders JKM (2000) Molecular evolution: dynamiuc combinatorial libraries, autocatalytic networks and the guest for molecular function. Curr Opin Chem Biol 4:270–279
8. Tarkanyi G, Jude H, Palinkas G, Stang PJ (2005) Dynamic NMR study of the hindered Pt-NBipyridine rotation in metal directed self-assembled macrocycles. Org Lett 7:4971–4973
9. Saur I, Scopelliti R, Severin K (2006) Utilisation of the self sorting processes to generate dynamic combinatorial libraries with new network topologies. Chem Eur J 12:1058–1066
10. Giuseppone N, Schmitt J-L, Schwartz E, Lehn J-M (2005) Transaminations under scandium triflate catalysis. Independent and constitutionally coupled reactions. J Am Chem Soc 127:5528–5539
11. Nguyen R, Allouche L, Eric Buhler E, Giuseppone N (2009) Dynamic combinatorial evolution within self-replicating supramolecular assemblies. Angew Chem Int Ed 48:1093–1096
12. Xu S, Giuseppone N (2008) Self-duplicating amplification in a dynamic combinatorial library. J Am Chem Soc 130:1826–1827
13. Sreenivasachary N, Lehn JM (2005) Gelation-driven component selection in the generation of constitutional dynamic hydrogels based on guanine-quartet formation. Proc Natl Acad Sci USA 102:5938–5943
14. Setnicka V, Urbanova M, Volka K, Nampally S, Lehn J-M (2006) Investigations of guanosine-quartet assemblies by vibrational and electronic circular dichroism spectroscopy, a novel approach for studying supramolecular entities. Chem Eur J 12:8735–8743
15. Baxter PNW, Lehn J-M, Rissanen K (1997) Generation of an equilibrating collection of circular inorganic copper(I) architectures and solid-state isolation of the dicopper helicate component. Chem Commun 1323–1324
16. Baxter PNW, Lehn J-M, Kneisel BO, Fenske D (1997) Self-assembly of a symmetric tetra-copper box-grid with guest trapping in the solid state. Chem Commun 2231–2232
17. Legrand YM, van der Lee A, Masquelez N, Rabu P, Barboiu M (2007) Temperature induced single crystal-to-single crystal transformations and structure directed effects on magnetic properties. Inorg Chem 46:9083–9089
18. Barboiu M, Dumitru F, Legrand Y-M, Petit E, van der Lee A (2009) Self-sorting of equili-brating metallosupramolecular DCLs via constitutional crystallization. Chem Commun 2192–2194
19. Legrand YM, Dumitru F, van der Lee A, Barboiu M (2009) Constitutional chirality – a driving force for self-sorting homochiral single-crystals from achiral components. Chem Commun 2667–2669
20. Legrand YM, van der Lee A, Barboiu M (2007) Self-optimizing charge transfer energy phenomena in metallosupramolecular complexes by dynamic constitutional self-sorting. Inorg Chem 46:9540–9547
21. Barboiu M, Petit E, van der Lee A, Vaughan G (2006) Constitutional self-selection of [2×2] homonuclear grids from a dynamic combinatorial library. Inorg Chem 45:484–486
22. Dumitru F, Petit E, van der Lee A, Barboiu M (2005) Homoduplex and heteroduplex complexes resulted from terpyridine-type ligands and Zn^{2+} metal ions. Eur J Inorg Chem 21:4255–4262
23. Barboiu M, Cerneaux S, van der Lee A, Vaughan G (2004) Ion-driven ATP pump by self-organized hybrid membrane materials. J Am Chem Soc 126:3545–3550

24. Cazacu A, Tong C, van der Lee A, Fyles TM, Barboiu M (2006) Columnar self-assembled ureido crown ethers: an example of ion-channel organization in lipid bilayers. J Am Chem Soc 128:9541–9548
25. Arnal-Hérault C, Barboiu M, Pasc A, Michau M, Perriat P, van der Lee A (2007) Constitutional self-organization of adenine-uracil-derived hybrid materials. Chem Eur J 13:6792–6800
26. Arnal-Hérault C, Banu A, Barboiu M, Michau M, van der Lee A (2007) Amplification and transcription of the dynamic supramolecular chirality of the guanine quadruplex. Angew Chem Int Ed Engl 46:4268–4272
27. Arnal-Hérault C, Pasc A, Michau M, Cot D, Petit E, Barboiu M (2007) Functional G-quartet macroscopic membrane films. Angew Chem Int Ed Engl 46:8409–8413
28. Michau M, Barboiu M, Caraballo R, Arnal-Hérault C, Perriat P, Van Der Lee A, Pasc A (2008) Ion-conduction pathways in self-organised ureidoarene-heteropolysiloxane hybrid membranes. Chem Eur J 14:1776–1783
29. Barboiu M (2010) Dynamic interactive systems: dynamic selection in hybrid organic-inorganic constitutional networks. Chem Commun (Camb) 46:7466–7476
30. Mihai S, Cazacu A, Arnal-Hérault C, Nasr G, Meffre A, van der Lee A, Barboiu M (2009) Supramolecular self-organization in constitutional hybrid materials. New J Chem 33:2335–2343
31. Barboiu M, Cazacu A, Michau M, Caraballo R, Arnal-Hérault C, Pasc-Banu A (2008) Functional organic- inorganic hybrid membranes. Chem Eng Proc 47:1044–1052
32. Mihai S Le, Duc Y, Cot D, Barboiu M (2010) Sol–gel selection of hybrid G-quadruplex architectures from dynamic supramolecular guanosine libraries. J Mater Chem 20:9443–9448
33. Barboiu M, Cazacu A, Mihai S, Legrand Y-M, Nasr G, Le Duc Y, Petit E, van der Lee A (2011) Dynamic constitutional hybrid materials-toward adaptive self-organized devices. Microp Mesop Mat 140:51–57
34. Barboiu M, Ruben M, Blasen G, Kyritsakas N, Chacko E, Dutta M, Radekovich O, Lenton K, Brook DJR, Lehn J-M (2006) Self-assembly, structure and solution dynamics of tetranuclear Zn^{2+} hydrazone [2×2] grid-type complexes. Eur J Inorg Chem 784–789
35. Cazacu A, Mihai S, Nasr G, van der Mahon E, Lee A, Meffre A, Barboiu M (2010) Lipophilic polyoxomolybdate nanocapsules in constitutional dynamic hybrid materials. Inorg Chim Acta 363:4214–4219
36. Barboiu M (2004) Supramolecular polymeric macrocyclic receptors – hybrid carrier vs channel transporters in bulk liquid membranes. J Incl Phenom Macrocycl Chem 49:133–137
37. Ramström O, Lehn J-M (2000) In situ generation and screening of a dynamic combinatorial carbohydrate library against concanavalin A. Chembiochem 1:41–48
38. Ramström O, Lohman S, Bunyapaiboonsri T, Lehn J-M (2004) Dynamic combinatorial carbohydrate libraries: probing the binding site of the concanavalin A lectin. Chem Eur J 10:1711–1715
39. Mahon E, Aastrup T, Barboiu M (2010) Dynamic glycovesicle systems for amplified QCM detection of carbohydrate-lectin multivalent biorecognition. Chem Commun 46:2441–2443
40. Mahon E, Aastrup T, Barboiu M (2010) Multivalent recognition of lectins by glyconanoparticle systems. Chem Commun 46:5491–5493
41. Bunyapaiboonsri T, Ramström O, Lohman S, Lehn J-M (2001) Dynamic deconvolution of a preequilibrated dynamic combinatorial library of acetylcholinesterase inhibitors. Chembiochem 2:438–444
42. Larsson R, Pei Z, Ramström O (2004) Catalytic self-screening of cholinesterase substrates from a dynamic combinatorial thioester library. Angew Chem Int Ed 43:3716–3718
43. Hochgurtel M, Kroth H, Piecha D, Hofmann MW, Nicolau C, Krause S, Schaaf O, Sonnenmoser G, Eliseev AV (2002) Target-induced formation of neuraminidase inhibitors from in vitro virtual combinatorial libraries. Proc Natl Acad Sci USA 99:3382–3387
44. Hochgurtel M, Biesinger R, Kroth H, Piecha D, Hofmann MW, Krause S, Schaaf O, Nicolau C, Eliseev AV (2003) Ketones as building blocks for dynamic combinatorial libraries: highly active neuraminidase inhibitors generated via selection pressure of the biological target. J Med Chem 46:356–358

45. Valade A, Urban D, Beau J-M (2007) Two galactosyltransferases' selection of different binders from the same uridine-based dynamic combinatorial library. J Comb Chem 9:1–4
46. Gerber-Lemaire S, Popowycz F, Rodriguez-Garcia E, Asenjo ATC, Robina I, Vogel P (2002) An efficient combinatorial method for the discovery of glycosidase inhibitors. Chembiochem 3:466–470
47. Whitney AM, Ladame S, Balasubramanian S (2004) Templated ligand assembly by using G-quadruplex DNA and dynamic covalent chemistry. Angew Chem Int Ed 43:1143–1146
48. Tsujita S, Tanada M, Kataoka T, Sasaki S (2007) Equilibrium shift by target DNA substrates for determination of DNA binding ligands. Bioorg Med Chem Lett 17:68–72
49. Huc I, Lehn J-M (1997) Virtual combinatorial libraries: dynamic generation of molecular and supramolecular diversity by self-assembly. Proc Natl Acad Sci USA 94:2106–2110
50. Nguyen R, Huc I (2001) Using and enzyme's active site to template inhibitors. Angew Chem Int Ed 40:1774–1776
51. Poulsen S-A, Bornaghi LF (2006) Fragment-based drug discovery of carbonic anhydrase II inhibitors by dynamic combinatorial chemistry utilizing cross metathesis. Bioorg Med Chem 14:3275–3284
52. Poulsen S-A (2006) Direct screening of a dynamic combinatorial library using mass spectrometry. J Am Soc Mass Spectrom 17:1074–1080
53. Poulsen S-A, Davis RA, Keys TG (2006) Screening natural product-based combinatorial library using FTICR mass spectrometry. Bioorg Med Chem 14:510–515
54. Wilson SR, Czarnik AW (eds) (1997) Combinatorial chemistry-synthesis and applications. Wiley, New York
55. Supuran CT (2008) Carbonic anhydrases: novel therapeutic applications for inhibitors and activators. Nat Rev Drug Discov 7:168–181
56. Supuran CT, Scozzafava A, Conway J (eds) (2004) Carbonic anhydrase – its inhibitors and activators. CRC Press, Boca Raton, pp 1–376, and references cited therein
57. Supuran CT, Winum J-Y (2009) Selectivity issues in the design of CA inhibitors. In: Supuran CT, Winum J-Y (eds) Drug design of zinc-enzyme inhibitors: functional, structural, and disease applications. Wiley, Hoboken
58. De Simone G (2009) X-Ray crystallography of CA inhibitors and its importance in drug design. In: Supuran CT, Winum J-Y (eds) Drug design of zinc-enzyme inhibitors: functional, structural, and disease applications. Wiley, Hoboken
59. Luca C, Barboiu M, Supuran CT (1991) Stability constant of complex inhibitors and their mechanism of action. Rev Roum Chim 36(9–10):1169–1173
60. Winum J-Y, Scozzafava A, Montero J-L, Supuran C-T (2008) Design of zinc binding functions for carbonic anhydrase inhibitors. Curr Pharm Des 14:615–621
61. Supuran CT, Scozzafava A, Casini A (2003) Carbonic anhydrase inhibitors. Med Res Rev 23:146–189
62. Barboiu M, Supuran CT, Menabuoni L, Scozzafava A, Mincione F, Briganti F, Mincione G (1999) Carbonic anhydrase inhibitors, synthesis of topically effective intraocular pressure lowering agents derived from 5-(aminoalkyl-carboxamido)-1,3,4-thiadiazole-2-sulfonamide. J Enz Inhib 15:23–46
63. Supuran CT, Barboiu M, Luca C, Pop E, Dinculescu A (1996) Carbonic anhydrase activators. Part 14. Syntheses of positively charged derivatives of 2-amino-5-(2-aminoethyl) and 2-amino-5-(2-aminopropyl)-1,3,4 thiadiazole and their interaction with isozyme II. Eur J Med Chem 31:597–606
64. Winum J-Y, Rami M, Scozzafava A, Montero J-L, Supuran C (2008) Carbonic anhydrase IX: a new druggable target for the design of antitumor agents. Med Res Rev 28:445–463
65. Nasr G, Petit E, Supuran CT, Winum JY, Barboiu M (2009) Carbonic anhydrase II-induced selection of inhibitors from a dynamic combinatorial library of Schiff's bases. Bioorg Med Chem Lett 19:6014–6017
66. Nasr G, Petit E, Vullo D, Winum JY, Supuran CT, Barboiu M (2009) Carbonic anhydrase-encoded dynamic constitutional libraries: towards the discovery of isozyme-specific inhibitors. J Med Chem 42:4853–4859

67. Giusepponne N, Lehn J-M (2006) Protonic and temperature modulation of constituent expression by component selection in a dynamic combinatorial library of imines. Chem Eur J 12:1715–1722
68. Giuseppone N, Lehn JM (2004) Constitutional dynamic self-sensing in a zinc(II)-polyiminofluorene system. J Am Chem Soc 126:11448–11449
69. Giuseppone N, Fucks G, Lehn JM (2006) Tunable fluorene-based dynamers through constitutional dynamic chemistry. Chem Eur J 12:1723–1735
70. Cazacu A, Legrand YM, Pasc A, Nasr G, Van der Lee A, Mahon E, Barboiu M (2009) Dynamic hybrid materials for constitutional self-instructed membranes. Proc Natl Acad Sci USA 106:8117–8122
71. Themed Issue: Recent progress in hybrid materials science. Chem Soc Rev (2011) issue 40
72. Barboiu M, Hovnanian N, Luca C, Cot L (1999) Functionalized derivatives of benzo-crown-ethers, V. Multiple molecular recognition of zwitterionic phenylalanine. Tetrahedron 55:9221–9232
73. Barboiu M, Guizard C, Luca C, Albu B, Hovnanian N, Palmeri J (1999) A new alternative to amino acid transport: facilitated transport of L-phenylalanine by hybrid siloxane membrane containing a fixed site macrocyclic complexant. J Memb Sci 161:193–206
74. Barboiu M, Guizard C, Luca C, Hovnanian N, Palmeri J, Cot L (2000) Facilitated transport of organics of biological interest II. Selective transport of organic acids by macrocyclic fixed site complexant membranes. J Memb Sci 174:277–286
75. Barboiu M, Guizard C, Hovnanian N, Palmeri J, Reibel C, Luca C, Cot L (2000) Facilitated transport of organics of biological interest I. A new alternative for the amino acids separations by fixed-site crown-ether polysiloxane membranes. J Memb Sci 172:91–103
76. Barboiu M, Guizard C, Hovnanian N, Cot L (2001) New molecular receptors for organics of biological interest for the facilitated transport in liquid and solid membranes. Sep Purif Technol 25:211–218
77. Guizard C, Bac A, Barboiu M, Hovnanian N (2001) Hybrid organic-inorganic membranes with specific transport properties. Applications in separation and sensors technologies. Sep Purif Technol 25:167–180
78. Davis JT, Spada GP (2007) Supramolecular architectures generated by self-assembly of guanosine derivatives. Chem Soc Rev 36:296–313
79. Davis JT (2004) G-quartets 40 years later: from 5'-GMP to molecular biology and supramolecular chemistry. Angew Chem Int Ed Engl 43:668–698

Top Curr Chem (2012) 322: 55–86
DOI: 10.1007/128_2011_203
© Springer-Verlag Berlin Heidelberg 2011
Published online: 2 August 2011

Dynamic Systemic Resolution

Morakot Sakulsombat, Yan Zhang, and Olof Ramström

Abstract Dynamic Systemic Resolution is a powerful technique for selecting optimal constituents from dynamic systems by applying selection pressures, either externally by addition of target entities, or internally within the system constraints. This concept is a subset of Constitutional Dynamic Chemistry, and the dynamic systems are generally based on reversible covalent interactions between a range of components where the systems are maintained under thermodynamic control. In the present chapter, the concept will be described in detail, and a range of examples will be given for both selection classes. For external pressure generation, target enzymes, in aqueous and/or organic solution, have been used to demonstrate the resolution processes. In a first example, a dynamic transthiolesterification system was generated in aqueous solution at neutral pH, and resolved by hydrolysis using serine hydrolases (cholinesterases). In organic solution, lipase-catalyzed acylation was chosen to demonstrate asymmetric resolution in different dynamic systems, generating chiral ester and amide structures. By use of such biocatalysts, the optimal constituents were selectively chosen and amplified from the dynamic systems in one-pot processes. In internal selection pressure resolution, self-transformation and crystallization-induced diastereomeric resolution have been successfully used to challenge dynamic systems. The technique was, for example, used to identify the best diastereomeric substrate from a large and varied dynamic system in a single resolution reaction.

Keywords Adaptivity · Asymmetric synthesis · Biocatalysis · Crystallization · Dynamic chemistry · Dynamic system · Enzyme · Resolution

---

M. Sakulsombat, Y. Zhang, and O. Ramström (✉)
Department of Chemistry, Royal Institute of Technology (KTH), Teknikringen 30,
10044 Stockholm, Sweden
e-mail: ramstrom@kth.se

Contents

1. Introduction 56
2. Resolutions Using External Selection Pressures 57
   - 2.1 Dynamic Systemic Resolution by External Selection Pressure in Aqueous Solution 58
   - 2.2 Dynamic Systemic Resolution by External Selection Pressure in Organic Solution 66
3. Dynamic Systemic Resolution by Internal Selection Pressures 77
4. Conclusion 83
References 83

1 Introduction

Constitutional Dynamic Chemistry (CDC) has been established as an efficient, pseudo-evolutive selection and identification concept relying on the generation of constitutional dynamic systems using covalent and supramolecular interconnections. These dynamic systems can, for example, be applied to identify ligands or inhibitors for various biomolecules and to generate dynamic hosts for different kinds of ligands. The establishment of the dynamic systems from its individual components is the central criterion of the concept, where all components can react/interact with each other under thermodynamic control. This generation can in principle be accomplished under uncatalyzed conditions, but the systems are generally subjected to catalytic action in order to increase the overall equilibration rate. The distribution of the component associations – constituents – depends on their respective stabilities under the system conditions. By changing the reaction parameters, or by installing selection pressures, the systems will spontaneously respond by amplifying the fittest distribution of constituents. Once selected constituents, and hence their underlying components, are removed (e.g., bound) or depleted from the equilibrating pool, the system will be forced to re-equilibrate until the selection process is completed.

Dynamic Systemic Resolution (DSR) is an intriguing subset of the CDC concept, where dynamic covalent systems are formed under thermodynamic control, and subsequently subjected to an irreversible resolution process to identify the optimal constituent(s) from the system (Scheme 1). In DSR, the dynamic system is applied to an irreversible secondary process that generates a selection pressure. During the resolution process, the optimal constituent (A_n–B_m in Scheme 1) is selectively recognized and resolved from the dynamic system. The equilibrium of the dynamic system is disturbed by the selective resolution, and forced to re-equilibrate until the resolution process reaches completion.

In recent years, a range of examples based on the DSR concept, using different systems and resolution formats, has been presented. Depending on the resolution process, DSR can be divided into two major classes: resolutions using external selection pressures and resolutions based on internal selection pressures within the system constraints. By external selection pressures, an entity – generally an

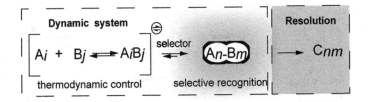

Scheme 1 Dynamic systemic resolution

enzyme – is used in catalytic amount to select and transform the optimal constituent from the dynamic system by an irreversible transformation. The resolution processes by enzymatic reactions can be performed in both aqueous and organic solution depending on the solubility of the dynamic system components/constituents and the target enzyme. The advantages of this technique are the abilities not only to identify the optimal constituent of the dynamic system but also to evaluate the enzyme specificities. On the other hand, internal selection pressures can be installed by self-transformation processes or phase-separation processes such as crystallization. These processes are dependent on the specific characteristics of the constituents and can be used to screen specific target entities from large and diverse dynamic systems in a one-pot process.

2 Resolutions Using External Selection Pressures

One of the major challenges in CDC is the application of proteins as target receptors in order to identify and amplify a matching ligand from a dynamic system. Generally, amplification of the fittest constituent is desired, and for this reason stoichiometric amounts of protein are generally required. In addition, many proteins are inherently unstable under conditions used for dynamic system generation. The first issue can however be addressed using kinetic resolution mechanisms, in principle only requiring catalytic amount of the target entities. For the second issue, the use of biocatalysts under mild conditions has been demonstrated, either in aqueous media or in organic solutions. This requires that the dynamic systems can be efficiently generated, and are sufficiently stable, under the conditions, and also that they are compatible with the enzyme-mediated reactions.

Hydrolase enzymes were initially chosen as target proteins to catalyze and amplify the optimal constituents from dynamic systems. Among the six enzyme classes, hydrolases (EC 3) are one of the most commonly used, both in bulk industrial processes as well as for laboratory scale reactions [1]. These enzymes do not require any cofactors to perform the reactions, and a large variety is commercially available. Hydrolases furthermore catalyze reactions for a broad range of substrates, e.g., hydrolysis of esters, amides, thiolesters, etc., often accompanied with high stereoselectivity. An example of hydrolases is the family of serine hydrolases, which employs

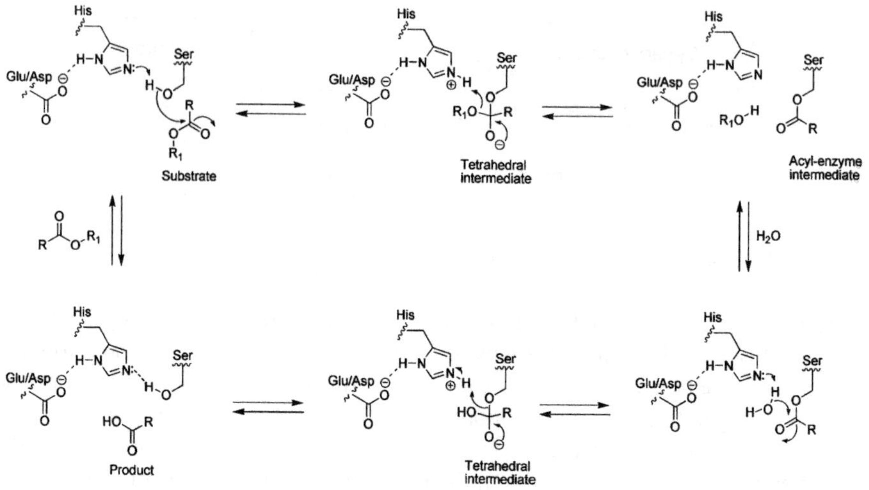

Scheme 2 Catalytic mechanism of serine hydrolases

three main amino acid residues – serine (Ser), histidine (His), and glutamate/aspartate (Glu/Asp) – in a catalytic triad as shown in Scheme 2.

The three amino acids in the triad play significant roles in the catalytic processes. The serine residue serves as the active nucleophile, activated by the other two residues. The carbonyl group of the incoming substrate, and the tetrahedral intermediate formed from the addition, are stabilized by H-bonding residues in the so-called oxyanion hole. The whole process proceeds in two stages, where the first stage generates an acyl-enzyme intermediate, while releasing a leaving group of the substrate. An incoming water molecule is subsequently activated in the second stage by the histidine and glutamate/aspartate residues, releasing the hydrolyzed product and restoring the serine residue. The catalytic triad is subsequently available for the next substrate.

2.1 Dynamic Systemic Resolution by External Selection Pressure in Aqueous Solution

An early example of DSR employed dynamic thiolester systems, generated from transthiolesterification reactions, combined with hydrolases in aqueous solution [2, 3]. The transthiolesterification reaction, of fundamental biological importance [4], is the reversible reaction between a thiolester and a thiol. This dynamic reaction is rapid and sufficiently stable under mild conditions in aqueous solution. Dynamic thiolester systems could thus efficiently be generated from a series of thiolester compounds and thiols. During equilibration, the hydrolase was applied as an external selection pressure to resolve the fittest constituents of the system, where the optimal thiolesters were continuously hydrolyzed to acid and thiol products. During

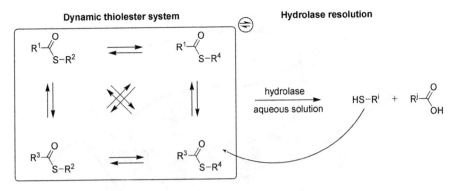

Scheme 3 Hydrolase resolution of dynamic thiolester system

the hydrolase resolution process, the hydrolyzed thiol product became in its turn part of the dynamic system and the equilibrium of the system was forced to produce the best thiolester until the reaction had reached completion as shown in Scheme 3.

2.1.1 Target Enzymes: Cholinesterases

Cholinesterases, e.g., acetylcholinesterase (AChE, EC 3.1.1.7) and butyrylcholinesterase (BChE, EC 3.1.1.8), are serine hydrolases that break down the neurotransmitter acetylcholine and other choline esters [5]. In the neurotransmission processes at the neuromuscular junction, the cationic neurotransmitter acetylcholine (ACh) is released from the presynaptic nerve, diffuses across the synapse and binds to the ACh receptor in the postsynaptic nerve (Fig. 1). Acetylcholinesterase is located between the synaptic nerves and functions as the terminator of impulse transmissions by hydrolysis of acetylcholine to acetic acid and choline as shown in Scheme 4. The process is very efficient, and the hydrolysis rate is close to diffusion controlled [6, 7].

In contrast to acetylcholinesterase, which is selective for acetylcholine, butyrylcholinesterase tolerates a wider variety of esters and is more active with butyryl- and propionylcholines than acetylcholine [7]. Structure-activity relationship studies have shown that different steric restrictions in the acyl pockets of AChE and BChE cause the difference in specificity to the acyl moiety of the substrate [6].

2.1.2 Dynamic Transthiolesterification Resolution

In order to evaluate the optimal substrate for cholinesterases, a variety of thiolesters and thiols were used to generate dynamic thiolester systems under mild reaction conditions, compatible with the target enzyme. The reversible thiolesterification reactions were performed in deuterated water at neutral pH, and the equilibration of the systems was followed by ^{1}H-NMR spectroscopy. Based on the structure of

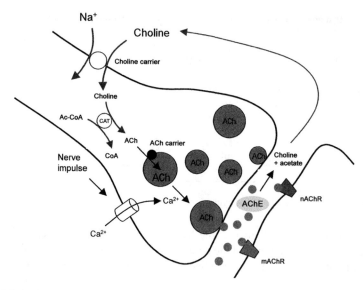

Fig. 1 Mechanism of acetylcholine action. Formed in the synapse, the compound is released by exocytosis into the synaptic cleft, where it is rapidly hydrolyzed by acetylcholinesterase

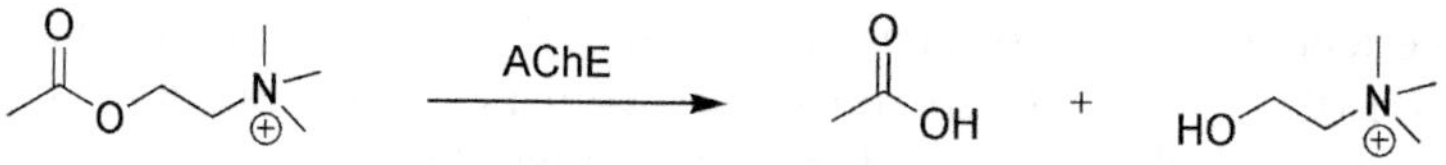

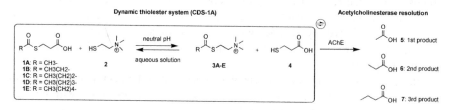

Scheme 4 Hydrolysis of acetylcholine by acetylcholinesterase

Scheme 5 Acetylcholinesterase resolution of CDS-1A

the neurotransmitter acetylcholine, the dynamic thiolester systems were mainly investigated regarding the performance from modifications of the acyl and thiol moieties.

The performances of the acyl groups were estimated from a dynamic thiolester system, generated from thiocholine **2** (choline analog) and five thiolesters **1A–E** as shown in Scheme 5. All thiolester compounds were prepared from 3-sulfanyl-propionic acid **4**, in order to keep all acyl components soluble in aqueous solution at neutral pH. After mixing one equivalent of each thiolester **1A–E** and five equivalents of thiocholine **2** in neutral deuterated buffer solution, the exchange system was monitored at 25 °C. The dynamic system generated 12 constituents/components in total; five initial thiolesters **1A–E**, five new thiolesters **3A–E**, thiocholine **2**, and new thiol **4**. The exchange rate of the dynamic system was rapid and reached equilibrium with a $t_{1/2}$ value of 50 min. The concentrations of all thiolesters **1A–E** and **3A–E** were relatively comparable, with a slightly higher ratio of thiolesters from sulfanylpropionic acid than those of thiocholine (3:2), and the system showed close to isoenergetic behavior. However, branched acyl thiolesters, *iso*-propanoyl, and *tert*-butanoyl, were also investigated, and compared to linear acyl thiolesters **1D–E**; the exchange rate of the dynamic thiolester system was in this case lower ($t_{1/2} = 110$ min) [3]. As expected, the branched acyl groups reduced the exchange rate of the dynamic system owing to steric hindrance from the branched acyl groups.

After the challenge of the dynamic thiolester system (CDS-1A) with acetylcholinesterase, the optimal constituents were immediately identified and hydrolyzed to acid and thiol products. The thiol product formed was simultaneously incorporated in the dynamic system to regenerate the optimal thiolester. During the acetylcholinesterase resolution process, two acid products, acetic and propionic acid, respectively, were mainly detected, with the acetyl ester hydrolyzed more rapidly ($t_{1/2} = 210$ min) than the propionyl ester ($t_{1/2} = 270$ min) as shown in Fig. 2. Only after significant hydrolysis of these two acyl species did the enzyme start to hydrolyze slowly the butyrate thiolester ($t_{1/2} = 1,100$ min), a lag phase possibly caused by inhibitory activities of the present thiolesters [8]. All the other acyl groups remained untouched by enzyme, a result which is in accordance with the known specificity of acetylcholinesterase.

The dynamic thiolester system was subsequently expanded by incorporating three additional thiols in order to probe the performances of the thiol moiety to the target enzyme as well as the enzyme selectivity. Three thiols, **8**, **10**, and **12**, having different adjacent functionalities, amine, sulfonic acid, and amide, respectively, were chosen and added to the dynamic thiolester system. The thiols were chosen due to their solubilities in aqueous media at neutral pH, and also for having similar exchange rates with acetylthiocholine. The kinetic and thermodynamic behavior of thiols/thiolesters are dependent on their structure and functionality,

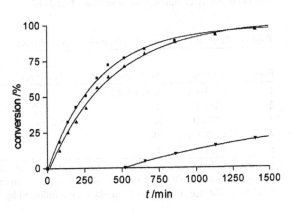

Fig. 2 Formation of acetic acid (*filled squares*), propionic acid (*filled triangles*) and butyric acid (*filled inverted triangles*) products in CDS-1A [2]. Copyright Wiley-VCH Verlag GmbH & Co. KGaA. Reproduced with permission

and some show lower exchange rates and/or generate less stable thiolesters compared to acetylthiocholine [3]. To avoid the formation of biased systems, only comparable thiols were included. The dynamic thiolester system CDS-1B was thus generated with equimolar amounts of five thiolesters **1A–E** and four thiols **2, 8, 10**, and **12** as shown in Scheme 6. For every thiol added, five additional thiolesters were formed, thus resulting in 25 thiolesters and five thiols in total. The formed dynamic thiolester system resulted in relatively comparable concentrations of all thiolesters and the system showed close to isoenergetic behavior.

Challenging the dynamic thiolester system CDS-1B with acetylcholinesterase resulted in similar product formations, but a slightly slower enzyme resolution process, than CDS-1A due to the larger size of CDS-1B. To probe the effect of the thiol moiety, dynamic thiolester systems, using only one thiol per system, were furthermore generated and applied to the enzyme resolution process. Thus, the additional CDS-1B, 1C, 1D, and 1E were prepared with equimolar amounts of five thiolesters **1A–E** and one equivalent of either thiol **2, 8, 10**, or **12** (Table 1). These systems contained ten thiolesters and two thiols. Challenging these systems with acetylcholinesterase resulted in half-lives of formation of the different hydrolysis products, acetic, propionic, and butyric acids, as shown in Table 1. These results indicate that only CDS-1A (Table 1, Entry 2), generated from five thiolesters **1A–E**

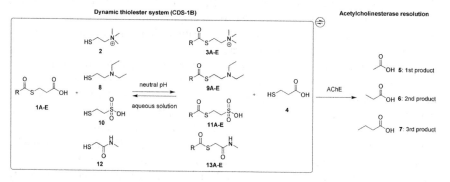

Scheme 6 Acetylcholinesterase resolution of CDS-1B

Table 1 Rates of formation of hydrolysis products in five CDSs

			$t_{1/2}$ (min)		
Entry	System	Additional thiol	Acetic acid	Propionic acid	Butyric acid
1	CDS-1B	2, 8, 10, 12	260	310	>1,800
2	CDS-1A	2	210	270	>1,500
3	CDS-1C	8	≈2,500	4,000	>>4,000
4	CDS-1D	10	–	–	–
5	CDS-1E	12	≈4,000	4,000	–

Dynamic thiolester systems were generated from one equivalent of each thiolesters **1A–E** and additional thiol in NaOD/D$_3$PO$_4$ buffer solution at pD 7.0 and then applied to acetylcholinesterase resolution. The rates of product formations were followed by ^{1}H-NMR spectroscopy

Dynamic Systemic Resolution

and thiocholine **2**, provided similar rates of product formation as CDS-1B (Table 1, Entry 1). The slightly shorter enzymatic resolution time required is a consequence of the smaller system size. For other dynamic systems CDS-1C–E (Table 1, Entries 3–5), the overall rates of product formation were considerably lower. By comparison with CDS-1A and 1C, where thiocholine **2** was replaced by thiol **8**, the acid product formations in CDS-1C (Table 1, Entry 3) were significantly slower than in CDS-1A (Table 1, Entry 2). This emphasized the enzyme specificity, where even similar thiols **2** and **8**, having adjacent ammonium and amine groups respectively, were investigated.

As expected, no hydrolysis products were formed when thiol **10** was used in the dynamic system CDS-1D (Table 1, Entry 4), and CDS-1E (Table 1, Entry 5) only very slowly produced acetic acid and propionic acid.

To test further the selectivity of the resolution process, six other hydrolase enzymes were examined under the same set of conditions as in dynamic thiolester system CDS-1A. These screening enzymes were butyrylcholinesterase (BChE, EC 3.1.1.8), horse liver esterase (HLE, EC 3.1.1.1), *Candida rugosa* lipase (CRL, EC 3.1.1.3), β-galactosidase (β-Gal, EC 3.2.1.23), trypsin (EC 3.4.21.4), and subtilisin Carlsberg (EC 3.4.21.62). The dynamic thiolester systems were exposed to each enzyme and the formation of hydrolysis products was followed by ^{1}H-NMR spectroscopy. The results are displayed in Table 2, showing that all hydrolases acting on carboxylic ester bonds (EC 3.1.1.X, Table 2, Entries 1–4) showed some hydrolysis, although the lipase from *Candida rugosa* (CRL, Table 2, Entry 4) only very modestly. In contrast to acetylcholinesterase (AChE, Table 2, Entry 1), butyrylcholinesterase (BChE, Table 2, Entry 2) acted on all acyl groups and hydrolyzed all groups in roughly the same time. The esterase from horse liver (HLE, Table 2, Entry 3), on the other hand, catalyzed the hydrolysis of the longer acyl chains slightly more efficiently than their shorter counterparts.

For the two proteases trypsin and subtilisin Carlsberg (Table 2, Entries 6 and 7), only the latter showed some activity under these conditions, also with some selectivity for the longer acyl chains similar to HLE. β-Galactosidase (β-Gal, Table 2, Entry 5), did not, however, show any activity, and control experiments with bovine serum albumin (BSA, Table 2, Entry 8) resulted in no hydrolysis products.

Table 2 Rates of product formations from enzymatic resolutions of dynamic thiolester system 1A

| Entry | Enzyme | Product yield (%)[a] | | | | |
		Acetic acid	Propionic acid	Butyric acid	Valeric acid	Caproic acid
1	AChE	50	45	–	–	–
2	BChE	37	42	44	44	43
3	HLE	16	19	20	23	31
4	CRL	–	<5	<5	<5	<5
5	β-Gal	–	–	–	–	–
6	Trypsin	–	–	–	–	–
7	Subtilisin	–	<5	<5	9	14
8	BSA	–	–	–	–	–

[a]$t = 210$ min ($t_{1/2}$ for acetic acid/AChE)

2.1.3 Tandem-Driven Dynamic Self-Inhibition of Acetylcholinesterase

To challenge dynamic systems with acetylcholinesterase further, the reversible transthiolesterification reaction was used to design and generate an in situ dynamic inhibitor system [9]. In this strategy, the target enzyme is directly associated with the dynamic one-pot tandem reaction, and the active enzyme inhibitor is exclusively formed through the enzyme action. From a pharmaceutical point of view, AChE is an important drug target for diseases like myasthenia gravis, glaucoma, and Alzheimer's disease. Various inhibitors of AChE have for this reason been synthesized and studied, among which *bis*-quaternary ligands such as decamethonium have been shown to be very potent as they are able to bridge two binding sites of AChE in close vicinity to each other. Also, the positively charged quaternary nitrogen ion interacts through cation-π-binding with the aromatic amino acid residues situated in the catalytic site [10–12].[1] In analogy, dithiolesters can be used to generate inhibitors of acetylcholinesterase. In the in situ dynamic dithiolester system, hydrolysis of acetylthiocholine by AChE leads to the product thiocholine, which undergoes reversible transthiolesterification reaction with the dithiolester, forming the actual inhibitors of acetylcholinesterase (Scheme 7).

In this self-inhibition study, three stealth inhibitors, compounds **14**, **15**, and **16**, were designed and used in the dynamic systems (Scheme 8). All structures were designed to contain an aromatic moiety and appended sulfonate groups in order to be easily soluble in aqueous solution. Dithiolesters **14** and **15** were functionalized in the *para*- and *meta*-positions, respectively, while compound **16** was monofunctionalized. These compounds were challenged with acetylcholinesterase in the dynamic system in neutral buffer solution, and the reactions were followed by

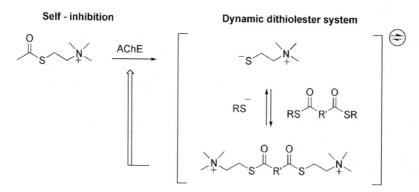

Scheme 7 Tandem-driven dynamic self-inhibition of acetylcholinesterase involving transthiolesterification

---

[1]Consult the ESTHER database for a wide range of inhibitors: bioweb.ensam.inra.fr/ESTHER/general? what = index.

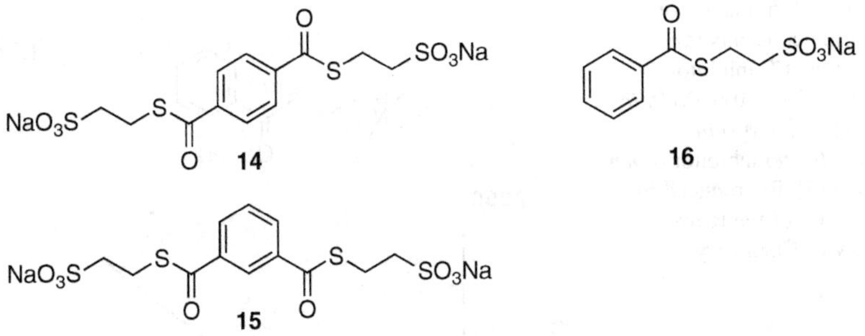

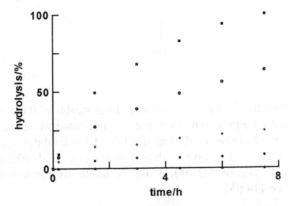

Scheme 8 Water-soluble non-inhibitory stealth inhibitors used in the dynamic system

Fig. 3 Demonstration of dynamic self-inhibition using 0.2 mM of compounds **14**, **15** and **16** (*filled circles*, *filled diamonds*, and *open circles*, respectively), together with 2 mM of ATCh and 0.2 mU of AChE. Control experiments in the absence of dithiolesters (*filled squares*), and with ATCh only (*filled triangle*) for comparison [9]. Reproduced by permission of the Royal Society of Chemistry

^{1}H-NMR spectroscopy. The resulting dynamic one-pot tandem system, generated from these compounds and acetylthiocholine, reached equilibrium in short time.

As expected, all the stealth inhibitors **14**, **15**, and **16** clearly showed inhibitory effects toward AChE as a result of the process. Figure 3 shows the kinetic hydrolysis reactions by acetylcholinesterase of the in situ dynamic thiolester system together with a control experiment without any inhibitors. Of the three compounds, terephthalic compound **14** proved to be the most effective, with only 9% of the substrate being hydrolyzed during the tested time frame. Its geometric isomer, compound **15**, also resulted in good inhibition activity in the dynamic system, although of lower efficiency than compound **14**, probably due to the more angled, isophthalic structure. The monothiolester **16**, on the other hand, demonstrated considerably lower inhibitory effect, although still reducing the degree of hydrolysis compared to the system with no inhibitor added, which emphasized the importance of bridging charges.

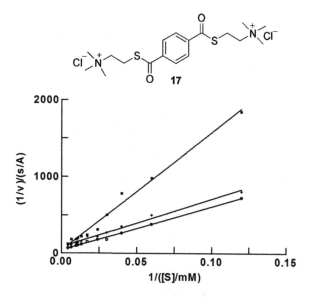

Fig. 4 Determination of inhibition constants for compound 17. Inhibitor concentrations: 100 nM (*filled squares*), 33 nM (*filled diamonds*), no inhibitor (*open circles*) [9]. Reproduced by permission of the Royal Society of Chemistry

Compound 17, the final product resulting from stealth inhibitor 14, was subsequently synthesized and its inhibitory potency evaluated using the spectrophotometric procedure of Ellman et al. (Fig. 4) [13]. Mixed inhibition of compound 17 was recorded, with a competitive inhibition constant (K_i) of 47 nM, and a noncompetitive constant (αK_i) of 103 nM, while the Michaelis constant for ATCh was determined to be 114 μM.

2.2 Dynamic Systemic Resolution by External Selection Pressure in Organic Solution

Organic building blocks for the synthesis of enantiomerically pure bioactive compounds are of high importance in academic laboratories and in industry. In organic synthesis, many transformations can be used to prepare chiral compounds, but efficient enantiomeric discrimination processes are strongly required. Enzymatic resolution of racemic compounds has in this respect become an increasingly attractive alternative to conventional chemical methods, due to its economic efficiency and benign environmental impact [14]. From this point of view, reversible covalent reactions, producing racemic intermediates, were chosen to generate dynamic systems in organic solvents. Mild reaction conditions, compatible with enzymatic resolution processes, were used to produce the dynamic systems under thermodynamic control. The dynamic systems were subsequently challenged by biomolecular selectors, able to recognize specific constituents. These constituents, containing stereogenic centers, could then be resolved by irreversible enzyme-catalyzed reactions under kinetic control and transformed to the corresponding

Dynamic Systemic Resolution

products in one-pot processes. Equilibration of the reversible reactions forced the dynamic systems to reproduce the optimal constituents until the reactions reached completion.

In synthetic organic chemistry, carbon–carbon bond forming reactions are fundamentally important for preparing optically pure compounds. One such reaction is the nitroaldol (Henry) reaction, where β-nitroalcohol adducts are formed by addition of nitroalkanes to carbonyl compounds. This reaction was first used to efficiently generate racemic dynamic systems in organic media [15]. Challenging these systems with enzymes subsequently led to the identification and asymmetric resolution of the optimal constituent from the dynamic systems in one-pot processes. Another traditional C–C-bond forming reaction, cyanohydrin formation, was further challenged by the resolution process. Dynamic cyanohydrin systems were generated from a cyanide source and carbonyl compounds, and the enzyme resolution process could be evaluated in situ. Both the nitroaldol and the cyanohydrin reactions are practically useful reactions since the products can be transformed into a variety of functional groups [16–22]. In these two types of dynamic systems, the enzyme resolution process was used to asymmetrically resolve chiral alcohol intermediates from a single dynamic process. However, a multiple dynamic covalent system based on the Strecker reaction, where transimination is combined with imine cyanation, could further be developed to generate chiral N-substituted α-aminonitrile adducts under mild reaction conditions. The dynamic aminonitrile system could in this case be resolved through a coupled process, where specific α-aminonitrile adducts were identified and catalyzed by the enzyme to give the corresponding amide product in high enantiomeric purities. The resolution of such dynamic aminonitrile system represents a breakthrough in DSR, because the continuous multi-exchange reactions were not only successfully used to generate dynamic systems, but could also be challenged by enzymes outside of their natural specificities.

2.2.1 Target Enzymes: Lipases

Among the hydrolases, the family of lipases (EC 3.1.1.3) holds a privileged position in synthetic protocols, due to their high stabilities under many operational conditions. This is especially the case for different reaction media, ranging from water, two phase systems composed of water and organic solvent, liquid substrates, to organic solvents with low water content [1]. Lipases are furthermore available from many sources, for example of fungal, bacterial and mammalian origin, and many different lipases are therefore commercially available in both free form and immobilized on solid support materials. To date, most applications are reported using such immobilized lipases, leading to higher stability, higher efficiency, and easier recovery from the reaction solution [23].

Lipases are commonly recognized as biocatalysts in hydrolysis and esterification reactions. The primary advantages of the used reactions are in asymmetric hydrolysis of chiral esters, as well as asymmetric esterification of a wide range of substrate structures [24, 25]. Current challenges in biocatalysis are to use enzymes beyond their normal catalytic reactions, for example, new substrate structures, new catalytic bond formation, etc. For example, some reports describe lipase-catalyzed amide formation reactions [26–28]. In DSR, lipase-catalyzed reactions were chosen to exert external selection pressures in resolution processes. Lipases were used to evaluate the best binder from different dynamic systems, including nitroaldol, cyanohydrin, and aminonitrile reactions, by transesterification and transamidation processes.

2.2.2 Dynamic Nitroaldol (Henry) Resolution

The nitroaldol (Henry) reaction, first described in 1859, is a carbon–carbon bond-forming reaction between an aldehyde or ketone and a nitroalkane, leading to a nitroalcohol adduct [29]. The nitroalcohol compounds, synthetically versatile functionalized structural motifs, can be transformed to many important functional groups, such as 1,2-amino alcohols and α-hydroxy carboxylic acids, common in chemical and biological structures [18, 20, 30, 31]. Because of their important structural transformations, new synthetic routes using transition metal catalysis and enzyme-catalyzed reactions have been developed to prepare enantiomerically pure nitroaldol adducts [32–34].

In the preparation of dynamic nitroaldol systems, different aldehydes and nitroalkanes were first evaluated for reversible nitroaldol reactions in the presence of base to avoid any side- or competitive reactions, and to investigate the rate of the reactions. ^{1}H-NMR spectroscopy was used to follow the reactions by comparison of the ratios of aldehyde and the nitroalcohols. Among various bases, triethylamine was chosen as catalyst because its reactions provided the fastest exchange reaction and proved compatible with the enzymatic reactions. Then, five benzaldehydes **18A–E** and 2-nitropropane **19** (Scheme 9) were chosen to study dynamic nitroaldol system (CDS-2) generation, because of their similar individual reactivity and product stabilities in the nitroaldol reaction. Ten nitroaldol adducts (±)-**20A–E** were generated under basic conditions under thermodynamic control, showing

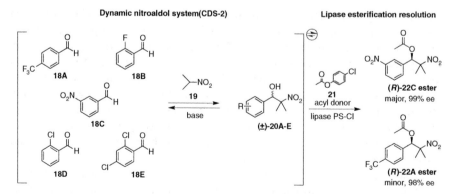

Scheme 9 Lipase-catalyzed resolution of CDS-2

close to isoenergetic behavior in the resulting dynamic system. Equimolar amounts of benzaldehydes **18A–E** and 2-nitropropane **19** in dry toluene were used.

In order to increase the exchange rate, ten equivalents of triethylamine were added, and the dynamic system was generated at 40 °C. Figure 5 shows ^{1}H-NMR spectra of the dynamic nitroaldol system at different reaction times. In the absence of any catalyst, none of the nitroalcohol adducts was observed, but addition of triethylamine resulted in efficient equilibrium formation (Fig. 5a). The aldehyde protons of compounds **18A–E** were easily followed (10.0–10.5 ppm), as well as the α-protons of 2-nitropropane **19** and adducts **20A–E** (4.5–6.5 ppm). The selected dynamic nitroaldol reaction proved to be stable without producing any side reactions within several days.

The resulting dynamic nitroaldol system was subsequently challenged with lipase-catalyzed transesterification reactions using different lipases and operational

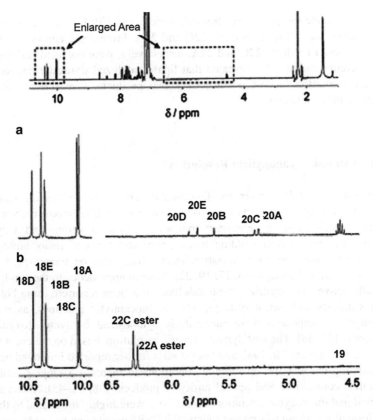

Fig. 5 ^{1}H-NMR spectra of (**a**) CDS-2 at equilibrium in the absence of lipase and acyl donor and (**b**) lipase-catalyzed esterification resolution of dynamic system after 14 days. Compound numbering as in Scheme 9. Modified from [15], copyright Wiley-VCH Verlag GmbH & Co. KGaA. Reproduced with permission

reaction conditions in order to obtain the best resolution process. Lipase PS-C I from *Burkholderia cepacia* (formerly *Pseudomonas cepacia*), immobilized on ceramic support, provided in this case the best resolution of the dynamic system. A variety of acyl donors were also evaluated where *p*-chlorophenyl acetate **21** proved the most useful in the resolution process because it did not lead to any side reactions, and resulted in complete reactions in reasonable time. As can be seen from the NMR spectra, only two products, **22C** and **22A**, were resolved from the dynamic nitroaldol system after 1 day. This pattern remained during the overall process (95% conversion) and ester products **22C** and **22A**, from corresponding 3-nitrophenyl and 4-trifluoromethylphenyl nitroalcohols **20C** and **20A**, were amplified as major and minor products, respectively. The results indicate that the lipase was specific for different nitroalcohol structures. The substitution patterns of the aromatic moiety of phenyl nitroalcohols **20A–E** showed that *meta*-substituted aromatic β-nitroalcohols were more efficiently transformed than *para*-substituted compounds. In addition, the *ortho*-substituted compound was not recognized by the lipase.

To investigate the lipase enantiomeric specificity, HPLC was used to monitor the enantiomeric ratios of products **22C** and **22A**. From these analyses, 99% and 98% *ee* of ester products **22C** and **22A**, respectively, were obtained and shown to have *R*-configuration. This revealed that lipase-catalyzed transesterification is an efficient resolution technique since only two products were selected from 20 β-nitroalcohol substrates.

2.2.3 Dynamic Cyanohydrin Resolution

To investigate DSR further by lipase-catalyzed transesterification reactions, dynamic cyanohydrin systems were used. The cyanohydrin carbon–carbon bond-forming reaction can be generated from an aldehyde or ketone and a cyanide source. The cyanohydrin building blocks, core structures of many biologically active compounds, are very versatile because they can be transformed into a range of functional groups [16, 17, 19, 22]. Several approaches to the synthesis of optically active cyanohydrin compounds have thus been reported, using both synthetic catalysts and biocatalysts [35–40]. In biocatalytic reactions, asymmetric cyanohydrin compounds were successfully synthesized by lyases (oxynitriles) and lipases [37, 38]. The first dynamic kinetic resolution, based on the cyanohydrin reaction, was reported in 1989 and later studies further reported improved methods by changing the enzyme operational conditions to increase a range of substrate structures, conversion, and optical purity of products [38, 41–45]. These studies revealed that the enzyme operational conditions were highly influential to the product formations. From this point of view, the DSR concept can be used to evaluate and match lipase performances to the selective dynamic cyanohydrin substrates, with respect to regio- and enantioselectivity, by changing the reaction parameters in one-pot processes.

Dynamic Systemic Resolution

To prepare dynamic cyanohydrin systems under mild conditions, a range of aldehyde compounds and cyanide sources was evaluated. As a result, benzaldehydes **23A–E** were selected due to their diverse substitution patterns and their inability to generate any side reactions. Even though there are many cyanide sources, acetone cyanohydrin **24** was chosen as cyanide source in presence of triethylamine base, resulting in smooth cyanide release. Dynamic cyanohydrin systems (CDS-3) were thus generated from one equivalent of each benzaldehyde **23A–E** and acetone cyanohydrin **24** in chloroform-*d* at room temperature (Scheme 10). One equivalent of triethylamine was added to accelerate the reversible cyanohydrin reactions and this amount was satisfactory to force the dynamic system to reach equilibrium even at low temperature.

^{1}H-NMR spectroscopy was used to study the dynamic cyanohydrin systems, following the aldehyde protons and the α-protons of the intermediates and ester products at different time intervals. Because of their similar structures, the α-protons of the cyanohydrin intermediates and ester products were detected in the same regions, 5.40–5.95 and 6.30–6.70 ppm, respectively, in the NMR spectra as shown in Fig. 6. The dynamic cyanohydrin system reached equilibrium in 3 h (Fig. 6a). As can be seen, cyanohydrin intermediates **25A** and **25C** were formed as major intermediates, while intermediates **25B, 25D**, and **25E** have similar ratios and were formed as minor intermediates in the dynamic system. The resulting dynamic system was proven to be stable without any side reactions within several days.

The dynamic cyanohydrin system was next challenged with lipase-catalyzed transesterification resolution using different operational conditions. Thus, different lipases, organic solvents, additives, and acyl donors were evaluated. Isopropenyl acetate **26** was chosen and used as acyl donor because its reaction produces acetone as by-product, which does not interfere in the reaction and the NMR spectra. Molecular sieve 4 Å was also added in the dynamic resolution process to control the water activity. The lipase preparation PS-C I was chosen in the resolution process since it expressed the highest lipase activities for both the substrate structure and the enantiomeric selectivities. Different organic solvents were also

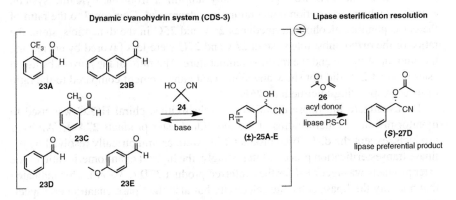

Scheme 10 Lipase-catalyzed esterification resolution of CDS-3

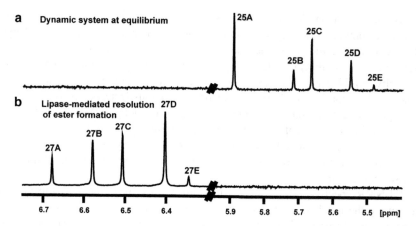

Fig. 6 ¹H-NMR spectra (enlarged areas) of (**a**) CDS-3 at equilibrium and (**b**) lipase-catalyzed transesterification resolution at 0 °C

probed, where the results using chloroform-*d* provided the highest lipase enantiomeric discrimination of the cyanohydrin substrates. In the subsequent lipase-catalyzed resolution of the dynamic cyanohydrin system, two parameters – the amount of enzyme and the reaction temperature – were further evaluated to reveal the lipase activities. The ratios of ester products **27A–E** were easily followed by NMR spectroscopy, and the results indicate that the ester **27D** was the preferred product, the ratio of which gradually increased when the amount of lipase was reduced (cf. Fig. 6b).

To improve the enzyme activities, decreasing the reaction temperature has been reported to enhance the enantiomeric discrimination [46–48]. Lower reaction temperatures were also applied to the lipase-catalyzed resolution of the dynamic cyanohydrin system. At 0 °C, the resolution process required a longer time, but the dynamic system was stable and showed similar intermediate ratios until the reaction was completed. The results (Fig. 6b) indicated that the non-substituted ester products **27B** and **27D** were preferentially amplified from the dynamic system, especially when the reaction temperature was decreased. Compared to the ratio of the corresponding alcohol intermediates **25A** and **25C** in the dynamic system, the ratios of the *ortho*-substituted esters **27A** and **27C** were less favored by decreasing the amount of lipase and the reaction temperature. The ratio of the *para*-substituted ester product **27E** did not show any significant enhancement compared to the ratio of its corresponding intermediate **25E**.

To investigate the lipase enantiomeric selectivities, chiral HPLC was used to monitor the enantiomeric ratios of the individual ester products **27A–E**. All ester products, except the disfavored product **27A**, were asymmetrically resolved by the lipase transesterification process. Interestingly, the highest enantiomeric ratio of the ester products was recorded for the preferred product **27D** (83% *ee*). This indicated that not only the lipase substrate selectivity but also the lipase enantiomeric specificity could be controlled in the dynamic cyanohydrin system.

2.2.4 Dynamic Aminonitrile (Strecker) Resolution

One of the key challenges in dynamic systemic chemistry is the exploration and development of new efficient dynamic covalent systems. To generate a dynamic system, a reversible single reaction type with a range of substrate structures is commonly used. On the other hand, very few examples of dynamic covalent systems using multiple exchange processes have been reported, especially examples where multiple exchange reactions are performed simultaneously and continuously communicate with each other during the process [49]. This is a considerable task that is of great interest, since multiple reactions in principle can be used to modify all substituents on a stereogenic carbon center in a one-pot process.

After surveying different possibilities, the aminonitrile (Strecker) reaction was chosen as design element for dynamic chiral systems [50]. The original Strecker reaction refers to the condensation of an aldehyde, an ammonium salt, and a cyanide source in water to obtain an α-aminonitrile product. This powerful carbon–carbon bond-forming reaction provides chiral α-branched amine structures, which are common substructures in many biologically active entities [51, 52]. Using this reaction, double dynamic covalent systems can be generated from transimination and cyanide addition reaction under thermodynamic control. The dynamic aminonitrile systems were subsequently challenged with lipase-catalyzed amidation, an uncommon catalytic activity for this family of enzymes, as well as specificities toward the different structures of the aminonitrile adducts.

The formation and the reversibility of aminonitriles were initially studied in order to generate efficient and stable dynamic aminonitrile systems and to evaluate reaction conditions that were compatible with the enzymatic resolution process. Imine starting material **28A**, prepared from its corresponding aldehyde and methylamine, was treated with trimethylsilyl cyanide (TMSCN) in the presence of acetic acid in chloroform-*d* to obtain aminonitrile product **29A** (Scheme 11a). Various cyanide sources were initially tested, where the reaction of the imine and TMSCN provided a suitable route for preparing aminonitrile compounds with no accompanying formation of side reactions. Following formation of aminonitrile **29A**, imine **28B**, carrying an ethylamine moiety, was added. The expected exchange reaction was not observed even after monitoring for several days. Different acidic and basic catalysts, for example, trifluoroacetic acid, triethylamine, DBU, etc., were screened but none of them provided any satisfactory results. Different Lewis acids were subsequently tested, and the exchange aminonitrile reaction of compounds **29A** and **28B** in the presence of zinc bromide proved efficient in inducing the desired product **29B**. The cyanide ion was thus released from compound **29A**, and was able to react with imine **28B** to form the corresponding product **29B**. The resulting exchange reaction was stable and rapid, and, most importantly, these reaction conditions were compatible with the enzymatic resolution.

In order to form double reversible reactions in a one-pot process, amine **D** was added to aminonitrile **29A**, followed by zinc bromide as catalyst as shown in Scheme 11b. The expected product **29D** was formed and detected by NMR

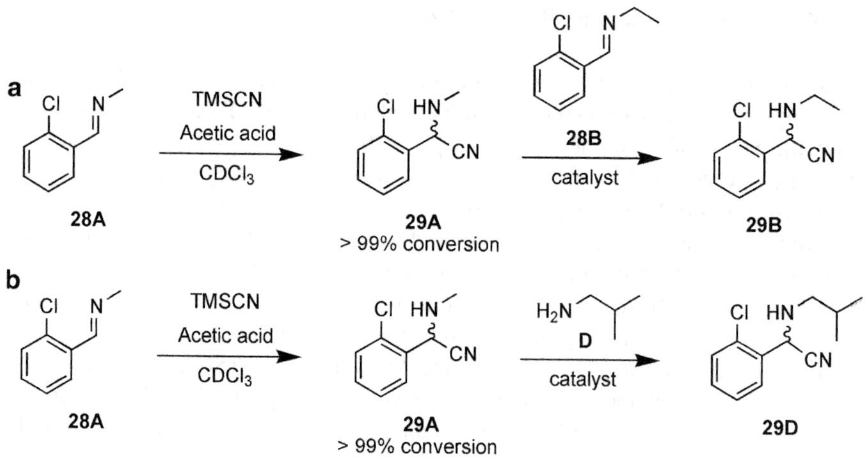

Scheme 11 Model reversible aminonitrile reactions: exchange reactions with (**a**) imine and (**b**) amine

spectroscopy. The results from this model reaction confirmed the double reversible reactions, transimination and cyanide addition, respectively. In this context, the aldehyde-free transimination [53] is an exchange reaction carried out in the presence of catalyst between imine and amine, without the involvement of water or aldehyde in the reaction. In this model reaction, aminonitrile **29A** underwent retro-Strecker reaction in the presence of zinc bromide to form imine **28A** and cyanide ion. The transimination between amine **D** and imine **28A** then formed the imine **28D**, which underwent cyanide addition to produce aminonitrile **29D**.

After thoroughly testing the reversibility of the double exchange reactions, dynamic aminonitrile systems (CDS-4B) were generated from the cyanide addition reactions of the dynamic transimination system (CDS-4A), formed from a series of imines **28A**, **31B**, and **34C**, and amine **D** as shown in Scheme 12. ^{1}H-NMR spectroscopy was used to monitor the formation of the dynamic aminonitrile systems at different reaction time intervals. Imine starting materials **28A**, **31B**, and **34C** (Fig. 7a), carrying different aromatic and amine moieties, were allowed to undergo transimination reactions with amine **D** to form 12 imine intermediates **28A–D**, **31A–D**, and **34A–D** (Fig. 7b). The resulting imine intermediates proved to have similar ratios in the dynamic transimination system. By addition of trimethylsilyl cyanide, acetic acid, and zinc bromide to the dynamic transimination system, 24 chiral aminonitrile intermediates **29A–D**, **32A–D** and **35A–D** were formed, showing close to isoenergetic behavior under thermodynamic control (Fig. 7c). By this double dynamic system based on carbon–carbon bond formation, three different aromatic imines could be subjected to the transimination reaction with another amine to form 12 imine intermediates. The addition of cyanide ion to these imines generated 24 new chiral building blocks, carrying three different

Dynamic Systemic Resolution

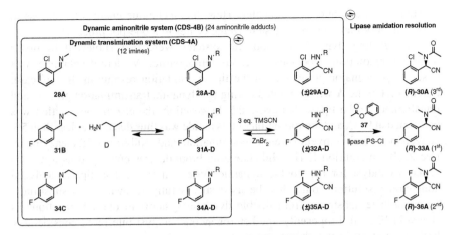

Scheme 12 Lipase-catalyzed amidation resolution of CDS-4A, CDS-4B conjugated with cyanide addition reaction

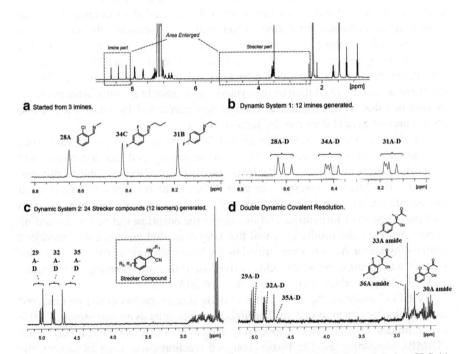

Fig. 7 ^{1}H-NMR spectra of dynamic aminonitrile system. (**a**) Initial imine signals before CDS-4A. (**b**) Twelve imine signals in CDS-4A. (**c**) α-Protons of 24 chiral aminonitriles in CDS-4B. (**d**) Methyl proton signals of three amide products from lipase-catalyzed amidation resolution of double dynamic covalent system. Adapted with permission from [50]. Copyright 2009 American Chemical Society

substituents at a single carbon center. Thus, this technique enables the access to multi-axis modifications on one stereogenic center in a one-pot process.

The resulting dynamic aminonitrile systems were first subjected to lipase mediated resolution processes at room temperature. N-Methylacetamide was observed as a major product from the lipase amidation resolution. In this case, free methylamine **A** was generated during the dynamic transimination process and transformed by the lipase. To avoid this by-reaction, the enzymatic reaction was performed at 0 °C, and the formation of this amide was thus detected at less than 5% conversion. To circumvent potential coordination, and inhibition of the enzyme by free Zn(II) in solution [54], solid-state zinc bromide was employed as a heterogeneous catalyst for the double dynamic system at 0 °C. The lipase-catalyzed amidation resolution could thus be used successfully to evaluate N-substituted α-aminonitrile substrates from double dynamic systems in one-pot reactions as shown in Fig. 7d. Proposedly, the heterogeneous catalyst interfered considerably less or not at all in the chemo-enzymatic reaction because the two processes are separated from each other. Moreover, the rate of the by-reaction was reduced due to strong chelation between the amine and zinc bromide in the heterogeneous system.

To obtain the best lipase-catalyzed amidation resolution, the lipase operational conditions, i.e., additives, lipase preparations, acyl donors, and solvents, were further evaluated. When molecular sieve 4 Å was added to control the water content in the enzymatic resolution, decompositions of aminonitrile intermediates were observed. Among a range of lipases, the resolution process by lipase PS-C I provided the highest conversion of amide products. Phenyl acetate **37** was chosen as acyl donor because its reaction led to marginal by-reactions. Thus, the lipase-catalyzed amidation resolution of the dynamic aminonitrile systems in the presence of zinc bromide as heterogeneous catalyst was performed by lipase PS-C I and phenyl acetate as acyl donor in dry toluene at 0 °C.

The double dynamic system comprised 24 chiral aminonitrile intermediates, carrying different substituents on the aromatic moiety and the N-position, and was subsequently challenged by the lipase-mediated amidation resolution process. Not only ^{1}H-NMR spectroscopy was used to follow the reactions; HPLC, through the integration of preparative Zorbax and chiral OD-H columns, was used to confirm the product formations and to analyze the enantioselectivities. According to both analyses, the results indicated that only N-methyl aminonitriles, produced from methylamine **A**, were transformed by the lipase resolution process. The major product was found to be amide **33A**, derived from its corresponding intermediate **32A**. The other products were amides **36A** and **30A**.

By HPLC analysis, the lipase enantiomeric discrimination could be observed from the double dynamic system. The highest enantioselectivity was around 5–10% ee for amides **33A** and **36A**, respectively. However, when tert-butyl methyl ether (TBME), commonly used in lipase-catalyzed reactions, was used as solvent, the lipase enantiomeric discrimination of the dynamic systems was much improved. High enantioselectivities were observed for all three final products: 90% ee (**33A**), 93% ee (**36A**), and 73% ee (**30A**). Thus, only three chiral amide products were resolved from dynamic system, containing 24 chiral intermediates, through lipase-catalyzed resolution. The dynamic resolution process demonstrated not only an uncommon catalytic amidation reaction but also structural and enantiomeric specificity toward the dynamic aminonitrile system.

3 Dynamic Systemic Resolution by Internal Selection Pressures

As mentioned, asymmetrically pure compounds are important for many applications, and many different strategies are pursued. However, in spite of many methods being developed, the classic resolution technique of diastereomeric crystallization is still preferentially used to prepare optically active pure compounds in bulk quantity. Crystallization is commonly used in the last purification steps for solid compounds because it is the most economic technique for purification and resolution. Attempts to achieve crystallization after completed reaction without workup and extraction is called a direct isolation process. This technique can be cost-effective even though the product yield obtained is lower. Special conditions may be needed in this case, and the diastereomers can be classified into two types: diastereomeric salts and covalent diastereomeric compounds, respectively. Diastereomeric salts can, for example, be used in the crystallization of a desired amine from its racemic mixture using a chiral acid. Covalent diastereomers can, on the other hand, be separated by chromatography, but are more difficult to prepare. Another advantage of crystallization is the possibility of combining in situ racemization reactions and diastereomeric formation reactions to get the desired pure compounds. This crystallization-induced resolution technique is still under development because of its requirements for optimized conditions [55, 56].

In this section, diastereomeric crystallization is presented as a driving force – or internal selection pressure – to resolve dynamic diastereomeric systems. The dynamic diastereomeric systems are generated from reversible covalent bond formation, leading to compounds carrying chiral carbon centers under thermodynamic control. The dynamic systems can represent more variety of the possible diastereomer adducts. The selective diastereomers, A_n–B_m, are subsequently chosen from the dynamic system by self-transformation and/or self-preferential crystallization. When the selective product C_{nm} is formed, the ratio of its corresponding diastereomer adducts A_n–B_m in the dynamic system will be decreased. The equilibrium in the dynamic system will force the reproduction of the intermediate until the resolution has reached completion. In the end, only one diastereomeric product C_{nm} is selectively crystallized and easily purified from the solution.

Dynamic nitroaldol systems, based on different benzaldehydes and nitroalkanes, were used to evaluate the concept [57]. To extend the size of the dynamic systems, the formation of diastereomeric nitroaldol adducts carrying two adjacent chiral carbons was one of the objectives. When the dynamic diastereomeric nitroaldol systems were formed, one of the starting materials was rapidly consumed and a white solid was formed during system equilibration. The solid compound was analyzed and further studied in mechanistic explanation and synthetic applications.

The dynamic diastereomeric nitroaldol system (CDS-5A) was formed from equimolar amounts of five benzaldehydes **18D–E** and **37A–C** and nitroethane **38** in chloroform-*d*. In the presence of triethylamine, 20 possible diastereomeric/enantiomeric nitroaldol adducts **39A–E** were formed under thermodynamic control as shown in Scheme 13. The formation of the dynamic systems was followed by ^{1}H-NMR spectroscopy. Figure 8 shows the aldehyde region of compounds **18D–E** and **37A–C** and the α-proton region of their intermediates and products. After 30 min reaction time, benzaldehyde **37C** was the most consumed starting material as shown in Fig. 8a. When the reaction was left longer and monitored for 24 h, 2-cyanobenzaldehyde **37C** was completely consumed and only product **40** was formed and detected in the NMR spectra (Fig. 8b).

After the reaction reached completion, the solid product was filtered off and identified by NMR spectroscopy and X-ray crystallography. By analyzing the data, the unexpected formation of amide product **40** from benzaldehyde **37C** and its corresponding intermediate **39C** was revealed. This represented a new transformation pathway, for which only a few related systems have been reported [58–60]. Scheme 14 displays the proposed mechanistic formation of amide product **40**, supported by density functional theory (DFT) calculations [61]. In the presence of triethylamine as catalyst, the key intermediate **39C** was formed from nitroaldol (Henry) reaction of 2-cyanobenzaldehyde **37C** and nitroethane **38**. The nitroaldol adduct **39C** then underwent internal cyclization (*5-exo-dig*) to give iminolactone **41**,

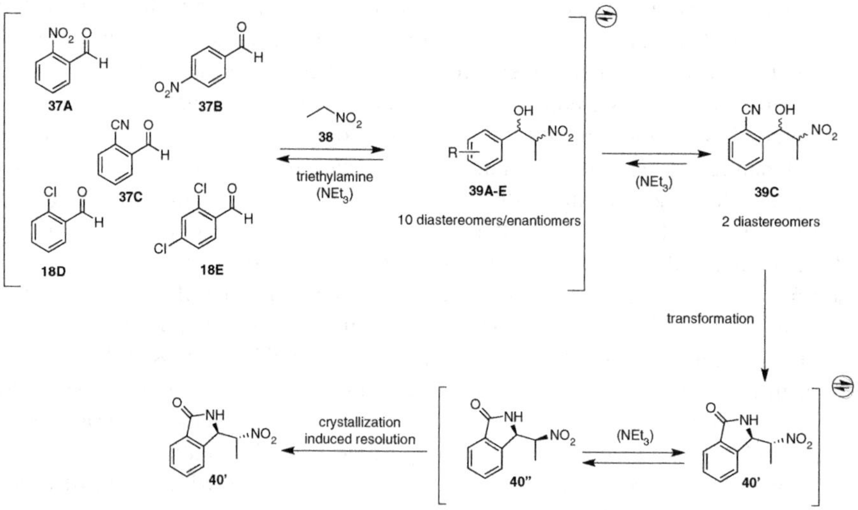

Scheme 13 Transformation–crystallization resolution of dynamic diastereomeric nitroaldol system CDS-5A

Dynamic Systemic Resolution

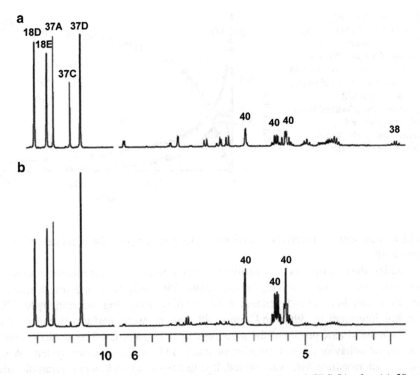

Fig. 8 NMR spectra of transformation-crystallization resolution of CDS-5A after (**a**) 30 min reaction time and (**b**) 24 h reaction time. Modified from [57], Reproduced by permission of the Royal Society of Chemistry

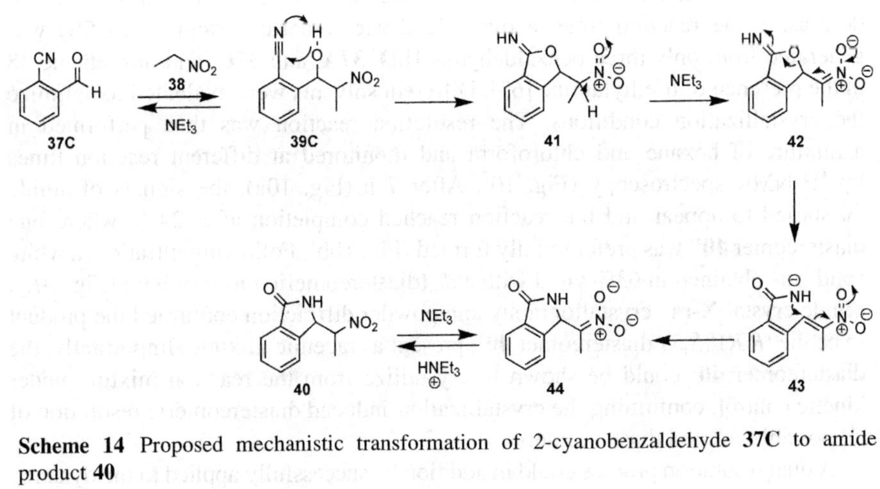

Scheme 14 Proposed mechanistic transformation of 2-cyanobenzaldehyde **37C** to amide product **40**

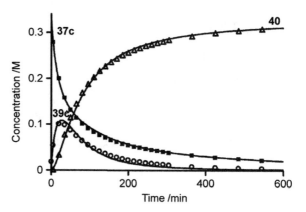

Fig. 9 Kinetic study of transformation from aldehyde 37C (*filled squares*), to nitroaldol adduct 39C (*open circles*) and amide product 40 (*open triangles*), respectively, followed by NMR spectroscopy. Modified from [57], Reproduced by permission of the Royal Society of Chemistry

which was intramolecularly transformed to product 44, the enolate anion of amide 40.

During the dynamic system formation, many possible diastereomeric nitroaldol adducts were formed but only intermediate 39C underwent the transformation reaction. Detailed time-dependent NMR studies, following benzaldehyde 37C, alcohol intermediate 39C, and amide 40 signals, were performed in order to evaluate the kinetic profile of the transformation (Fig. 9) [62]. The results revealed the kinetic behavior of the formation of amide 40 from the dynamic system. When alcohol intermediate 39C was formed, it simultaneously underwent intramolecular cyclization and rearrangement. The dynamic system was at the same time forced to reequilibrate until benzaldehyde 37C was completely consumed.

According to the structure of amide 40, four possible diastereomers can be formed by cyclization and rearrangement. The results from X-ray crystallographic analysis of the solid product from the dynamic system however indicated only one major diastereomer, 40'. In order to investigate the crystallization phenomenon and decreasing the reaction time, a dynamic diastereomeric system (CDS-5B) was generated from only three benzaldehydes 18D, 37A, and 37C, and nitroethane 38 in the presence of triethylamine [63]. Different solvents were evaluated to optimize the crystallization conditions. The resolution reaction was thus performed in a mixture of hexane and chloroform and monitored at different reaction times by ^{1}H-NMR spectroscopy (Fig. 10). After 7 h (Fig. 10a), the signals of amide 40 started to appear and the reaction reached completion after 24 h, where one diastereomer 40" was preferentially formed (Fig. 10b). Following filtration, a white solid was obtained in 63% yield with a *dr* (diastereomeric ratio) of 93:3 (Fig. 10c). Single crystal X-ray crystallography and powder diffraction confirmed the product to be the (*R*,*R*)/(*S*,*S*)-diastereomer 40', present as racemic mixture. Importantly, the diastereomer 40' could be shown to crystallize from the reaction mixture under kinetic control, confirming the crystallization-induced diastereomeric resolution of all possible formed diastereomers.

A dual resolution process could in addition be successfully applied to the dynamic diastereomeric nitroaldol system. By selective transformation of intermediate 39C,

Dynamic Systemic Resolution

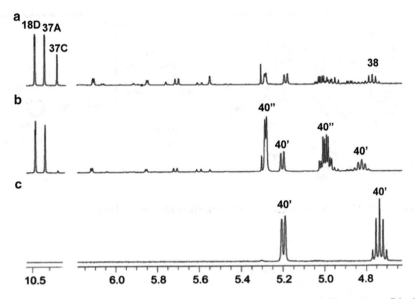

Fig. 10 NMR spectra of transformation-crystallization resolution of CDS-5B at (**a**) $t = 7$ h, (**b**) completed resolution, and (**c**) filtered crystalline precipitate after resolution. Adapted with permission from [63]. Copyright 2008 American Chemical Society

iminolactone **41** was formed and rearranged to yield diastereomeric amides **40**. These amides subsequently underwent crystallization-induced diastereomeric resolution to give only one diastereomer **40'**. The obtained amide products, 3-substituted isoindolinones, are the core structure of several biologically active natural products, such as related alkaloids lennoxamine, magallanesine, nuevamine, and chilenine, as well as the pharmaceutical compounds, including the blood pressure lowering drug chlorthalidone and other candidates [64–73]. A number of synthetic routes to 3-isoindolinones have been described, although only very few addressing stereoselective pathways [74–76].

The dynamic crystallization-induced diastereomeric resolution was further investigated in order to develop practical approaches to obtain one diastereomer by exploiting more diversity of the diastereomeric nitroaldol adducts in the dynamic system [77]. A larger and more diverse dynamic nitroaldol system (CDS-5C) was generated by equimolar amounts of nine different benzaldehydes, nitroethane **38**, and triethylamine was used as catalyst as shown in Scheme 15. A total of 36 nitroaldol diastereomers were formed under thermodynamic control in chloroform-d, and the reaction was followed by ^{1}H-NMR spectroscopy.

Figure 11 shows the NMR spectra of the dynamic system at different time intervals. After addition of triethylamine, the system was monitored by following the aldehyde protons at $t = t_0$ (Fig. 11b) and at 17 h (Fig. 11c). Only the signal of benzaldehyde **45I** was significantly consumed. Interestingly, the signal of the corresponding nitroaldol intermediate **46I** did not increase, but formation of a crystalline solid could instead be observed. After complete consumption of

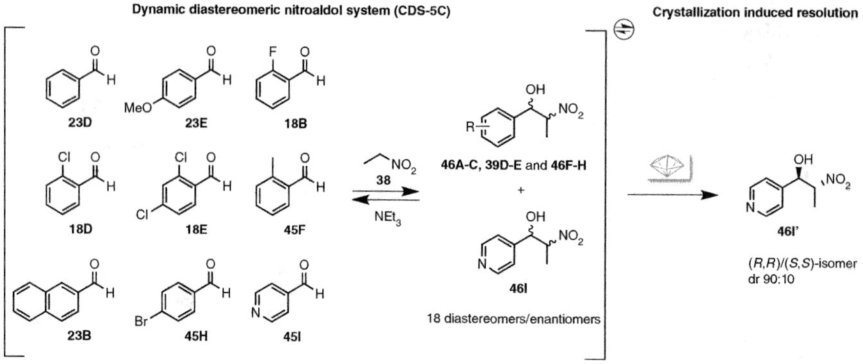

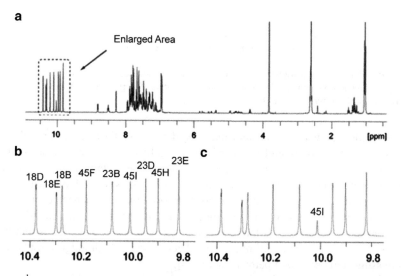

Scheme 15 Crystallization-induced diastereomeric resolution of CDS-5C

Fig. 11 ^{1}H-NMR spectra of CDS-5C: (**a**) full spectrum after 17 h reaction time; aldehyde region after (**b**) addition of base ($t = t_0$); (**c**) 17 h reaction time. Modified from [77], Copyright Wiley-VCH Verlag GmbH & Co. KGaA. Reproduced with permission

aldehyde **45I**, the reaction mixture was filtered, and the solid component was identified as pyridine β-nitroalcohol **46I**. Supported by X-ray crystallography, product **46I** was obtained in a *dr* of 90:10 to be (*R*,*R*)/(*S*,*S*)-isomer **46I'**, present as racemic mixture.

The results from the analysis indicated that nitroaldol intermediate **46I'** was resolved by crystallization from the dynamic system (36 possible diastereomers). This shows the advantage of DSR using diastereoselective crystallization technique to identify the optimal constituent from large and diverse dynamic systems in one-pot processes. After the resolution process, the reaction conditions of the single nitroaldol reaction of 4-pyridinecarboxyaldehyde **45I** was further optimized in

Dynamic Systemic Resolution

order to improve the conversion and diastereomeric ratio of its corresponding product **46I'**. When the concentration of the reaction was increased tenfold, crystalline product **46I'** was obtained in very high yield with a *dr* of 96:4.

4 Conclusion

The above examples demonstrate the DSR concept as a useful approach to generate and interrogate simultaneously complex systems for different applications. A range of reversible reactions, in particular carbon–carbon bond-formation transformations, was used to demonstrate dynamic system formation in both organic and aqueous solutions. By applying selection pressures, the optimal constituents were subsequently selected and amplified from the dynamic system by irreversible processes under kinetic control. The DSR technique can be used not only for identification purposes, but also for evaluation of the specificities of selection pressures in one-pot processes. The nature of the selection pressure applied leads to two fundamentally different classes: external selection pressures, exemplified by enzyme-catalyzed resolution, and internal selection pressures, exemplified by transformation- and/or crystallization-induced resolution. Future endeavors in this area include, for example, the exploration of more complex dynamic systems, multiple resolution schemes, and variable systemic control.

References

1. Gais HJ, Leuchtenberger W (1995) In: Drauz K, Waldmann H (eds) Enzyme catalysis in organic synthesis: a comprehensive handbook. Wiley-VCH, Weinheim, Section B.1, pp 165–363
2. Larsson R, Pei Z, Ramström O (2004) Catalytic self-screening of cholinesterase substrates from a dynamic combinatorial thiolester library. Angew Chem Int Ed 116:3802–3804
3. Larsson R, Ramström O (2006) Dynamic combinatorial thiolester libraries for efficient catalytic self-screening of hydrolase substrates. Eur J Org Chem 285–291
4. Kielly WW, Bradley LB (1954) Glutathinone thiolesterase. J Biol Chem 206:327–333
5. Kaplay SS (1976) Acetylcholinesterase and butyrylcholinesterase of developing human brain. Biol Neonate 28:65–73
6. Järv J (1984) Stereochemical aspects of cholinesterase catalysis. Bioorg Chem 12:259–278
7. Main AR, Soucie WG, Buxton IL, Arinc E (1974) The purification of cholinesterase from horse serum. Biochem J 143:733–744
8. Cho SJ, Garsia ML, Bier J, Tropsha A (1996) Structure-based alignment and comparative molecular field analysis of acetylcholinesterase inhibitors. J Med Chem 39:5064–5071
9. Zhang Y, Angelin M, Larsson R, Albers A, Simons A, Ramström O (2010) Tandem driven dynamic self-inhibition of acetylcholinesterase. Chem Commun 46:8457–8459
10. Cuillou C, Mary A, Renko DZ, Gras E, Thal C (2000) Potent acetylcholinesterase inhibitors: design, synthesis and structure–activity relationships of alkylene linked bis-galanthamine and galanthamine-galanthaminium salts. Bioorg Med Chem Lett 10:637–639

11. Harel M, Schalk I, Ehretsabatier L, Bouet F, Goeldner M, Hirth C, Axelsen PH, Silman I, Sussman JL (1993) Quaternary ligand binding to aromatic residues in the active-site gorge of acetylcholinesterase. Proc Natl Acad Sci USA 90:9031–9035
12. Wong M, Greenblatt HM, Dvir H, Carlier PR, Han YF, Pang YP, Silman I, Sussman JL (2003) Acetylcholinesterase complexed with bivalent ligands related to huperzine A: experimental evidence for species-dependent protein–ligand complementarity. J Am Chem Soc 125: 363–373
13. Ellman CL, Courtney KD, Andres V Jr, Featherstone RM (1961) A new and rapid colorimetric determination of acetylcholinesterase activity. Biochem Pharmacol 7:88–90
14. Breuer M, Ditrich K, Habicher T, Hauer B, Kesseler M, Sturmer R, Zelinski T (2004) Industrial methods for the production of optically active intermediates. Angew Chem Int Ed 43:788–824
15. Vongvilai P, Angelin M, Larsson R, Ramström O (2007) Dynamic combinatorial resolution: direct asymmetric lipase-mediated screening of a dynamic nitroaldol library. Angew Chem Int Ed 119:966–968
16. Brussee J, Roos EC, Van der Gen A (1988) Bioorganic synthesis of optically active cyanohydrins and acyloins. Tetrahedron Lett 29:4485–4488
17. Effenberger F, Gutterer B, Jäger J (1997) Stereoselective synthesis of (1R)- and (1R, 2 S)-1-aryl-2-alkylamino alcohols from (R)-cyanohydrins. Tetrahedron Asymmetry 8:459–467
18. Kordik CP, Reitz AB (1999) Pharmacological treatment of obesity: therapeutic strategies. J Med Chem 42:181–201
19. Krieble VL, Wieland WA (1921) Properties of oxynitrilase. J Am Chem Soc 43:164–175
20. Ono N (2001) The nitro group in organic synthesis. Wiley-VCH, Weinheim
21. Schales O, Graefe AH (1952) Arylnitroalkenes: a new group of antibacterial agents. J Am Chem Soc 74:4486–4490
22. Stelzer U, Effenberger F (1993) Preparation of (S)-2-fluoronitriles. Tetrahedron Asymmetry 4:161–164
23. Seebach D (1990) Organische synthese – wohin? Angew Chem 102:1363–1409
24. Faber K (1992) Biotransformations in organic chemistry. Springer, Berlin
25. Klibanov AM (1990) Asymmetric transformations catalyzed by enzymes in organic solvents. Acc Chem Res 23:114–120
26. Inada Y, Nishimura H, Takahasshi K, Yoshimoto T, Ranjan Saha A, Saito Y (1984) Ester synthesis catalyzed by glycol-modified gly-modified lipase in benzene. Biochem Biophys Res Commun 122:845–850
27. Matos JR, Blair WJ, Wong CH (1987) Lipase catalyzed synthesis of peptides: preparation of a penicillin G precursor and other. Biotechnol Lett 9:233–236
28. Zaks A, Klibanov M (1985) Enzyme-catalyzed processes in organic solvents. Proc Natl Acad Sci USA 82:3192–3196
29. Henry L (1895) C R Hebd Séances Acad Sci 120:1265
30. Luzzio FA (2001) The Henry reaction: recent examples. Tetrahedron Lett 57:915–945
31. Rosini G (1991) In: Trost BM, Fleming I, Heathcock CH (eds) Comprehensive organic synthesis, vol 2. Pergamon, Oxford, p 321
32. Dalko PI, Moisan L (2001) Enantioselective organocatalysis. Angew Chem Int Ed 40: 3726–3748
33. Dalko PI, Moisan L (2004) In the golden age of organocatalysis. Angew Chem Int Ed 43:5138–5175
34. Vongvilai P, Larsson R, Ramström O (2008) Direct asymmetric dynamic kinetic resolution by combined lipase catalysis and nitroaldol (Henry) reaction. Adv Synth Catal 350:448–452
35. Belokon YN, Blacker AJ, Clutterbuck LA, Hogg D, North M, Reeve C (2006) An asymmetric, chemo-enzymatic synthesis of O-acetylcyanohydrins. Eur J Org Chem 4609–4617
36. Greingl H, Schwab H, Fechter M (2008) The synthesis of chiral cyanohydrins by oxynitrilases. Trends Biotechnol 18:252–256

Dynamic Systemic Resolution

37. Holt J, Hanefeld U (2009) Enantioselective enzyme-catalysed synthesis of cyanohydrins. Curr Org Synth 6:15–37
38. Johnson DV, Zabelinskaja-Mackova AA, Griengl H (2000) Oxynitrilases for asymmetric C–C bond formation. Curr Opin Chem Biol 4:103–109
39. North M, Usanov DL, Young C (2008) Lewis acid catalyzed asymmetric cyanohydrin synthesis. Chem Rev 108:5146–5226
40. Seoane G (2000) Enzymatic C-C bond-forming reactions in organic synthesis. Curr Org Chem 4:283–304
41. Inagaki M, Hiratake J, Nishioka T, Oda J (1989) Kinetic resolution of racemic benzaldehyde cyanohydrin via stereoselective acetylation catalyzed by lipase in organic solvent. Bull Inst Chem Res, Kyoto Univ 67:132–135
42. Inagaki M, Hiratake J, Nishioka T, Oda J (1991) Lipase-catalyzed kinetic resolution with in situ racemization: one-pot synthesis of optically active cyanohydrin acetates from aldehydes. J Am Chem Soc 113:9360–9361
43. Inagaki M, Hatanaka A, Mimura M, Hiratake J, Nishioka T, Oda J (1992) One-pot synthesis of optically active cyanohydrin acetates from aldehydes via quinidine-catalyzed transhydrocyanation coupled with lipase-catalyzed kinetic resolution in organic solvent. Bull Chem Soc Jpn 65:111–120
44. Inagaki M, Hiratake J, Nishioka T, Oda J (1992) One-pot synthesis of optically active cyanohydrin acetates from aldehydes via lipase-catalyzed kinetic resolution coupled with in situ formation and racemization of cyanohydrins. J Org Chem 57:5643–5649
45. Xu Q, Geng X, Chen P (2008) Kinetic resolution of cyanohydrins via enantioselective acylation catalyzed by lipase PS-30. Tetrahedron Lett 49:6440–6441
46. Sakai T (2004) Rational strategies for highly enantioselective lipase-catalyzed kinetic resolutions of very bulky chiral compounds: substrate design and high-temperature biocatalysis. Tetrahedron Asymmery 15:2765–2770
47. Sakai T, Hayashi K, Yano F, Takami M, Ino M, Korenaga T, Ema T (2003) Enhancement of the efficiency of the low temperature method for kinetic resolution of primary alcohols by optimizing the organic bridges in porous ceramic-immobilized lipase. Bull Chem Soc Jpn 76:1441–1446
48. Sakai T, Kishimoto T, Tanaka Y, Ema T, Utaka M (1998) Low-temperature method for enhancement of enantioselectivity in the lipase-catalyzed kinetic resolutions of solketal and some chiral alcohols. Tetrahedron Lett 39:7881–7884
49. Leclair J, Vial L, Otto S, Sanders JKM (2005) Expanding diversity in dynamic combinatorial libraries: simultaneous exchange of disulfide and thioester linkages. Chem Commun 1959–1961
50. Vongvilai P, Ramström O (2009) Dynamic asymmetric multicomponent resolution: lipase-mediated amidation of a double dynamic system. J Am Chem Soc 131:14419–14425
51. Gröger H (2003) Catalytic enantioselective Strecker reactions and analogous syntheses. Chem Rev 103:2795–2828
52. Merino P, Marqeslopez E, Tejero T, Herrera R (2009) Organocatalyzed Strecker reactions. Tetrahedron 65:1219–1234
53. Giuseppone N, Schmitt JL, Schwartz E, Lehn JM (2005) Scandium(III) catalysis of transimination reactions. Independent and constitutionally coupled reversible processes. J Am Chem Soc 127:5528–5539
54. Ema T, Jittani M, Furuie K, Utaka M, Sakai T (2002) 5-[4-(1-Hydroxyethyl)phenyl]-10,15,20-triphenylporphyrin as a probe of the transition-state conformation in hydrolase-catalyzed enantioselective transesterifications. J Org Chem 67:2144–2151
55. Anderson NG (2005) Developing processes for crystallization-induced asymmetric transformation. Org Proc Res Dev 9:800–813
56. Ebbers EJ, Ariaans GJA, Houbiers JPM, Bruggink A, Zwanenburg B (1997) Controlled racemization of optically active organic compounds: prospects for asymmetric transformation. ChemInform 53:9417–9476

57. Angelin M, Vongvilai P, Fischer A, Ramström O (2008) Tandem driven dynamic combinatorial resolution via Henry-iminolactone rearrangement. Chem Commun 768–770
58. Sato R, Ohmori M, Kaitani F, Kurosawa A, Senzaki T, Goto T, Saito M (1988) Synthesis of isoindoles. Acid- or base-induced cyclization of 2-cyanobenzaldehyde with alcohols. Bull Chem Soc Jpn 61:2481–2485
59. Sato R, Senzaki T, Goto T, Saito M (1984) Novel synthesis of 3-(N-substituted amino)-1-isoindolenones from 2-cyanobenzaldehyde with amines. Chem Lett 13:1599–1602
60. Song YS, Lee CH, Lee KJ (2003) Application of Baylis–Hillman methodology in a new synthesis of 3-oxo-2,3-dihydro-1 H-isoindoles. J Heterocycl Chem 40:939–941
61. Angelin M, Rahm M, Fischer A, Ramström O (2010) Diastereoselective one-pot tandem synthesis of 3-substituted isoindolinones: a mechanistic investigation. J Org Chem 75:5882–5887
62. Hoops S, Sahle S, Gauges R, Lee C, Pahle J, Simus N, Singhal M, Xu L, Mendes P, Kummer M (2006) COPASI – a COmplex PAthway SImulator. Bioinformatics 22:3067–3074
63. Angelin M, Fischer A, Ramström O (2008) Crystallization-induced secondary selection from a tandem driven dynamic combinatorial resolution process. J Org Chem 73:3593–3595
64. Belliotti TR, Brink WA, Kesten SR, Rubin JR, Wustrow DJ, Zoski KT, Whetzel SZ, Corbin AE, Pugsley TA, Heffner TG, Wise LD (1998) Isoindolinone enantiomers having affinity for the dopamine D4 receptor. Bioorg Med Chem Lett 8:1499–1502
65. Carter BL, Ernst ME, Cohen JD (2004) Hydrochlorothiazide versus chlorthalidone: evidence supporting their interchangeability. Hypertension 43:4–9
66. Fajardo V, Elango V, Cassels BK, Shamma M (1982) Chilenine: an isoindolobenzazepine alkaloid. Tetrahedron Lett 23:39–42
67. Fuwa H, Sasaki M (2007) An efficient method for the synthesis of enol ethers and enecarbamates. Total syntheses of isoindolobenzazepine alkaloids, lennoxamine and chilenine. ChemInform 5:1849–1853
68. Moreau A, Courure A, Deniau E, Grandclaudon P (2005) Construction of the six- and five-membered aza-heterocyclic units of the isoindoloisoquinolone nucleus by parham-type cyclization sequences – total synthesis of Nuevamine. Eur J Org Chem 3437–3442
69. Salvettei A, Ghiadoni L (2006) Thiazide diuretics in the treatment of hypertension: an update. J Am Soc Nephrol 17:S25–S29
70. Stuk TL, Assink BK, Bates RC, Erdman DT Jr, Fedij V, Jennings SM, Lassig JA, Smith RJ, Smith TL (2003) An efficient and cost-effective synthesis of pagoclone. Org Process Res Dev 7:851–855
71. Valencia E, Freyer AJ, Shamma M, Fajardo V (1984) (±)-Nuevamine, an isoindoloisoquinoline alkaloid, and (±)-lennoxamine, an isoindolobenzazepine. Tetrahedron Lett 25:599–602
72. Valencia E, Fajardo V, Freyer AJ, Shamma M (1985) Magallanesine: an isoindolobenzazocine alkaloid. Tetrahedron Lett 26:993–996
73. Yoneda R, Sakamoto Y, Oketo Y, Harusawa S, Kurihara T (1996) An efficient synthesis of magallanesine using [1,2]-Meisenheimer rearrangement and Heck cyclization. Tetrahedron 52:14563–14576
74. Chen MD, He MZ, Zhou X, Huang LQ, Ruan YP, Huang PQ (2005) Studies on the diastereoselective reductive alkylation of (R)-phenylglycinol derived phthalimide: observation of stereoelectronic effects. Tetrahedron 61:1335–1344
75. Deniau E, Couture A, Grandclaudon P (2008) A conceptually new approach to the asymmetric synthesis of 3-aryl and alkyl poly-substituted isoindolinones. Tetrahedron Asymmetry 19:2735–2740
76. Lamblin M, Couture A, Deniau E, Grandclaudon P (2008) Alternative and complementary approaches to the asymmetric synthesis of C3 substituted NH free or N-substituted isoindolin-1-ones. Tetrahedron Asymmetry 19:111–123
77. Angelin M, Vongvilai P, Fischer A, Ramström O (2010) Crystallization driven asymmetric synthesis of pyridine β-nitroalcohols via discovery-oriented self-resolution of a dynamic system. Eur J Org Chem 6315–6318

Top Curr Chem (2012) 322: 87–106
DOI: 10.1007/128_2011_198
© Springer-Verlag Berlin Heidelberg 2011
Published online: 5 July 2011

Dynamic Combinatorial Self-Replicating Systems

Emilie Moulin and Nicolas Giuseppone

Abstract Thanks to their intrinsic network topologies, dynamic combinatorial libraries (DCLs) represent new tools for investigating fundamental aspects related to self-organization and adaptation processes. Very recently the first examples integrating self-replication features within DCLs have pushed even further the idea of implementing dynamic combinatorial chemistry (DCC) towards minimal systems capable of self-construction and/or evolution. Indeed, feedback loop processes – in particular in the form of autocatalytic reactions – are keystones to build dynamic supersystems which could possibly approach the roots of "Darwinian" evolvability at mesoscale. This topic of current interest also shows significant potentialities beyond its fundamental character, because truly smart and autonomous materials for the future will have to respond to changes of their environment by selecting and by exponentially amplifying their fittest constituents.

Keywords Chemical evolution · Dynamic combinatorial chemistry · Self-replication · Systems chemistry

Contents

1 Introduction 88
2 Dynamic Combinatorial Self-Replication of Small Molecules 90
3 Dynamic Combinatorial Self-Replication of Large Self-Assemblies 95
4 Conclusions and Perspectives 102
References 103

E. Moulin and N. Giuseppone (✉)
SAMS research group – icFRC, University of Strasbourg – Institut Charles Sadron, CNRS, 23 rue du Loess, BP 84047, 67034 Strasbourg cedex 2, France
e-mail: giuseppone@unistra.fr

1 Introduction

Dynamic combinatorial chemistry (DCC) is founded on the study and the construction of mixtures of discrete constituents which are produced by reversible molecular or supramolecular associations [1, 2]. The composition of a dynamic combinatorial library (DCL) is thermodynamically driven and, as such, is able to adapt itself to any parameter that – permanently or transiently – modifies its constitution/energy potential surface [3, 4]. Thus, in the presence of various internal or external parameters, the involved equilibria can be displaced toward the amplification of given products through an adaptation process that will occur through an in situ screening of these species. A schematic representation using Emil Fisher lock-and-key metaphora can be used to illustrate these concepts (Fig. 1).

In this figure, the process can be divided into three steps: (1) mixing of the building blocks capable of reversibly interacting with one another by molecular or supramolecular processes, (2) formation of the library made of all the possible combinations of assembly between the building blocks, and (3) interaction of the library members with a target and, subsequently, amplification/selection. Without the receptor, the proportion between the products will reflect kinetic and thermodynamic parameters that are associated with the basic units and with their connections. These proportions will be modified in the presence of the target if one interaction, at least, exists between this latter and, at least, one of the library members. Thus, the differential analysis of the relative proportions between these two experiments, with and without target, permits the determination of the compound that preferentially interacts with the receptor [5]. This approach has been broadly extended during the last few years and has resulted in the rich development of a very attractive domain of modern chemistry because, first of all, it associates combinatorial features together with the spontaneous self-organization of molecules in response to perturbations. A number of scientific fields have been coupled with this seminal concept, such as the synthesis of bioactive products and drug discovery [6–8], the discovery of catalysts [9, 10], the synthesis of coordination complexes [11–13], the design of supramolecular systems for sensing [14] or, more recently, the synthesis of dynamic covalent [15, 16] or non-covalent polymers [17],

Fig. 1 Schematic representation of the basic principles of DCC in the framework of a receptor/substrate interaction

liquid crystals [18, 19], membranes [20, 21], or responsive surfaces [22], which have resulted in new opportunities for implementations in material science [23].

The Darwinian Theory describes the evolution of biological species and links it to their "information gaining" properties. Although the origin of life is quite a controversial issue, it is commonly accepted, based on a statement of Oparin [24], that an evolving system is capable of three fundamental features: (1) to metabolize, (2) to self-replicate, and (3) to undergo mutations. Using this definition, self-replication represents one of the keystones which allow distinguishing between non-living and living systems. At the molecular level, because the complementary base-pairing in DNA transfers genetic information by a replication process, nucleic acids are candidates to have been the first reproducing molecules. In particular, the finding that RNA can act itself as an enzyme-like catalyst [25, 26] – or even as a "self-replicating ribozyme" [27] – supports Kühn [28], Crick [29], Orgel [30], Eigen and Shuster [31], and Joyce [32], in their hypothesis of an original RNA world that have existed before translation. By pushing even further these ideas to a *minimal* chemical framework, the simplest three-step model depicted in Fig. 2 can be used to conceptualize the process of a molecular template directed replication [33].

In this cycle, the main feature is that template **T** can bind to itself by complementary supramolecular binding units, thus autocatalytically increasing its concentration during the course of the reaction. In a first step, **T** reversibly binds its constituents **A** and **B** to yield a termolecular complex **M**. This complex pays the entropy penalty that brings the two precursors into close proximity, which facilitates the formation of a covalent bond between them, whether or not this one is assisted by an enzyme [34]. In a second step, termolecular complex **M** is

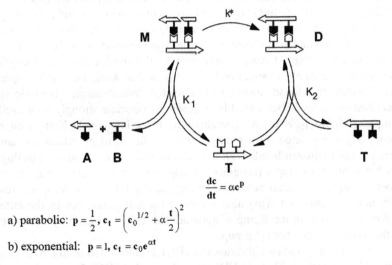

Fig. 2 Schematic representation of a minimal self-replicating system; rate equation for parabolic and exponential growth. (Reproduced from [33])

irreversibly transformed into duplex **D**. Supramolecular dissociation of **D** gives two template molecules, doubling the quantity of **T** for each replication cycle. This minimal self-replicating cycle given in Fig. 2 has been used for the development of non-enzymatic systems based on nucleotidic and non-nucleotidic precursors, such as peptides [35] and other small molecules as templates [36, 37]. Current implementations are supported by advanced analytical techniques and theoretical approaches [38] based either on autocatalytic, cross-catalytic, or collective catalytic pathways for template formation. A major drawback of these systems comes from the stability of duplex **D** which leads to product inhibition [39], namely leading to parabolic growth instead of exponential growth, this latter being the prerequisite for selection in the Darwinian sense [40, 41].

Very recently, the idea of coupling the yet separated fields of DCC and self-replicating systems was envisaged as a topic of interest. Indeed, the over expression of a compound by making copies of itself from a pool of reshuffling constituents, and in a series of competing equilibria, would be of significance for scientists who are interested in various minimal "self-amplifying" systems in their quest to bridge the gap with a possible origin of life.

2 Dynamic Combinatorial Self-Replication of Small Molecules

The first scientific article combining these two concepts and describing an experimental proof that DCC can be linked to self-replicating systems was published by Xu and Giuseppone [42]. In order to design such a system, they envisioned several building blocks capable of (1) reversible covalent associations and (2) displaying – or not displaying – complementary supramolecular units in order to produce – or not produce – a template with self-recognition properties (Fig. 3). The key molecule was chosen as imine Al_1Am_1, synthesized by the condensation of aldehyde Al_1 (derived from a Kemp's imide) and adenosine amine Am_1. This self-complementary dynamic compound, inspired by a related "non-dynamic" isosteric system described by Julius Rebek [43, 44], was able to associate strongly with itself into complex $[Al_1Am_1]_2$. While Al_1 contains a free imide bond on the Kemp recognition group, Al_2 is protected by a methyl group on the nitrogen which prevents the formation of hydrogen bonds with the adenine moiety. Al_3 is also an analogue of Al_1 but without a Kemp's recognition group and with an acetyl group instead, in order to display a similar activation energy as Al_1 for the condensation reaction with amines. Am_1 and Am_2 also present close structures, but in the latter the hydrogen bonds with the Kemp's imide are sterically restricted by the protection of the adenine with a benzyl group.

They then compared two libraries, one (**DCL1**) constituted of Al_2, Al_3, Am_1, and Am_2 (15 mM each at 22 °C in $CDCl_3$), and the other (**DCL2**) constituted of Al_1, Al_2, Al_3, Am_1, and Am_2 (15 mM each at 22 °C in $CDCl_3$). Both libraries were

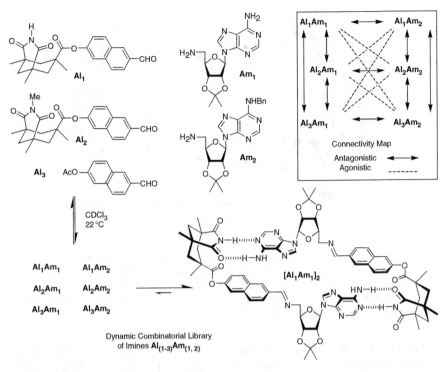

Fig. 3 Representation of the 11 library members obtained by mixing the 3 aldehydes (**Al₁–Al₃**) and the 2 amines (**Am₁**, **Am₂**) in deuterated chloroform. In this combinatorial set, one imine, namely **Al₁Am₁**, can self-assemble through hydrogen bonds and produce homodimer [**Al₁Am₁**]₂. In the frame (*middle right*), the connectivity map describes the antagonistic and the agonistic constitutional relationships between the six imines of the library. (Reproduced from [42])

shown to equilibrate under thermodynamic control, i.e., without local pseudo-minima on the hypersurface of energy/constitution and, by comparing the "non-self-duplicating" **DCL1** and the "self-duplicating" **DCL2**, conclusions were made on the three following features. First, the expression of the self-duplicator **Al₁Am₁** (9.03 mM) is increased by +83% compared to the theoretical statistical distribution (4.94 mM). Second, this amplification of the self-duplicator is realized by the takeover of the resources of its direct competitors, i.e., the imines having antagonistic connectivity with **Al₁Am₁** (namely **Al₁Am₂**, **Al₂Am₁**, and **Al₃Am₁** (3.0 mM each); see frame in Fig. 3). Thus, the amplification of the self-duplicator **Al₁Am₁** compared to its direct competitors reaches a value of +200%. And last, the agonistic connectivity between the self-duplicator **Al₁Am₁** together with **Al₂Am₂** and **Al₃Am₂** leads to a small increase of the last two products compared to the statistical distribution (5.7 mM; +14%).

The authors also studied the kinetic behavior of the 11 member library **DCL2** by plotting the concentration of the constituents as a function of time (Fig. 4). This kinetic evolution is biased compared to the reference "non-self-duplicating" **DCL1**

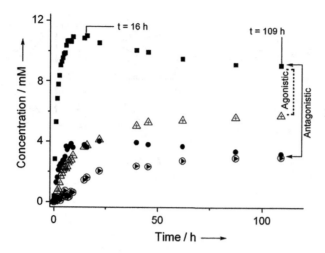

Fig. 4 Evolution as a function of time of the concentration of the 6 products $Al_{(1-3)}Am_{(1,2)}$ in the 11-member dynamic combinatorial library (**DCL2**) described in Fig. 3. Concentrations were determined by ^{1}H NMR and products are represented as follows: (*filled squares*) **Al$_1$Am$_1$**, (*filled circles*) **Al$_1$Am$_2$**, (*open circles*) **Al$_2$Am$_1$**, (*plus signs*) **Al$_2$Am$_2$**, (*filled triangles*) **Al$_3$Am$_1$**, and (*open triangles*) **Al$_3$Am$_2$**. (Reproduced from [42])

involving **Al$_2$**, **Al$_3$**, **Am$_1$**, and **Am$_2$**, in which the four imines display very similar initial rates in the competition experiment ($\approx 4.5 \; 10^{-1}$ mM h^{-1}). In the duplicating **DCL2**, the self-duplicator **Al$_1$Am$_1$** is produced with a V_0 of 59×10^{-1} mM h^{-1}, which is about 60 times faster than the condensation of **Al$_2$Am$_1$**, and **Al$_3$Am$_1$** (0.96×10^{-1} mM h^{-1}), 13 times faster than **Al$_2$Am$_2$** and **Al$_3$Am$_2$** (4.6×10^{-1} mM h^{-1}) and 6 times faster than **Al$_1$Am$_2$** (9.6×10^{-1} mM h^{-1}). These differential rates lead to a maximum (+200%) of amplification (kinetic amplification) of the self-duplicator (10.98 mM) compared to its immediate competitor **Al$_1$Am$_2$** – and of +160% compared to the average concentration of the six imines – at $t = 16$ h.

This example clearly demonstrates that it is possible to self-amplify one product in a DCL, namely the one that can self-complementarily direct its own formation. It was shown that the expression of the components in the library evolves along both kinetic and thermodynamic biases that both lead to the amplification of the best duplicator. Importantly, because of the double reversibility of the system (supramolecular H-bonds and molecular imine condensation), the competition is ruled not only by the differential rates of formation of the components, but also by the possible takeover of the building blocks of the antagonistic competitors, thus leading to the decrease of their absolute concentration. Such a system illustrates the spontaneous screening and selection of the most efficient self-duplicator by the destruction of the entities which are not (or less, such as **Al$_1$Am$_2$**) able to duplicate themselves.

In such self-replicating DCLs, because all the components are linked by thermodynamics, the equilibrium is always reached at infinite times, which necessitates highly differential variations of free energies between library members to reach high amplifications. To improve the replication efficiency at longer time, one might consider using a non-reversible autocatalytic loop that is fed by a DCL that serve as a reservoir of reshuffling starting materials. In principle, the system being out of equilibrium, a full conversion of the replicator can be reached, although the system cannot go back and readapt afterwards.

The first example of this nature was published by Sadownik and Philp [45], who reported an efficient synthetic replicator based on 1,3-dipolar cycloaddition between a nitrone and a maleimide. They constructed their library from two aldehydes, with only one having an amidopyridine recognition site (Fig. 5).

In this exchange pool, the presence of 4-fluoroaniline leads to the formation of imines **1** and **4**, and the presence of 4-fluorophenylhydroxylamine leads to the condensation in imines **2** and **3**. While **1** and **4** do not react with maleimides **5a,b**, nitrones **2** and **3** can react irreversibly with them, leading to a product pool constituted of *cis*- and *trans*-cycloadducts **6** and **7**. Importantly, *trans*-**7b** is able to catalyze its own formation because of the binding of nitrone **3** and maleimide **5b** to the product which forms ternary complex [**3**·**5b**·*trans*-**7b**] and accelerates the non-

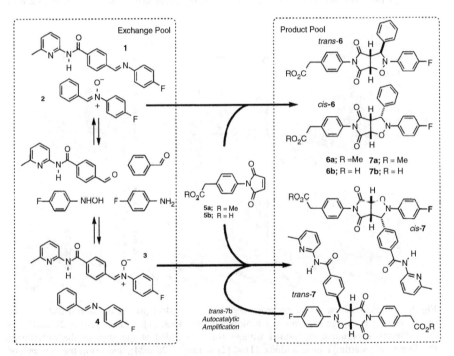

Fig. 5 A pool of compounds containing imines **1** and **3** and nitrones **2** and **4** can exchange freely in CD$_2$Cl$_2$ saturated with *p*-toluenesulfonic acid monohydrate at 273 K. Material can be transferred irreversibly to a pool of products, present in the same solution, that cannot be interconverted or returned to the exchange pool, through reaction of nitrones **2** or **3** with an appropriate maleimide (**5a** or **5b**). When maleimide **5b** is used as the dipolarophile, replicator *trans*-**7b** is formed in the product pool and this species can act as a catalyst for its own formation. (Reproduced from [45])

templated cycloaddition by 100 times with a diastereomeric ratio > 50:1. After checking the full reversibility in the exchange pool, the authors performed a control experiment where maleimide was modified in its methyl ester form (**5a**), thus precluding its binding to aminopyridine recognition sites of **1** and **3** by hydrogen bondings. The concentrations of all species in both exchange and product pools were determined by deconvolution of the appropriate resonances signals in ^{19}F NMR spectra, and are presented after three different reaction times in Fig. 6a.

In the product pool, after 16 h of reaction, the total conversion through cycloaddition is only 21% and the relative distribution of the adducts shows small deviations compared to a perfectly non-selective reaction. This result illustrates that simply coupling an irreversible reaction to the DCL generates no bias in either the exchange or product pools. However, when using maleimide **5b**, that recovers its binding potential with the aminopyridine recognition site of **3**, because it bears a free carboxylic acid pending group, ternary complex [**3**·**5b**·*trans*-**7b**] is generated and the pools' compositions are strongly modified after 16 h (Fig. 6b). Overall conversion is now 48% and **7b** is present with a concentration of 7.7 mM (nearly 80% of the product pool) and with a *trans*-**7b** over *cis*-**7b** ratio of 21:1. In the exchange pool the effect is also demonstrated by comparing the two nitrones **2** and

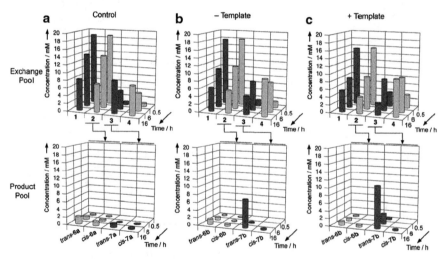

Fig. 6 The composition of the exchange pool and the product pool at 0.5 h, 6 h and 16 h for (**a**) the control exchange experiment (starting concentrations: [**1**] = [**2**] = [**5a**] = 20 mM), (**b**) the exchange experiment performed without *trans*-**7b** added at the start of the experiment ("– Template"; starting concentrations: [**1**] = [**2**] = [**5b**] = 20 mM), and (**c**) in the presence of *trans*-**7b** added at the start of the experiment ("+ Template"; starting concentrations: [**1**] = [**2**] = [**5b**] = 20 mm, [*trans*-**7b**] = 2 mM). All experiments were performed in CD$_2$Cl$_2$ saturated with *p*-toluenesulfonic acid monohydrate at 273 K. *Dark gray* cylinders represent compounds bearing an amidopyridine recognition site, *light gray* cylinders represent compounds bearing no recognition site. Where no cylinder is shown, the concentration of that compound is below the limit of detection (<50 μM). (Reproduced from [45])

3 which are affected differently by the presence or the absence of **5b** (Fig. 6a,b). Finally, the autocatalytic effect was confirmed by the classical "seeding" experiment with **7b** that proved univocally the presence of a replicating loop (Fig. 6c).

3 Dynamic Combinatorial Self-Replication of Large Self-Assemblies

Very recently, in the quest for a strong autocatalytic behavior within an all-equilibrated system, Giuseppone and coworkers extended the DCC approaches towards self-replicating large self-assemblies [46]. Indeed, as stated above, a common problem in self-replication of small molecules is related to the inhibition by the product that avoids reaching an exponential growth of the replicator. Thus, the idea of exploiting the supramolecular instabilities of larger self-assemblies under shearing forces might be of interest to enforce the system to grow and divide without generating a thermodynamically trapped binary complex. As pioneered by Luisi in purely kinetically controlled systems, structures such as micelles [47] or vesicles [48] can provide an efficient matrix with the ability to self-replicate by taking advantage of the growth/division cycles of surfactant self-assemblies (Fig. 7) [49, 50].

Such particular self-replicating systems, sometimes also named as self-reproducing, display a particular case of autocatalysis, namely *autopoiesis*. This concept appeared in the mid-1970s when Maturana and Varela submitted that living systems are essentially characterized by their aptitude to organize continuously the generation of their own components, thus maintaining the very network process that produces them [51]. The minimal criteria defining autopoiesis should verify whether (1) the system has a semipermeable boundary (2) that is produced within the system and (3) that encompasses reactions which regenerate the components of the system. While the definition of life is controversial and is more popularly defined by self-replication according to the prebiotic RNA world view, autopoiesis remains at least a complementary approach and defines a very interesting conceptual framework that encompasses collective properties such as self-assembly, self-organization, and emergence [52, 53].

In their investigations to implement DCC towards self-replicating large self-assemblies [44], Giuseppone and coworkers designed a new type of very simple amphiphilic molecular objects that, because of their reversible connections and through molecular recombination, allow the production of various types (in size and shape) of micellar self-assemblies in water. These objects were constructed by using the reversible connection of a single imine bond between hydrophilic and hydrophobic blocks, thus leading to dynamic amphiphilic block (the so-called dynablocks). The individual condensations of aliphatic, benzylic, aromatic, and hydroxyl amines **1–8** (having various lengths of polyethylene glycol units) with the *p*-substituted benzaldehyde **A** (having a hydrophobic tail of eight carbons) led to the

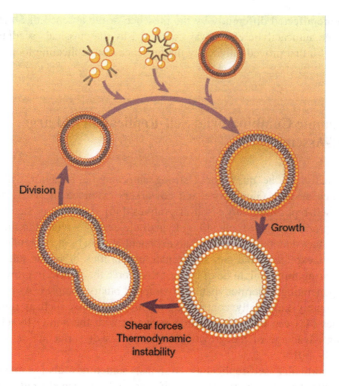

Fig. 7 Mode of vesicle growth and division by increasing concentration of surfactants. (Reproduced from [50])

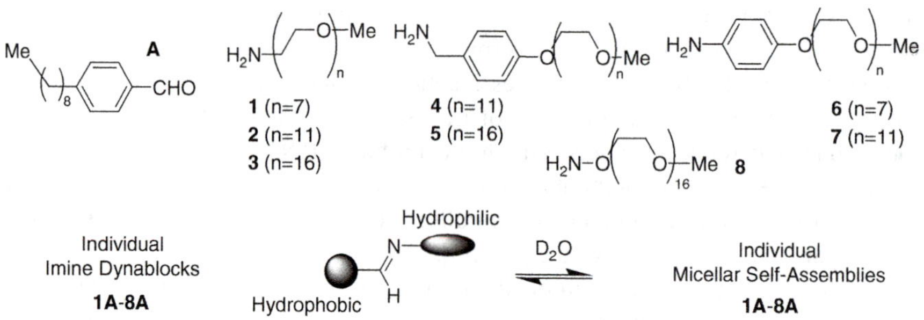

Fig. 8 Chemical structures of dynablocks. The individual reactions of hydrophilic amines **1–8** with hydrophobic aldehyde **A** lead to imines **1A–8A**. (Reproduced from [46])

formation of dynablocks **1A–8A** (having different hydrophilic/hydrophobic radii ($r_{H/h}$) related to surfactant shape parameters) (Fig. 8). The micellar structures were confirmed for **1A**, **4A**, **5A**, **7A**, and **8A** by using the DOSY NMR technique with which the diffusion of the imine self-assemblies in water can be correlated to their

Dynamic Combinatorial Self-Replicating Systems

hydrodynamic radii. The results indicate the formation of micellar self-assemblies having hydrodynamic radii comprised between 5 and 7.1 nm, and which vary with the reverse of $r_{H/h}$. For instance, dynablock **4A** ($r_{H/h} = 2.1$) displays a hydrodynamic radius of 6.9 nm, while **5A** ($r_{H/h} = 2.9$) shows a smaller hydrodynamic radius of 6.5 nm. The evolution from cylindrical to spherical micelles, depending on $r_{H/h}$, was proved by light and neutron scattering experiments for compounds **6A** and **8A**.

In this study, another asset of the DOSY NMR technique was to discriminate, within mixtures, the components displaying different diffusion coefficients [54]. For example, when studying dynablock **7A** in water, the correlation between the resonance signals in the first dimension with the diffusion values in the second dimension led to the following conclusions. The imine dynablock diffuses with a rate of 28 $\mu m^2 s^{-1}$, corresponding to a micellar object of 14.2 nm in diameter which contains in its core all the remaining free aldehyde while the free amine stays outside of the micellar system with a diffusion of 250 $\mu m^2 s^{-1}$. This location of the free aldehyde within the boundary of the structure is the first requirement to access an autopoietic behavior.

The authors then turned to kinetic and thermodynamic studies of the formation of compound **7A** by mixing **7** and **A** directly in deuterated water (equimolar ratio of 50 mM). The condensation of the product, measured by ^1H NMR, revealed a sigmoid concentration-time profile, characteristic of an autocatalytic system. To check the origin of the autocatalytic process, they set up the same experiment but in the presence of increasing initial amounts of preformed micelles **7A** (seeding). The progressive loss of the sigmoid shape clearly indicated that the micelle catalyses its own formation through the condensation of **7** and **A** with a V_{max} of 72×10^{-1} mmol h^{-1} (autocatalytic efficiency $\varepsilon \approx 80$). For low quantities of catalyst (5.1×10^{-4} mmol $< x < 51 \times 10^{-4}$ mmol), the plot of $\log(V_0)$ against log (initial micellar concentration) revealed a linear dependence of rate on catalyst concentration, demonstrating that the rate of the uncatalyzed reaction is comparatively negligible. In addition, the corresponding slope of 0.8 indicated a mainly exponential growth in this system.

Finally, two competition experiments were set up between **1A** and **7A** by mixing **1, 7,** and **A** (50 mM each) and the corresponding concentration-time profiles were determined by ^1H NMR (Fig. 9a,b).

In the first experiment, the beginning of the competition was performed in deuterated acetonitrile showing a clear domination of **1A** ($V_0 = 15$ mM h^{-1}, and $c = 31$ mM at equilibrium) over **7A** ($V_0 = 2.1$ mM h^{-1}, and $c = 16$ mM). Then the acetonitrile was exchanged for D_2O, keeping a constant concentration of 50 mM. The concentration time profile revealed a dramatic evolution of the selectivity in favor of **7A**, because of the formation of the most stable micellar self-assembly. The formation of **7A** ($V_0 = 15$ mM h^{-1}, and $c_{eq} = 32$ mM) was here achieved by the destruction of its competitor **1A** ($V_0 = -15$ mM h^{-1}, and $c_{eq} = 4.5$ mM). The second competition was performed directly in deuterated water from **1, 7,** and **A** (50 mM each). The sigmoid concentration-time profile indicated a highly selective self-replicating process in favor of **7A**, **1A** being formed in quantities

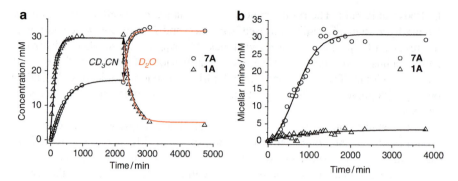

Fig. 9 Molecular selection in coupled equilibria through the self-replication of a specific mesostructure. (**a**) Concentration of imines **1A** and **7A** vs time starting from an equimolar mixture of **1**, **7**, and **A** ($c = 50$ mM each) in CD$_3$CN and, after reaching the thermodynamic equilibrium, by changing the solvent to pure D$_2$O. (**b**) Concentration of imines **1A** and **7A** vs time starting from an equimolar mixture of **1**, **7**, and **A** in D$_2$O ($c = 50$ mM each). (Reproduced from [46])

always inferior to 5 mM and reaching a concentration at the equilibrium similar to that in Fig. 9a. Moreover, the half-time time reaction for this experiment (720 min) was twice the half-time of the formation of neat **7A** (340 min), illustrating the competition between the coupled equilibria. The final sizes, measured by DOSY NMR, of the micellar self-assemblies produced from these two competitions were similar to one another (7.2 nm) but also near equal to the structure of neat **7A** (7.1 nm).

In sum, this work described a general concept for the synergistic relationships which exist at two length scales within a self-replicating DCL which could potentially be enlarged to a number of related systems presenting other structures in their self-assemblies (Fig. 10).

The molecular constituents compete at the subnanometer scale for the reversible production of dynablocks having different $r_{H/h}$. This ratio, together with the stacking effect, mainly determines the formation and the thermodynamic stability of the bounded structures at the tens of nanometer scale. Then, in a first *autocatalytic loop*, these self-assemblies are able to generate their own formation by increasing the rate of the dynablock condensation and they entirely fulfil the required characteristics of a minimal autopoietic process. Moreover, in a second *thermodynamic loop*, the self-assemblies discriminate between the incorporated dynablocks and thus favor the preferential synthesis of their own blocks. Such a system, combining cooperative processes at different length scales in networks of equilibria and displaying autocatalysis within DCLs, is of interest for the understanding of the emergence of self-organizing collective properties but also for the design of responsive systems.

As illustrated in the previous system, the importance of the growth/division processes is crucial because it avoids product inhibition that is a major drawback often encountered in the chemistry of small replicators. This phenomenon was nicely exploited further by Otto and coworkers who have recently developed a mechanosensitive self-replicating system by modulating the shearing forces applied

Dynamic Combinatorial Self-Replicating Systems

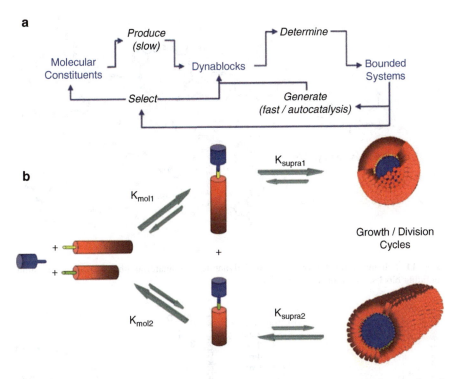

Fig. 10 General statement. (**a**) Synergistic constitutional relationships observed at two length scales within a model minimal self-replicating DCL described in (**b**). For the sake of clarity, the growth/division cycles of micellar structures are not represented. (Reproduced from [46])

on supramolecular assemblies [55]. In this work, they reported the results obtained from a small dynamic combinatorial library made of self-binding building blocks **1** (Fig. 11).

An alternating peptide sequence made of hydrophobic (leucine) and hydrophilic (lysine) amino acids, and which is known to self-assemble in β-sheet structures, was linked to a phenyl dithiol organizing unit, capable of reacting with itself by oxidative disulfide formation in the presence of oxygen (**1**). Such a reaction being reversible at neutral pH, different macrocyclic structures can in principle be obtained with variations in their sizes. For the purpose of further combinatorial studies, two other compounds were synthesized, which do not feature a peptide chain (**2**), or which present a modified peptide chain (**3**). In a control experiment using **2** as starting material, in a borate buffer at pH 8, the final composition analyzed by HPLC revealed the major presence of cyclic trimer and tetramer with a stable ratio over a period of 15 days. In contrast, if the same experiment is performed from **1**, and although the same behavior is observed during the first 4 days, the composition suddenly and dramatically changes from the small cyclic units mostly towards the larger heptamer. In addition, this change was shown to depend on the agitation, as without mechanical stirring the mixture continues to

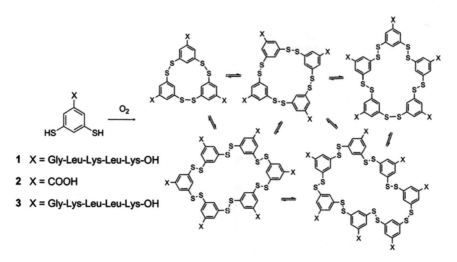

Fig. 11 Schematic illustration of a small dynamic combinatorial library made from dithiol building blocks. (Reproduced from [55])

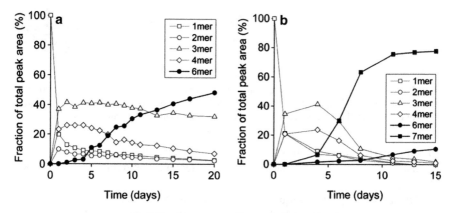

Fig. 12 Evolution of the product distribution with time upon agitating a solution of **1** (3.8 mM) by (**a**) shaking (500 rpm) or (**b**) stirring (1,200 rpm). (Reproduced from [55])

express mostly cyclic trimer and tetramer over a period of 7.5 months. Even more remarkably, the replacement of the mechanical stirring by shaking leads to major formation of the hexamer. In both cases the sigmoid growth of the formation of hexamer and heptamer suggested a self-replicating process (Fig. 12a,b).

To probe the self-replicating properties of the system, seeding experiments were performed by adding either preformed hexamer or heptamer to solutions of **1** which were then respectively shaken or stirred for several days. The results clearly indicated the induced formation of the corresponding macrocyles. In addition, the solutions of both hexamer and heptamer were characterized by cryo-TEM revealing the presence of 1–2 μm length fibers with diameters in the range of 4.7–4.9 nm

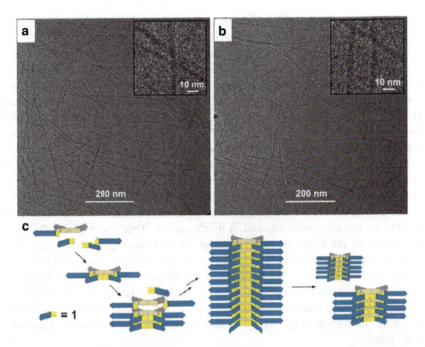

Fig. 13 Cryo-TEM images of (**a**) 1_6, (**b**) 1_7 (*inset*: magnified view of representative fibers), and (**c**) schematic representation of the proposed formation of fibers of 1_6. The benzenedithiol core of building block **1** is shown in *yellow* and the peptide chain in *blue*. Stacks of hexamer are held together by the assembly of the peptide chains into elongated cross-β sheets. (Reproduced from [55])

[corresponding to a single macrocycle with an extended conformation of the lateral peptide chains (Fig. 13a,b)]. Circular dichroism confirmed the presence of the β-sheet that held the supramolecular structures for the hexamer and the heptamer, a result consistent with C=O absorbance bands by FTIR spectroscopy as well as with fluorescence spectroscopy using thioflavine T that is commonly used to characterize amyloid fibers. Interestingly, compound **3** with its modified peptide chain produced a mixture dominated by the cyclic trimer and circular dichroism revealed the formation of a random coil peptide rather than a β-sheet structure.

Based on these spectroscopic evidences, Otto and coworkers proposed a model for the self-organization process, and for its mechanosensitivity, which is depicted in Fig. 13c. The mechanism starts with a nucleation phase showing a linear growth of the fibers occurring at their two ends. When reaching a critical size, the fibers break and increase the number of "active" ends, thus explaining the sigmoid growth of the time-concentration profile. To confirm this assumption, the length distribution as a function of time was statistically determined using cryo-TEM images, indicating an increased degree of fragmentation upon agitation. While hexamer and heptamer were similarly fragmented by stirring, the fragmentation was superior for the hexamer when shaken. Here the authors proposed that, although shaking leads to a lower stress than produced by stirring, the larger fraction of the sample touched in the first conditions improves in fine the fragmentation efficiency.

Finally, competition experiments were set up between hexamer and heptamer for building block **1** under various conditions of agitation, with the particularity to skip the nucleation phase by adding equal amounts of hexamer and heptamer to a solution containing trimer and tetramer. It was shown that the heptamer is able to grow even in the absence of agitation, while the hexamer is not in the same conditions. Heptamer also grows faster than the hexamer under stirring, which can only compete – with a same rate of formation – under shaking. Thus, the result of this competition is linked to the ability of the fibers to break, and to generate a higher number of "active" ends with nonlinear growth. One might thus assume that the "breakability" of the fibers is related to the number of β-sheets involved in the fibers, which makes the heptamer more difficult to break in less efficient shaking conditions, but also more efficient in its elongation. This also explains the greater propensity of the hexamer to grow faster upon shaking despite a lower rate of elongation of the individual fibers. The strong specificity in the elongation of the fibers – for their own constituting cyclic disulfides – shifts the system quite far away from the statistical distribution, and the authors here consider amplification as a result of a kinetic trap. One might prefer to call it a low thermodynamic minimum as all the individual steps are by essence reversible in the present experimental conditions.

4 Conclusions and Perspectives

In a couple of years, a series of proofs of principle have shown that the blossoming domain of DCC can be coupled to the field of self-replicating systems, and some reviews already report on that subject [56, 57]. These advances might have a significant impact in systems chemistry [58] because they bring new tools for studying fundamental problems related to self-organization processes and emergence of order from chaos. The selection process in such DCLs combines kinetic and thermodynamic rules, which do not simply rely on the amplification of the fittest components, but also on the depletion of their direct competitors, a statement that can be seen as a sort of "molecular ecology." Moreover, when fully reversible, these amplifications are transient and can in principle be reversed when external conditions modify the best replicator (in a putative event that could be considered as a mutation). Even more interestingly, the communication between different length scales, such as the molecular and the mesoscopic ones, leads to the superimposition of constitutional and structural networks, which both interfere with one another. Indeed, this approach is not limited to small molecules, but also to larger self-assemblies that constitute the functional entities in the chemistry of materials. In particular, the design of responsive materials, capable of "smartly" interacting with their environment, will play an increasing role in the future of nanosciences, and

Dynamic combinatorial self-replicating systems might become crucial to develop the next generation of responsive autonomous devices.

Acknowledgments We wish to thank the European Research Council (ERC StG n° 257099), the Agence Nationale de la Recherche (ANR-09-BLAN-034-02), the CNRS, the icFRC, and the University of Strasbourg. We also acknowledge ESF-COST action on Systems Chemistry (CM0703).

References

1. Lehn J-M (1999) Dynamic combinatorial chemistry and virtual combinatorial libraries. Chem Eur J 5:2455–2463
2. Corbett PT, Leclaire J, Vial L, West KR, Wietor J-L, Sanders JKM, Otto S (2006) Dynamic combinatorial chemistry. Chem Rev 106:3652–3711
3. Lehn J-M (2002) Toward self-organization and complex matter. Science 295:2400–2403
4. Lehn J-M (2002) Toward complex matter: supramolecular chemistry and self-organization. Proc Nat Acad Sci USA 99:4763–4768
5. Ludlow RF, Otto S (2010) The impact of the size of dynamic combinatorial libraries on the detectability of molecular recognition induced amplification. J Am Chem Soc 132:5984–5986
6. Ramström O, Lehn J-M (2002) Drug discovery by dynamic combinatorial libraries. Nat Rev Drug Discov 1:26–36
7. Gareiss PC, Sobczak K, McNaughton BR, Palde PB, Thornton CA, Miller BL (2008) Dynamic combinatorial selection of molecules capable of inhibiting the (CUG) repeat RNA-MBNL1 interaction in vitro: discovery of lead compounds targeting myotonic dystrophy (DM1). J Am Chem Soc 130:16254–16261
8. Herrmann A (2009) Dynamic mixtures and combinatorial libraries: imines as probes for molecular evolution at the interface between chemistry and biology. Org Biomol Chem 7:3195–3204
9. Brisig B, Sanders JKM, Otto S (2003) Selection and amplification of a catalyst from a dynamic combinatorial library. Angew Chem Int Ed 42:1270–1273
10. Gasparini G, Dal Molin M, Prins LJ (2010) Dynamic approaches towards catalyst discovery. Eur J Org Chem 2429–2440
11. Hasenknopf B, Lehn J-M, Kneisel BO, Baum G, Fenske D (1996) Self-assembly of a circular double helicate. Angew Chem Int Ed 35:1838–1840
12. Severin K (2004) The advantage of being virtual Target-induced adaptation and selection in dynamic combinatorial libraries. Chem Eur J 10:2565–2580
13. Legrand Y-M, Van Der Lee A, Barboiu M (2007) Self-optimizing charge-transfer energy phenomena in metallosupramolecular complexes by dynamic constitutional self-sorting. Inorg Chem 46:9540–9547
14. Buryak A, Severin K (2005) Dynamic combinatorial libraries of dye complexes as sensors. Angew Chem Int Ed 44:7935–7938
15. Giuseppone N, Lehn J-M (2004) Constitutional dynamic self-sensing in a zinc(II) polyimino-fluorenes system. J Am Chem Soc 126:11448–11449
16. Giuseppone N, Fuks G, Lehn J-M (2006) Tunable fluorene-based dynamers through constitutional dynamic chemistry. Chem Eur J 12:1723–1735
17. Sreenivasachary N, Lehn J-M (2005) Gelation-driven component selection in the generation of constitutional dynamic hydrogels based on guanine-quartet formation. Proc Nat Acad Sci USA 102:5938–5943
18. Giuseppone N, Lehn J-M (2006) Electric-field modulation of component exchange in constitutional dynamic liquid crystals. Angew Chem Int Ed 45:4619–4624

19. Herrmann A, Giuseppone N, Lehn J-M (2009) Electric-field triggered controlled release of bioacive volatiles from imine-based liquid crystalline phases. Chem Eur J 15:117–124
20. Cazacu A, Legrand Y-M, Pasc A, Nasr G, Van Der Lee A, Mahon E, Barboiu M (2009) Dynamic hybrid materials for constitutional self-instructed membranes. Proc Nat Acad Sci USA 106:8117–8122
21. Barboiu M (2010) Dynamic interactive systems: dynamic selection in hybrid organic-inorganic constitutional networks. Chem Commun 46:7466–7476
22. Tauk L, Schröder AP, Decher G, Giuseppone N (2009) Hierarchical functional gradients of pH-responsive self-assembled monolayers using dynamic covalent chemistry on surfaces. Nat Chem 1:649–656
23. Lehn J-M (2007) From supramolecular chemistry towards constitutional dynamic chemistry and adaptive chemistry. Chem Soc Rev 36:151–160
24. Oparin A (1924) The origin of life. Pabochii, Moscow
25. Sharp PA (1985) On the origin of RNA splicing and introns. Cell 42:397–400
26. Cech TR (1986) A model for the RNA-catalyzed replication of RNA. Proc Nat Acad Sci USA 83:4360–4363
27. Paul N, Joyce GF (2002) A self-replicating ligase ribozyme. Proc Nat Acad Sci USA 99:12733–12740
28. Kuhn H, Waser J (1981) Molecular self-organization and origin of life. Angew Chem 93:495–515
29. Crick FH (1968) The origin of the genetic code. J Mol Biol 38:367–379
30. Orgel LE (1968) Evolution for the genetic apparatus. J Mol Biol 38:381–393
31. Eingen M, Schuster P (1979) The hypercycle: a principle of natural self-organization. Springer, Berlin
32. Joyce GF (1989) RNA evolution and the origins of life. Nature 338:217–224
33. Patzke V, von Kiedrowski G (2007) Self replicating systems. Arkivoc v 293–310
34. Von Kiedrowski G (1986) A self-replicating hexadeoxynucleotide. Angew Chem 98:932–934
35. Lee DH, Granja JR, Martinez JA, Severin K, Ghadiri MR (1996) A self-replicating peptide. Nature 382:525–528
36. Luther A, Brandsch R, von Kiedrowski G (1998) Surface-promoted replication and exponential amplification of DNA analogs. Nature 396:245–248
37. Vidonne A, Philp D (2009) Making molecules make themselves – the chemistry of artificial replicators. Eur J Org Chem 593–610
38. Dieckmann A, Beniken S, Lorenz CD, Doltsinis NL, Von Kiedrowski G (2011) Elucidating the origin of diastereoselectivity in a self-replicating system: selfishness versus altruism. Chem Eur J 17:468–480
39. Sievers D, von Kiedrowski G (1994) Self-replication of complementary nucleotide-based oligomers. Nature 369:221–224
40. Szathmary E, Gladkih I (1989) Sub-exponential growth and coexistence of non-enzymatically replicating templates. J Theor Biol 138:55–58
41. Issac R, Chmielewski J (2002) Approaching exponential growth with a self-replicating peptide. J Am Chem Soc 124:6808–6809
42. Xu S, Giuseppone N (2008) Self-duplicating amplification in a dynamic combinatorial library. J Am Chem Soc 130:1826–1827
43. Tjivikua T, Ballester P, Rebek J Jr (1990) Self-replicating system. J Am Chem Soc 112:1249–1250
44. Reinhoudt DN, Rudkevich DM, de Jong F (1996) Kinetic analysis of the Rebek self-replicating system: is there a controversy? J Am Chem Soc 118:6880–6889
45. Sadownik JW, Philp D (2008) A simple synthetic replicator amplifies itself from a dynamic reagent pool. Angew Chem Int Ed 47:9965–9970
46. Nguyen R, Allouche L, Buhler E, Giuseppone N (2009) Dynamic combinatorial evolution within self-replicating supramolecular assemblies. Angew Chem Int Ed 48:1093–1096

47. Bachmann PA, Walde P, Luisi PL, Lang J (1991) Self-replicating micelles: aqueous micelles and enzymatically driven reactions in reverse micelles. J Am Chem Soc 113:8204–8209
48. Zepik HH, Blöchliger E, Luisi PL (2001) A chemical model for homeostasis. Angew Chem Int Ed 40:199–202
49. Bachmann PA, Luisi PL, Lang J (1992) Autocatalytic self-replicating micelles as models for prebiotic structures. Nature 357:57–59
50. Szostak JW, Bartel DP, Luisi PL (2001) Synthesizing life. Nature 409:387–390
51. Varela FG, Maturana HR, Uribe R (1974) Autopoiesis: the organization of living systems, its characterization and a model. Biosystems 5:187–196
52. Menger FM (1991) Aggregate organic molecules with collective properties. Angew Chem Int Ed Engl 30:1086–1099
53. Mann S (2008) Life as a nanoscale phenomenon. Angew Chem Int Ed 47:5306–5320
54. Giuseppone N, Schmitt J-L, Allouche L, Lehn J-M (2008) DOSY NMR experiments as a tool for the analysis of constitutional and motional dynamic processes: implementation for the driven evolution of dynamic combinatorial libraries of helical strands. Angew Chem Int Ed 47:2235–2239
55. Carnall JMA, Waudby CA, Belenguer AM, Stuart MCA, Peyralans JJ-P, Otto S (2010) Mechanosensitive self-replication driven by self-organization. Science 327:1502–1506
56. Peyralans JJ-P, Otto S (2009) Recent highlights in systems chemistry. Curr Opin Chem Biol 13:705–713
57. Del Amo V, Philp D (2010) Integrating replication-based selection strategies in dynamic covalent systems. Chem Eur J 16:13304–13318
58. Hunt RAR, Otto S (2011) Dynamic combinatorial libraries: new opportunities in systems chemistry. Chem Commun 47:847–858

Top Curr Chem (2012) 322: 107–138
DOI: 10.1007/128_2011_200
© Springer-Verlag Berlin Heidelberg 2011
Published online: 19 July 2011

DCC in the Development of Nucleic Acid Targeted and Nucleic Acid Inspired Structures

Benjamin L. Miller

Abstract Nucleic acids were one of the first biological targets explored with DCC, and research into the application has continued to yield novel and useful structures for sequence- and structure-selective recognition of oligonucleotides. This chapter reviews major developments in DNA- and RNA-targeted DCC, including methods under development for the conversion of DCC-derived lead compounds into probe molecules suitable for studies in vitro and in vivo. Innovative applications of DCC for the discovery of new materials based on nucleic acids and new methods for the modification of nucleic acid structure and function are also discussed.

Keywords DNA · G-quadruplex · Nucleic acid recognition · Resin-bound DCC · RNA · Transition metal complex

Contents

1 Introduction: DNA and RNA Recognition: The Problem and the Promise 108
2 Origins of DNA-Targeted DCC: Initial Miller Group Experiments 109
3 Sequence Versus Structure Targeting: The Example
   of G-Quadruplexes 111
4 Developing Methods for Working with Large Libraries: The RBDCC Concept 116
5 Using DCC to Address the "Grand Challenge" of RNA Recognition 119
   5.1 Initial Experiments: Cu-Salicylamides 119
   5.2 Expanding the Library: Identifying Sequence-Selective RNA Binders with
      RBDCC 120
6 Recognition of Nucleosides and Sequestration of Nucleic-Acid Binders 125
7 Beyond Nature: Towards Novel Materials and Nucleic Acids with Altered Properties . 128
   7.1 Generation of Novel Materials with DCC 128
   7.2 Dynamic Decoration of Nucleic Acids 130
   7.3 Nucleic Acid Cross-Linking 132
   7.4 Dynamic SNP Analysis 132
8 Conclusions 133
References 134

B.L. Miller (✉)
Department of Dermatology, University of Rochester Medical Center, Rochester, NY 14642, USA
e-mail: Benjamin_miller@URMC.Rochester.edu

1 Introduction: DNA and RNA Recognition: The Problem and the Promise

Recognition of nucleic acids in a sequence-selective manner is central to a host of cellular processes, ranging from replication to recombination to regulation of gene expression. In addition to the proteins and other biopolymers engaged in such recognition in the normal life of a cell, Nature has evolved a plethora of small molecule structures able to bind nucleic acids and alter biology. Chemists have found inspiration in these nucleic acid-binding natural products, both as exquisite examples of Nature's structural creativity and as gateways to understanding recognition processes. Selected examples of the rich diversity of nucleic acid-binding natural products are shown in Fig. 1. These range from RNA-binding aminoglycosides (kanamycin A) and DNA-binding polypyrroles (distamycin) to more exotic structures such as the bisintercalating peptide macrocycles (triostin A, TANDEM) and DNA-cleaving enediyne antibiotics (dynemicin A).

Despite the breadth of structural types found in oligonucleotide targeted natural products, however, the complete set of "interesting" target sequences (both from the perspective of basic biology and medicinal applications) far outstrips the number of known high-affinity, high-selectivity binders. Therefore, chemists have long sought new ways to facilitate the design and discovery of novel active oligonucleotide-binding compounds. From the earliest days of the development of Dynamic Combinatorial Chemistry (DCC), the inherent modularity of DNA and RNA singled these out as particularly attractive targets for application of the technique. This chapter will review the various ways in which researchers have applied DCC to the discovery of novel DNA- and RNA-binding compounds. As we shall also discuss, DCC and DCC-like processes have also been a mainstay of

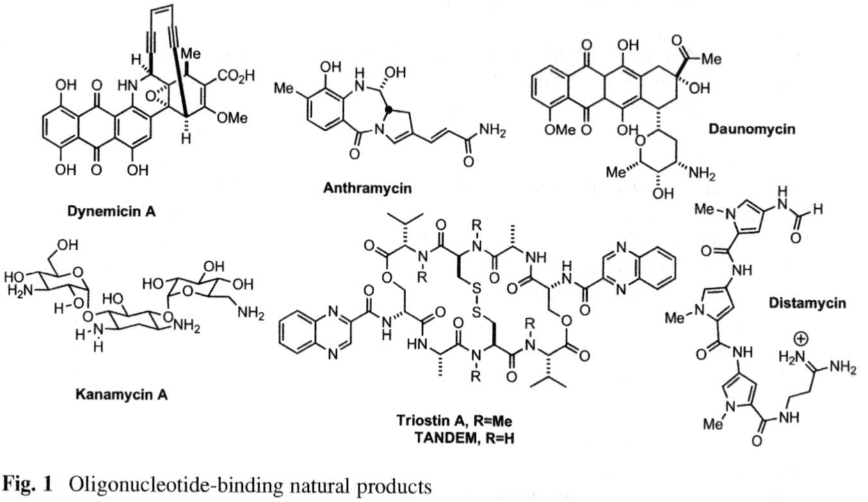

Fig. 1 Oligonucleotide-binding natural products

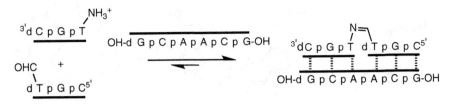

Fig. 2 DNA-templated synthesis of imine-linked oligonucleotides

studies related to understanding the origins of life, and the generation of novel nucleic acid-templated materials and DNA-based nanostructures.

The use of DNA as a template for the synthesis of non-natural materials predates the promulgation of DCC as such as a concept. For example, work carried out in the early 1990s by Wu and Orgel examined the ability of DNA hairpins incorporating an unpaired segment at the 5′ end to template the addition of nucleotides via a nonenzymatic, irreversible process [1]. In a contemporaneous report [2], and in further studies published in 1997 [3], Lynn and colleagues demonstrated that a *reversible* covalent reaction could also be employed to stitch together oligonucleotides assembled on a DNA template. The basic premise of the process is shown in Fig. 2. Incubation of 5′-amino and 3′-aldehyde functionalized DNA trimers with a hexanucleotide results in duplex formation, followed by proximity-enhanced production of the imine. Reduction of the product imine-linked oligonucleotide with potassium cyanoborohydride was used as a method for obtaining turnover of the template and amplification of the amine-linked product. To allow for analysis of the kinetics of the reaction (including templating, covalent linkage, and turnover) by NMR, the authors employed ^{15}N labeled oligonucleotides in their 1997 work. This allowed them to observe that incorporation of the reducing step transformed what was previously a third-order kinetic process into pseudo-first order. Somewhat surprisingly, the authors also found that the reaction was not product limited (although this would presumably change as the reaction progressed to completion). The authors noted prophetically with regard to the development of nucleic acid-targeted DCC in their 1992 communication that "the introduction of a reversible step therefore holds promise for enhancing thermodynamic control in template-directed synthesis."

2 Origins of DNA-Targeted DCC: Initial Miller Group Experiments

In 1995, the author of this chapter began the process of interviewing for academic positions, with a goal of focusing on nucleic acid targeted DCC (although it was not called that at the time) as a viable area around which to build a new research group. Our strategy was informed and inspired by several literature studies, including the

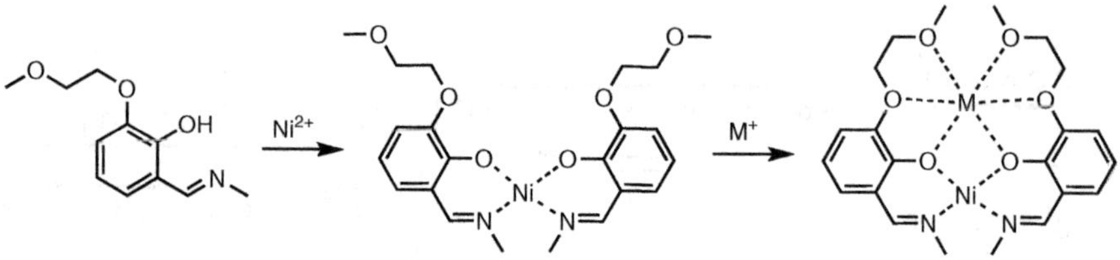

Fig. 3 A self-assembling ionophore. Formation of a Ni(II)–salicylaldimine complex pre-organizes a crown ether-like cation binding site

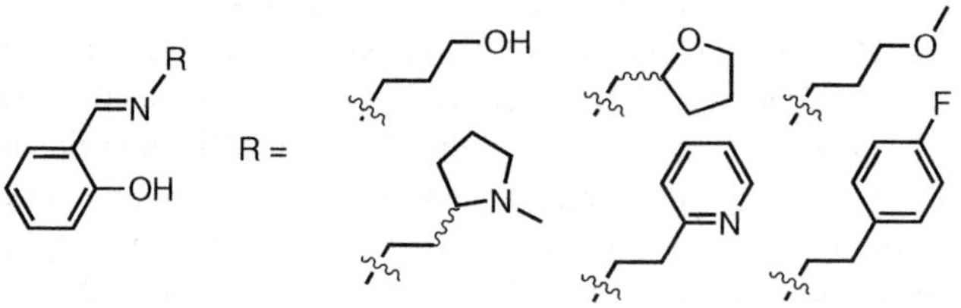

Fig. 4 Salicylaldimines used in the first DNA-targeted DCL

Orgel and Lynn work described above, as well as a plethora of innovative approaches to self-assembled structures pioneered by the Lehn group and others. Of particular importance in our thinking was a 1989 study by the Schepartz group, in which a self assembled nickel(II)salicylaldimine complex was used to form a binding site for alkali metal ions (Fig. 3) [4]. This "self-assembled ionophore" (as the paper was titled) led this author to begin considering (1) whether such structures could be used to bind more complex targets – specifically, nucleic acids; (2) if one could generate combinatorial libraries of such complexes, and (3) if the use of *labile* metal complexes in such a library would provide selection and amplification processes in accord with Le Chatelier's principle.

Thus, in the approach that took shape, a resin immobilized DNA duplex would be used as the target against which a mixture of in situ assembled metal salicylaldimine complexes would be allowed to evolve. In the first iteration of this approach, six pre formed salicylaldimines (Fig. 4) were dissolved in a buffered zinc chloride solution, and incubated with a commercial poly dT resin to which poly dA had been allowed to hybridize [5]. HPLC analysis of the material eluted from the resin following hydrolysis and derivatization with 2-naphthoyl chloride revealed that the salicylaldimine derived from 1-methyl-2-pyrrolidineethanamine was that most depleted from the library, suggesting that it (and by presumption complexes formed from it) had the highest affinity for the DNA. As this was the only member of the library carrying a positive charge at neutral pH, it was not a surprising result, but we were nonetheless pleased that the DCC experiment provided results consistent with chemical intuition. Titration experiments confirmed that the salicylaldimine formed from this amine bound oligo d[A·T] DNA with an approximately 1 μM dissociation constant (K_D) in the presence of zinc.

Of course, imines are themselves labile at neutral pH, and therefore in principle one should be able to obtain an identical result by simply mixing salicylaldehyde, amines, and zinc in buffer in the presence of DNA. In a second report in 1999, we examined exactly this possibility [6]. Once again as expected, the pyrrolidine was the amine most retained by the DNA resin in the presence of salicylaldehyde and zinc. An inherent complication of this strategy, however, is that the labile library constituents made analysis and understanding of the results of library selection difficult. Furthermore, while useful as a proof of concept study, there is no clear pathway from the metal complexes identified in these studies to compounds suitable for use in structural studies or in vivo. As we will discuss later, consideration of such issues has been an important aspect of the design of more recent libraries targeting RNA recognition.

3 Sequence Versus Structure Targeting: The Example of G-Quadruplexes

Much like the search for protein-binding compounds requires consideration of tertiary structure in addition to peptide sequence, thinking about selective nucleic acid binding solely in the context of primary sequence is often an oversimplification. In the context of DNA recognition, a particularly important example of this is the case of G-quadruplexes. As their name implies, these structures consist of stacked tetrads of guanosine bases, typically ordered around a monovalent cation. One representative structure, a quadruplex derived from human telomeric DNA, is shown in Fig. 5; many variants of this motif formed by either parallel or antiparallel DNA strands have been observed.

Because of their importance in gene regulation, [8] and in particular given their known involvement in the expression of oncogenes including c-kit [9] and c-myc [10], G-quadruplex DNA has been of broad interest to the molecular recognition community as a potential drug target [11]. This includes DCC researchers, who have demonstrated several successful approaches to G-quadruplex recognition. In general, these approaches build on a strategy of employing both planar aromatic groups, given their potential ability to pi-stack with the guanosine bases making up the G-tetrad, and charged groups able to interact with the anionic sugar-phosphate backbone.

The Balasubramanian group has been particularly active in this area. Their first proof of concept [12], appearing in 2004, involved a small disulfide library formed from a peptide (1) previously shown to interact with G-quadruplexes, and an acridine derivative (2). Using glutathione as a catalyst for disulfide exchange, the library was allowed to evolve in the presence of the human telomeric G-quadruplex sequence 5′-biotin(GTTAGG)$_5$. The 5′ biotin group was incorporated such that after halting disulfide exchange (facile at pH > 7.1 [13]) by acidification of the solution, compounds bound to the DNA were readily separated by incubation of the mixture with streptavidin-coated beads. As the authors had anticipated,

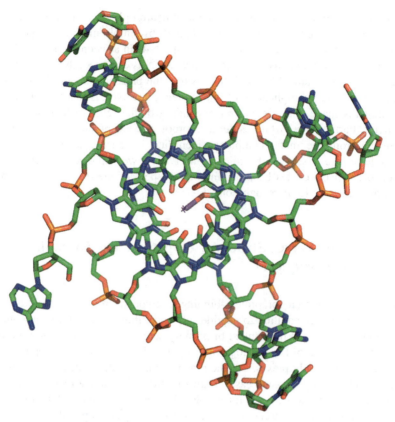

Fig. 5 Top view of a representative G-quadruplex X-ray crystal structure (human telomeric DNA; PDB ID 1KF1 [7]. Guanosine bases forming tetrads are at the center of the structure in this view

heterodimeric disulfide **1–2** was amplified (fourfold relative to the equilibrium library composition obtained in the absence of quadruplex DNA). Surprisingly, however, homodimeric **1–1** was also amplified to a significant extent (fivefold). Surface plasmon resonance experiments confirmed that **1–2** and **1–1** bound with dissociation constants (K_D) of 30.0 and 22.5 μM, respectively. No detectable binding of **2–2** was observed, consistent with its lack of selection in DCL experiments.

Balasubramanian and colleagues next examined the ability of G-quadruplex DNA to amplify structures selected from a library of polyamides based on the structure of distamycin (**3**), a well-studied DNA minor groove binder [14]. Libraries constructed from equimolar mixtures of monomers **4–6** were allowed to undergo selection in the presence of either a G quadruplex derived from human telomeric DNA, or a duplex DNA sequence taken from the human oncogene *c-kit*. As in the 2004 work described above, glutathione was added as a catalyst for exchange, and as a method for simplifying analysis; the thought behind the latter being that an excess of glutathione would favor the formation of simple glutathione-monomer conjugates in the absence of DNA target. This hypothesis was found to be experimentally valid, as amplification was most readily detected in the libraries containing the highest concentration of glutathione.

Amplification of hetero- and homodisulfides derived from **5** to **6** was found to be much stronger in the presence of duplex DNA than quadruplex DNA, an observation the authors interpreted as resulting from library members having higher affinity for the duplex target. The shortest disulfide (**4–4**) was not amplified, and thermal melting experiments confirmed that it had minimal affinity to either target. Compounds **5–5** and **5–6** both increased the melting temperature of target quadruplexes, consistent with binding by these disulfides.

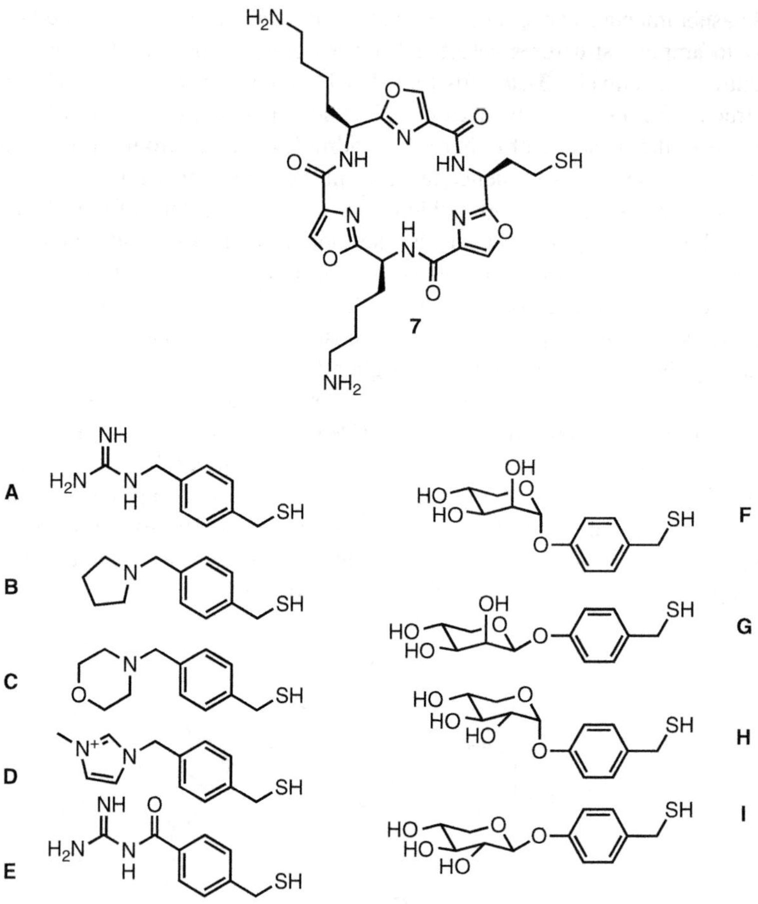

Fig. 6 Oxazole-peptide macrocycle libraries targeting G-quadruplex recognition

Most recently, the Balasubramanian and Sanders groups have used DCC to identify compounds able to distinguish between different G-quadruplex structures (an important task, since as we have already described how many G-quadruplexes are involved in "normal" cellular function as well as in disease states) [9]. Building on an oxazole-peptide macrocyclic G-quadruplex binder designed by the Balasubramanian group [15], the authors generated two libraries in which scaffold **7** was allowed to undergo equilibration with a series of thiols (Fig. 6), consisting of para-benzylic thiols functionalized with either positively charged amines (**A–E**) or simple sugars (**F–I**). As in the pyrrole-based library described above, reduced and oxidized glutathione were added to the library mixture as exchange catalysts and to simplify detection of the amplified species. Consistent with prior experiments, libraries allowed to equilibrate in the absence of DNA template primarily resulted in mixtures consisting of the disulfides formed from each starting library thiol and glutathione, with no strong self-selection. However, equilibration of **7** with **A–E** in the presence of the c-Kit21 quadruplex DNA resulted in strong (≥2,000%) amplification of **7-E** and **7-A**, with lesser amplification of **7-D** and **7-B**. Affinities measured for **7-E** ($K_D = 6.6 \pm 1$ μM) and **7-A** ($K_D = 10.9 \pm 1.9$ μM) were consistent with the observed relative amplification levels. Selection carried out in the presence of the c-Myc22 quadruplex reversed the relative amplification levels for **7-A** and **7-E**, in accord with subsequently measured dissociation constants (**7-A**: $K_D = 6.8 \pm 1.4$ μM; **7-E**: $K_D = 9.8 \pm 0.2$ μM). While the levels of selectivity obtained for the two quadruplexes are somewhat modest, it is nonetheless highly significant that these can be readily observed via the DCC experiment. Selectivity for quadruplex structures relative to duplex DNA was very high, as allowing the library to evolve in the presence of a duplex target caused no change in the relative concentration of library members. Similar experiments conducted with a library generated from **7** and **F–I** also revealed sequence-dependent differences in amplification factors.

An alternative arm-and-scaffold approach to DCC-based identification of G-quadruplex binders was reported in 2008 by Nielsen and Ulven [16]. Here, mono-, di-, and tritopic scaffolds **8**, **9**, and **10** were combined with side chain disulfides **J–L** (Fig. 7), the molar ratio of each scaffold being adjusted to reduce

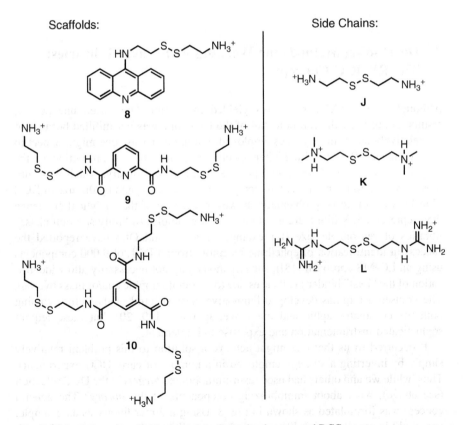

Fig. 7 Side-chain and scaffold approach to G-quadruplex targeted DCC

the possibility of "topicity dependent" bias (i.e., **8:9:10** = 3:6:10). Each side chain was present in 200-fold excess relative to **8**, and ethanethiol was added to initiate library exchange. Highlighting both the challenges and opportunities inherent in DCL analysis, the authors found it necessary to monitor multiple absorbance wavelengths in the HPLC trace in order to differentiate co-eluting library members. Acridine derivative **8-L** was found to have the highest amplification factor. In contrast, library selection conducted with an unstructured oligonucleotide resulted in no amplification. Confirmation of the affinity of **8-L** was obtained via thermal melting experiments, which demonstrated the ability of **8-L** to increase the melting temperature of $5'$-d[$G_3(T_2AG_3)_3$]-$3'$ by 12.8 °C.

Considered together, these four studies ably demonstrate the ability of DCC to rapidly yield lead structures for interesting targets. Despite the relatively small size of the libraries examined in each case, the diversity of structures accessible by the methodology is remarkable. G-quadruplex RNA has also been described [17], and therefore it is likely that future discovery efforts will also expand to targeting these sequences. Finally, as we will discuss later, G-quartets have served as scaffolds for the generation of novel materials via dynamic methods.

4 Developing Methods for Working with Large Libraries: The RBDCC Concept

Although early DCC experiments yielded interesting (and often unexpected) results, in our view the full potential of the technique was not fulfilled because of the relatively small library sizes employed. Put another way, one might expect to obtain higher affinity, with higher specificity compounds in proportion to the proportion of chemical structural space represented in a library, which is in turn loosely related to the numerical diversity of the library. Expanding the size of DCC libraries was felt to be problematic, however, because of the analytical challenge they represented. While there are a now a few examples of "fully solution phase" libraries of reasonable size (for example, Ludlow and Otto have reported the successful identification of ephedrine receptors from a DCL of 9,000 components using an LC-MS approach [18]), such methods may not necessarily allow identification of the "best" binder in all cases due to chromatographic and/or mass overlap. The Poulsen group has developed innovative mass spectral methods for avoiding both the chromatographic and mass overlap issues [19, 20], but these require sophisticated instrumentation and expertise to implement.

It occurred to us that one might achieve a solution to this problem relatively simply by inverting a strategy employed in a number of early DCC experiments. Thus, while we and others had used resin-immobilized *targets* in the DCC selection (see above), what about immobilizing components of the *library*? The general concept was formulated as shown in Fig. 8. Using a dimer library as an example, one could immobilize each library member at a discrete location on a grid, much like an oligonucleotide or protein microarray. Subjecting this "DCC chip" to a solution mixture containing an equivalent set of library monomers in solution, along with a fluorophore-tagged target, would allow for generation of the full dimer library, its evolution in the presence of the target biomolecule, and capture of the target on the chip surface by the highest affinity binder. One could then simply identify components of the winning compound (or compounds) by imaging the chip using a standard microarray scanner. For example, the illustration shown in Fig. 8 would indicate that monomers 1 and 2 would be the components warranting further study, and either homodimers 1–1 and 2–2 or heterodimer 1–2 the highest affinity compound.

We set out to test the concept in the context of DNA recognition by synthesizing a library of compounds based loosely on the structure of the bisintercalating antibiotics (such as triostin A, shown at the beginning of this chapter in Fig. 1). In addition to their proven ability to bind DNA with high affinity and significant selectivity, these natural product bisintercalators are inherently primed for mimicry using a DCC approach, as they contain both disulfide and ester groups as potential library scrambling functionality. The nucleic acid compatibility of disulfide exchange led us to focus on this as the central diversity generating reaction for proof of concept experiments for RBDCC. As such, the general library design is shown in Fig. 9.

Fig. 8 Immobilized DCC

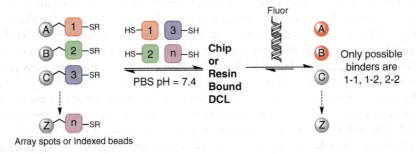

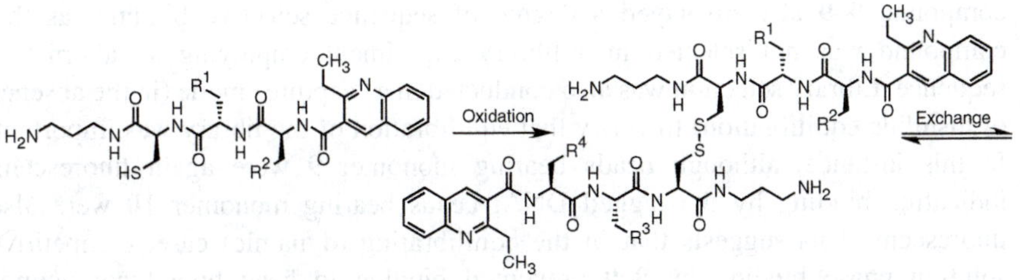

Fig. 9 Design of proof-of-concept RBDCC library targeting DNA. R^1 and R^2 represent amino acid side chains varied in the library. Only solution-phase dimer formation is shown

Incorporating a quinoline as a putative intercalator also allowed us to take advantage of recently developed synthetic methodology from our laboratory [21].

Synthesis of the library was accomplished using standard solid-phase peptide methods. Library monomers were immobilized via their terminal amine group in a microarray format on aldehyde-treated glass slides using a robotic pin spotter. In control experiments, disulfide formation and exchange was readily observable with a solution-phase peptide labeled with rhodamine. To our chagrin, however, screening the library against a fluorescently tagged DNA target provided no hits. Was this a problem with the library design, or an inherent failing of solid-phase supported DCC? Further analysis revealed that the answer was "neither," but rather an error in implementation. If one considers the design of an experiment of this type, the ability to visually detect binding to an immobilized library member depends on immobilized material outcompeting compound in solution. In the case of a microarrayed experiment, the amount of material in any array spot is vanishingly small relative to material in solution, unless solution volume is drastically minimized. Thus, the initial experiments we conducted were not able to overcome this problem.

On further consideration of the experimental setup, we realized that the use of resin beads for monomer immobilization would provide sufficient material to outcompete solution binding. Of course, one would lose the advantage of monomer indexing provided by the array, but still retain the advantage of phase separation, with the added benefit of being able to use resin beads with a "solution-like" (i.e., polyethylene glycol-based) matrix. We therefore set out to test this possibility using a semi-parallel library format, again employing the compound structure shown in Fig. 9. Nine vials were set up, each with a single type of monomer immobilized on resin beads. As a control experiment, labeled DNA was added to the mixture after first adding thiopropanol to yield disulfides. No binding was observed to any of the beads, demonstrating that the monomers alone had minimal DNA-binding ability. Next, an identical set of vessels containing bead-bound monomers were set up, and a mixture of solution phase monomers and labeled DNA was added to each vial. The library was then allowed to evolve under standard disulfide exchange conditions. After halting equilibration and rinsing the beads, strong fluorescence was readily detected in one tube. This suggested that compound **9–9**, formed by dimerization of resin bound **9** and solution **9**, was the highest affinity library member. Indeed, equilibrium dialysis of **9–9** revealed a dissociation constant of 2.8 µM, while other (unselected) library members had lower affinity. Surprisingly, compound **9–9** also displayed a degree of sequence selective binding, as this compound was not selected in a library experiment employing an alternative sequence. Library selection was also conducted in a "capture" mode (in the absence of disulfide equilibration) to verify that equilibration of the library was important. In this instance, although beads bearing monomer **9** were again fluorescent, indicating binding by the tagged DNA, beads bearing monomer **10** were also fluorescent. This suggests that in the equilibrating (dynamic) case, competitive solution phase binding by **9–9** prevented binding to bead bound compounds incorporating **10**. Completion of this study set the stage for later efforts targeting RNA with much larger RBDCC libraries, as described below. As we shall see, the competition between bead-bound and solution-phase compounds in an RBDCC library is a reproducible and highly useful phenomenon.

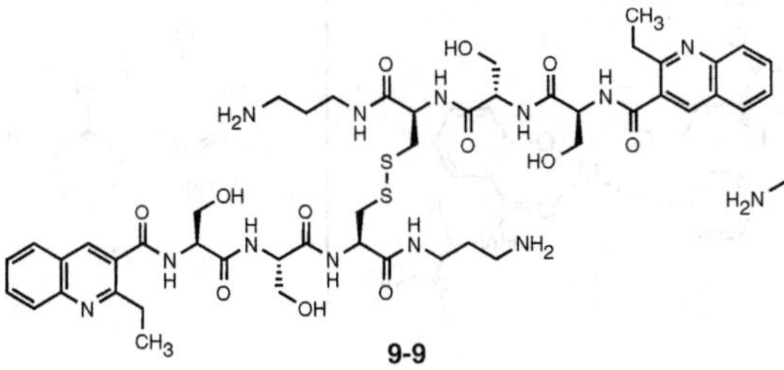

9-9

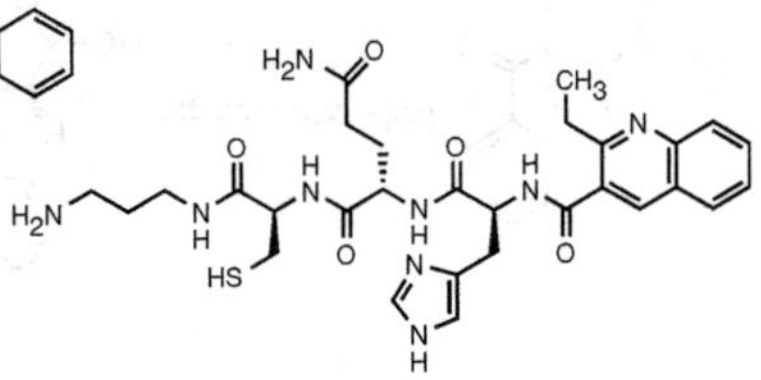

10

5 Using DCC to Address the "Grand Challenge" of RNA Recognition

The central dogma of molecular biology, in which RNA's role in the cell is simply one of messenger, obscures the complexity and richness of the ways in which RNA influences biological processes. Recent research is changing this myopic view, however, as entirely new classes of RNAs (for example, noncoding RNA [22], a subset of which includes micro RNAs [23]) as well as structural motifs within messenger RNAs capable of regulating translation [24] are being discovered. In tandem, the importance of RNA as a target for probe and pharmaceutical development is increasingly recognized [25–27]. Unfortunately, unlike DNA, there is as yet no "code" mapping RNA sequence to high-affinity synthetic binders. (One of the most well-known recognition "codes" for DNA is that developed by the Dervan group. For lead references, see [28].) The set of known RNA binding natural products is also relatively unhelpful in addressing this problem. The aminoglycosides, perhaps the most widely studied RNA binding natural products [26], have only limited binding selectivity, although synthetic variants of aminoglycosides are beginning to show promise [29]. A growing set of research groups is beginning to attempt to address this problem using an innovative range of molecular design and combinatorial strategies. Given the importance of the field and the relative lack of understanding with regard to which strategies might be successful, we hypothesized that DCC might be a particularly useful method for rapidly identifying novel RNA-binding structures.

5.1 Initial Experiments: Cu-Salicylamides

Our early efforts to target RNA, as with our group's DNA targeted work described above, focused on the use of transition metal complexes as library constituents.

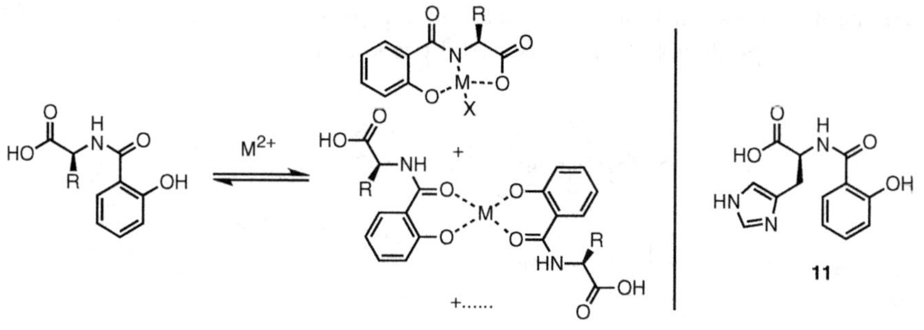

Fig. 10 RNA-targeted metal-salicylamide DCL

A small library of salicylamides was synthesized, in anticipation that these could form equilibrating mixtures of 1:1 and 2:1 complexes with a metal ion such as copper(II) (Fig. 10) [30]. Screening of this library was accomplished using an equilibrium dialysis strategy, in which library constituents and copper were allowed to equilibrate across a dialysis membrane with a target RNA derived from the Group I intron of *Pseudomonas carinii*, a common pathogen afflicting immuno-compromised patients [31]. Library analysis revealed that histidine salicylamide **11** underwent strongest selection. Subsequent binding titrations confirmed an apparent dissociation constant of 0.2 µM, with the primary binding species most likely existing as a 1:1 copper:salicylamide complex.

5.2 Expanding the Library: Identifying Sequence-Selective RNA Binders with RBDCC

While this metal-salicylamide DCL served as a useful first demonstration of RNA targeted DCC, progress towards the application of DCC for the identification of highly sequence selective compounds capable of serving as leads for further development as useful biological probes awaited the development of methods for the production and analysis of libraries capable of accessing greater numerical and structural space. The advent of RBDCC promised to make that possible. Thus far, our group has used RBDCC successfully to target two biomedically relevant RNA sequences: an RNA stemloop implicated in regulating frameshifting in HIV (**12**) [32], and a pathogenic triplet repeat RNA (CUG^{exp}, **13**) believed to be the causative agent of type 1 myotonic dystrophy (DM1) [33], the most common form of adult-onset muscular dystrophy.

```
              G-C
              U U
              C-G
              G-C
       C A    U U
       A A    C-G
       C-G    G-C
       C-G    U U
       C-G    C-G
       U-A    G-C
       U-A    U U
       C-G    C-G
       C-G    G-C
       G-C    U U
    5' G-C    C-G
              G-C
        12    C-G
              C-G
              C
              5'

              13
```

Although considerable NMR structural information about the HIV-1 frameshift-stimulating RNA stemloop (HIV-1 FSS RNA, **12**) was available, the flexibility of RNA coupled with the relatively early stage of understanding about atomic-level factors governing RNA recognition has made it generally resistant to structure-guided design approaches. Therefore, to approach the problem of sequence-selective RNA binding, expansion of the DNA-targeted RBDCC library described above to larger numerical and structural space was deemed necessary. This was accomplished by (1) expanding the peptide segment of each monomer from two amino acids (with a variable residue and conserved cysteine) to three (two variable amino acids), (2) allowing the position of the cysteine to vary, and (3) including two heterocyclic "cap" groups rather than one (Fig. 11). Amino acids were selected for the variable positions such that a broad range of functional group types was accessed, while providing uniquely defined monomer mass. The position of the cysteine was encoded by the size of the resin beads on which each monomer was synthesized. Thus, 150 solid-supported monomers could be allowed to exchange with 150 solution-phase monomers (in this case, an identical set to those on the resin beads) to provide a total theoretical library diversity of 11,325 unique

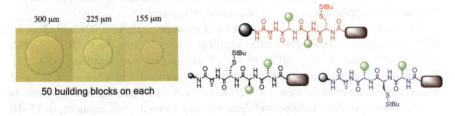

Fig. 11 RNA-targeted RBDCC. Cysteine position was encoded by resin size, while variable amino acids were selected to provide unique monomer mass

compounds after disulfide exchange-mediated equilibration. Bead-bound monomers were linked to the resin via a photocleavable moiety (*ortho*-nitrophenyl glycine), allowing simple post-screening cleavage and analysis via mass spectrometry.

After first verifying that bead-bound monomers had minimal affinity for target RNA **12**, the entire 11,325-compound library was screened with fluorescently-labeled HIV-1 FSS RNA as a pool. Fluorescent beads were then removed, cleaved, and cleaved material analyzed. As the stoichiometry of the library screen was set up such that bead-bound monomers were in excess (a choice made so as to favor dimmer formation on bead rather than in solution), mass spectrometry revealed only the identity of component monomers rather than the full structure of selected compounds. Nevertheless, this provided a substantial simplification of the library in a single step: three monomers were identified as being selected in replicate experiments, potentially representing six unique compounds, or 0.05% of the library (ignoring terminus differentiation due to resin attachment; taking this into consideration raises the number of compounds to a still-low 9).

Confirmation of the highest affinity binder was obtained via a secondary screen. Interestingly, this once again confirmed the value of RBDCC as an inherently competitive binding selection. As shown in Fig. 12, beads bearing **14** were found to capture target RNA, as evidenced by fluorescence, when allowed to undergo disulfide exchange with solution-phase **15**. The converse experiment, in which bead-bound **15** underwent exchange with solution-phase **14**, did not capture target RNA. This would lead one to conclude that although the **14–15** had measurable affinity for the HIV-1 FSS, it was less strong than that of the **14–14** homodimer, as in the second experiment this homodimer would be able to form in solution, and thereby compete away all of the available target RNA. Indeed, the exchange experiment consisting of both bead-bound and solution-phase **14** provided the greatest fluorescence, consistent with this interpretation.

Resynthesis of the selected compounds allowed verification of their relative affinities. Compound **14–14** was found to have a binding constant K_D of 4.1 ± 2.4 μM to the HIV-1 FSS RNA by surface plasmon resonance (SPR; compound was immobilized on-chip), while **14–15** had a much lower affinity ($K_D > 90$ μM; binding occurred, but did not saturate within the limits of the assay). Subsequent fully solution-phase fluorescence assays indicated a higher affinity for **14–14** ($K_D = 350 \pm 110$ nM), as one would anticipate given the sterically more accessible solution-phase ligand. Binding of **14–14** could not be competed away by the presence of a large (>tenfold) excess of total yeast tRNA, indicating a significant degree of sequence selectivity.

As a test of the RBDCC concept to yield sequence-selective RNA binders for different targets, we next screened an identical library for binding to CUG^{exp} RNA (**13**). In this case, a very different set of monomers was identified (Fig. 13). Secondary screening indicated the selection was not as strongly biased in favor of a single molecule as in the case of the HIV-1 FSS. Resynthesis and binding analysis of all of the possible hetero- and homodimers formed from monomers **15–18** confirmed that several of the selected compounds were able to bind CUG^{exp} RNA with low micromolar affinity. Sequence selectivity was found to be high, as

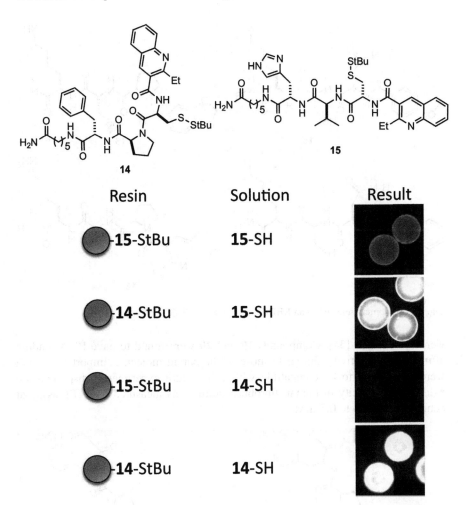

Fig. 12 Secondary screen of HIV FSS RNA-targeted RBDCC

compounds identified in this screen had little to no affinity for the HIV-1 FSS RNA (and vice versa). Importantly, several of the compounds were able to compete for RNA binding with MBNL1, a splicing factor known to bind CUG^{exp} RNA with high affinity. As this represented the *first* example of a synthetic non-nucleic acid based binder to this important RNA target, it highlighted the promise of DCC to rapidly provide useful lead compounds.

With these early successes in hand, a key question arising is that of whether RNA-targeted DCC yields compounds that can serve as relevant leads for further medicinal chemistry studies? Our group is currently engaged in efforts to demonstrate that the answer to this question is in the affirmative. For example, building on the efforts with the HIV-1 FSS RNA detailed above, non-reducible analogs incorporating a hydrocarbon (olefin or saturated alkane) bioisostere for the disulfide

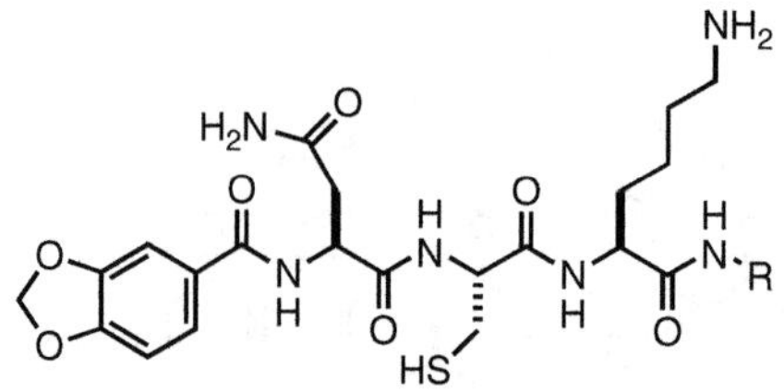

15

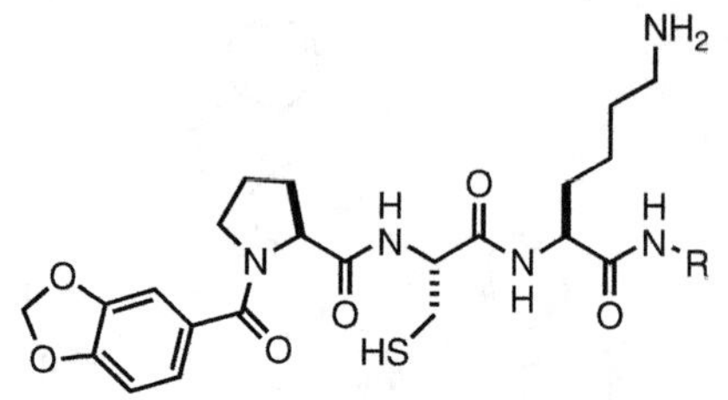

16

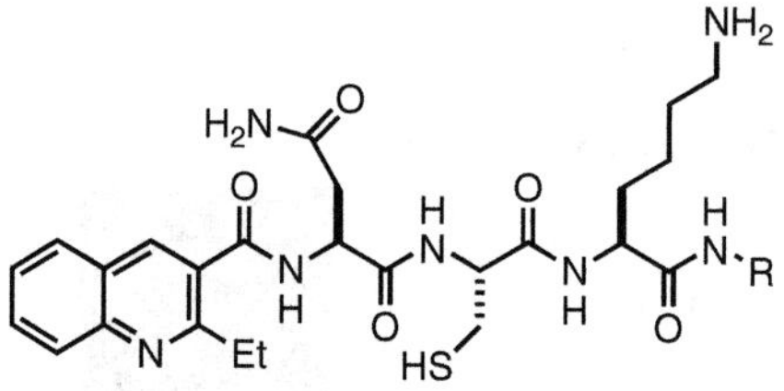

17

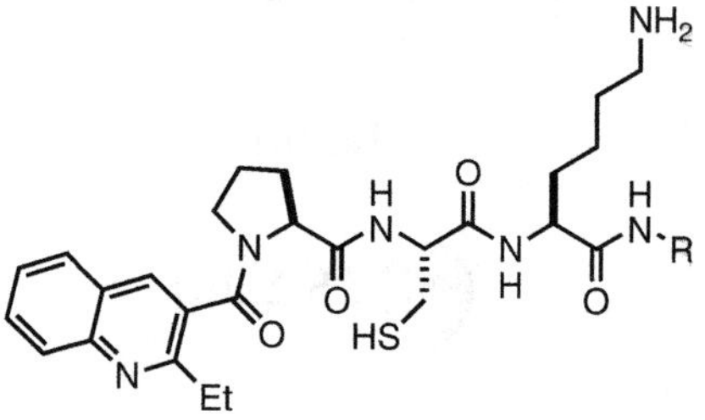

18

Fig. 13 Monomers selected in an RBDCC screen with (CUG^{exp}) RNA **13**

were synthesized [34]. Compounds **19** and **20** were found to have RNA-binding affinity and selectivity similar to those of the parent molecule. Importantly, both were also found to be compatible with cell culture experiments, displaying no evidence of toxicity in human fibroblast cultures as measured by MTT assay at concentrations up to 0.5 mM.

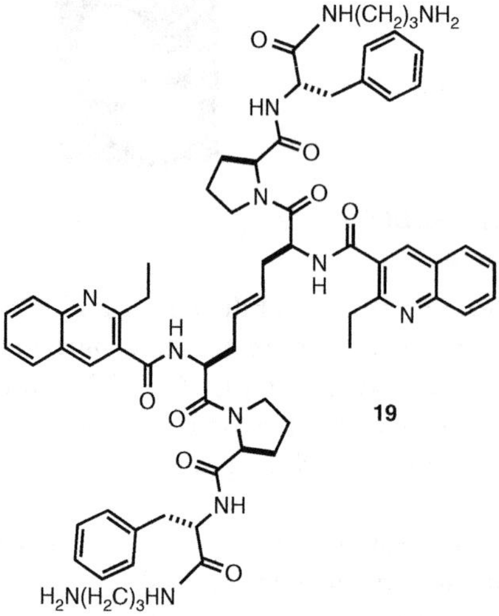

19

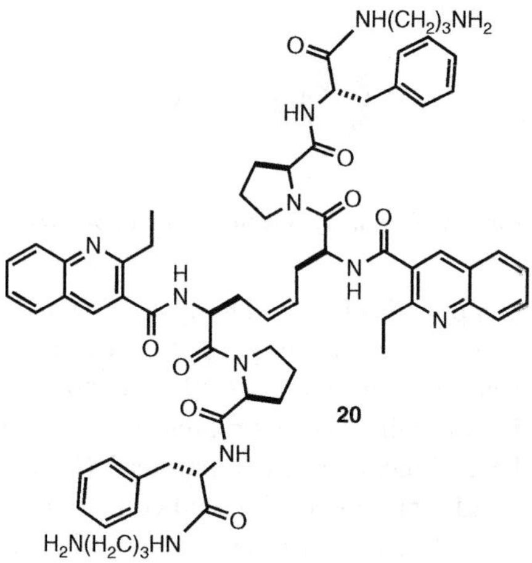

20

6 Recognition of Nucleosides and Sequestration of Nucleic-Acid Binders

Recognition of non-polymeric nucleic acids (found in the cell as cofactors and biosynthetic intermediates, for example) is also an important endeavor; two contemporaneous studies illustrate complementary aspects of the problem. Dynamic modification of a nucleotide *guest* has been reported by Abell, Ciulli, and colleagues, as a method for probing the adenosine binding pocket of pantothenate (vitamin B12) synthetase from *Mycobacterium tuberculosis* [35]. In this report, the authors set out to examine the ability of the enzyme to select disulfide analogs of a pantoyladenylate intermediate (**21**, Fig. 14). Beginning with 5′-deoxy-5′-thioadenosine (**22**), a library was formed via incorporation of thiols **23–30**. These were chosen as potentially facilitating charge-charge interactions with residues in the phosphate binding site of the enzyme (**23, 24, 27, 28**), or as hydrophobic groups capable of occupying the pantoate pocket (**25, 26, 29, 30**). Library equilibration was allowed to proceed in a glutathione redox buffer (in tris(hydroxymethol)aminomethane (Tris) buffer, pH 8.5), by analogy to protocols developed in Balasubramanian's G-quadruplex work described above [9]. Following acid quenching of disulfide exchange, HPLC analysis of the library revealed that the disulfide undergoing the most amplification was that incorporating benzylthiol **26**. Interestingly, free thiol **22** (R = H) was also strongly amplified. Binding of the **22–26** disulfide to the enzyme was confirmed by isothermal titration calorimetry (ITC; $K_D = 210 \pm 10 \mu M$). Free thiol **22** (R = H) also bound, albeit less well ($K_D = 380 \pm 30 \mu M$). An X-ray crystal structure of the **22–26** disulfide bound to the enzyme confirmed that the benzyl group bound within the pantoate pocket, as anticipated by the authors' initial hypothesis. As the authors suggest, this study serves as proof-of-concept for the development of novel

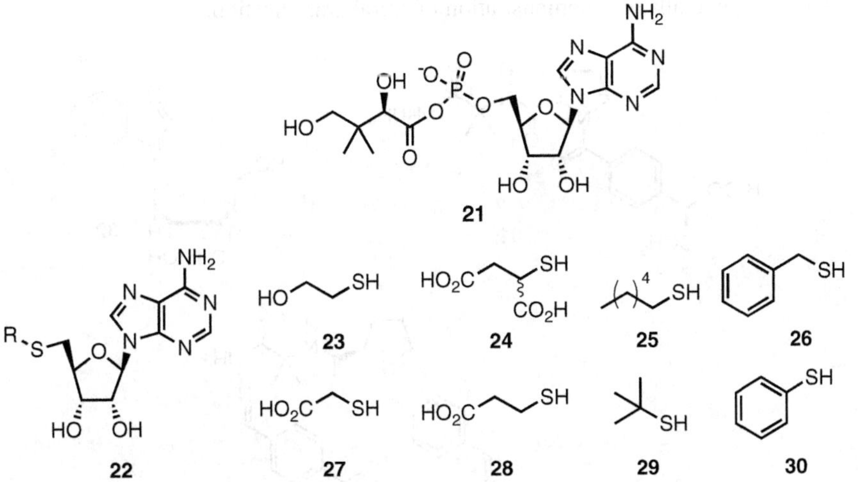

Fig. 14 Thioadenosine-derived library used for probing the adenosine-binding pocket of pantothenate synthetase

nucleotide-based enzyme inhibitors, or analogs of a broad range of biosynthetic intermediates and nucleotide cofactors.

21

Conversely, Gagné and coworkers have described efforts targeting the use of DCC to evolve *hosts* for nucleic acids. In particular, their libraries provide interesting examples of chiral amplification [36]. Racemic **31** was allowed to equilibrate via TFA-initiated hydrazone exchange in the presence or absence of adenosine (**32**). In the absence of template, the library mixture was determined to consist of a 44:29:24:3 ratio of dimers:trimers:tetramers:hexamers; in the presence of a fivefold excess of **32** this changed to a 56:21:23:0 ratio. Of particular interest, laser polarimetry coupled HPLC showed the appearance of a signal (indicative of the presence of a chiral material) coincident with the retention time of the dimers for the adenosine-containing library. This was subsequently determined to result from the selective production of the (S,S) configuration of **33** in preference to (R,R) or (S,R) **33**. Increasing the amount of adenosine to a tenfold excess allowed amplification of a tetramer or tetramers as well, although no chiral amplification was observed [37]. Switching to cytidine or 2-thiocytidine changed the course of library evolution; in this case a laser polarimetry signal was observed for dimers, trimers, and tetramers. Although binding affinities were modest in all cases, this study nevertheless serves as an important starting point for the development of nucleic acid-binding receptors, and more generally as a demonstration of chiral amplification.

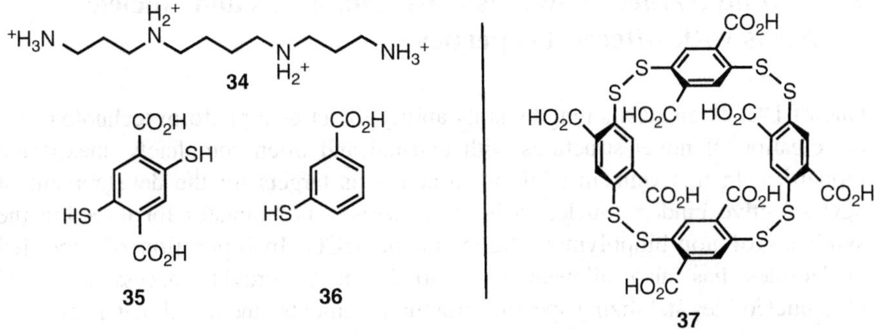

Fig. 15 A two-building-block DCL constructed for the identification of spermine receptors

A third variant on the theme of nucleic acid recognition is that of altering the properties of DNA via the identification of receptors capable of sequestering naturally occurring DNA binders. For example, spermine (**34**, Fig. 15) is a naturally occurring polyamine employed by the cell as a nonselective DNA and RNA binder. Among its many roles in the cell, spermine uniformly binds DNA and neutralizes its net negative charge. Thus, spermine is an essential component of the process of compacting DNA into nuclear chromatin. Otto and coworkers have employed DCC for the identification of high affinity spermine receptors, potentially yielding a synthetic reagent for interfering with a number of spermine-facilitated DNA processes [38].

Identification of the spermine receptor was accomplished using a library constructed from only two building blocks. Compounds **35** and **36** were chosen based on their ability to form both linear and cyclic oligomers via disulfide chemistry. Carboxylate functionality was included in each monomer both to provide ion-based recognition of spermine ammonium groups, as well as to ensure that the selected complex remained soluble in aqueous solution. Analysis of a library produced in the absence of spermine indicated that a linear tetramer consisting of two molecules of **35** capped by two molecules of **36** was the primary species produced. In contrast, a selection carried out in the presence of spermine resulted in the amplification of a peak subsequently shown to be the cyclic tetramer **37**. A **36–36** dimer was also found to be amplified under these conditions, as one would expect given the conversion of most of the other library material (**35–37**). Isothermal titration calorimetry (ITC) and NMR were employed to verify complexation between **37** and **34**; the binding constant (K_D) for the **37•34** complex was found to be 22 nM. As this is a stronger interaction than that reported by others for the binding of spermine to DNA, the authors examined whether **37** could competitively remove spermine from calf thymus DNA. Indeed, changes in the circular dichroism spectrum provided evidence this was successfully accomplished.

7 Beyond Nature: Towards Novel Materials and Nucleic Acids with Altered Properties

One of DCC's greatest strengths is its ability to act as a platform technology for the creation of novel structures with unusual and often completely unexpected properties. In this vein, in addition to acting as targets for the development of new selective binders, nucleic acids have served as templates for directing the synthesis of non-biopolymer oligomers via DCC. Incorporation of modified nucleotides has also allowed DCC to begin to provide access to novel oligonucleotides stabilizing specific structural elements, and novel materials.

7.1 Generation of Novel Materials with DCC

In an example of the latter class, the Lehn group has examined the ability of G-quartet derived DCLs to yield new types of hydrogels [39]. As we have discussed above, guanine-rich oligonucleotides readily form quadruplex structures. However, it is also known that guanine derivatives are able to self-assemble in the presence of metal ions to form tetrads even when no sugar-phosphate backbone is present [40, 41]. Testing this in the context of a dynamic system, a DCL was formed (Fig. 16) in which library members best able to contribute to self-assembly would undergo selection via phase segregation, as the columnar assemblies of G-tetrads form hydrogel matrices. After initially testing the ability of a series of aldehydes (39–45) to promote gel formation via the acyl hydrazone structure 38, a library consisting of aldehydes 43, 45, and hydrazides derived from guanosine and serine was examined. Gelation-driven selection strongly favored condensation of guanosine hydrazide with 45. In subsequent work, small-angle neutron scattering

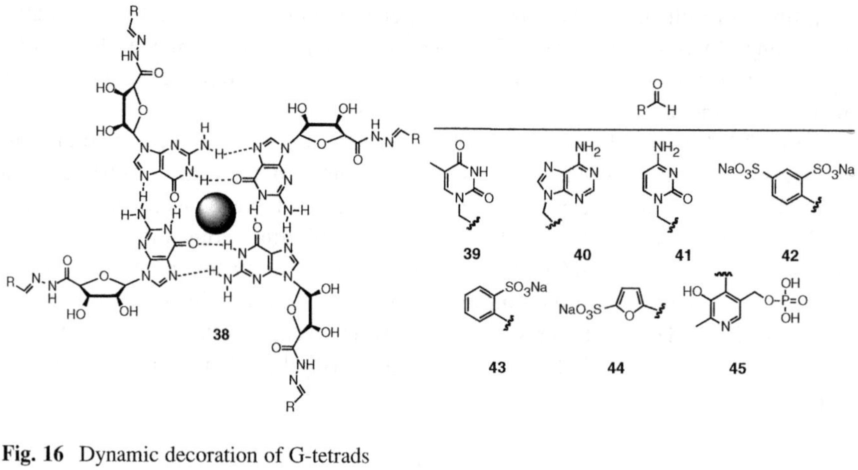

Fig. 16 Dynamic decoration of G-tetrads

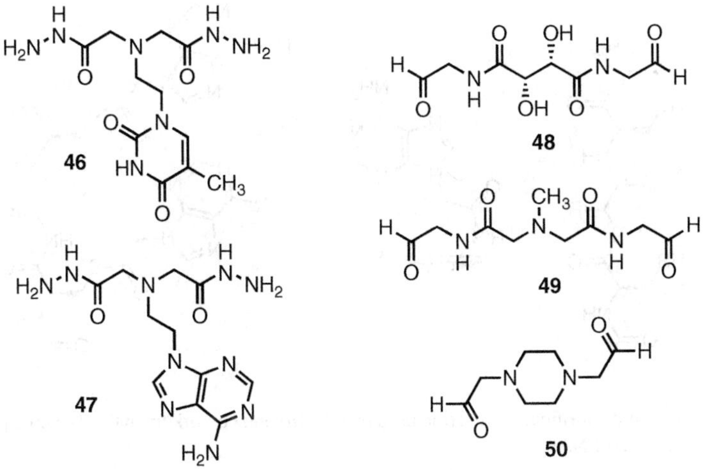

Fig. 17 Components used in the production of DNA-inspired dynamic polymers (DyNAs)

experiments were employed to examine the structure of the gels formed from the decorated G-quartet scaffold [42]. In addition to observing structural differences as a function of the central cation (thicker fibers were observed for Na^+ than K^+), a strong dependence on the identity of the decorating aldehyde was also found. Structures assembled from the guanosine hydrazide alone were compact, and characterized by crosslinking between columnar aggregates, while those decorated with a pyridoxal phosphate derivative (**45**) formed isolated columns with more widely spaced tetrads.

In an alternative strategy, Lehn and colleagues have explored the integration of nucleobase-bearing backbones into dynamically constructed polymers, or "DyNAs" [43]. After initial experiments involving the condensation of hydrazide **46** and tartrate derivative **48** (Fig. 17) resulted in precipitation, the authors shifted their focus to materials incorporating aldehydes **49** and **50**, as both would be charged at pH 6, and likely to promote solubility. This was indeed the case, and polymers ranging in average molecular weight from 1880 (**47** + **50**) to 23,920 (**46** + **49**) were formed. Increasing concentration was found to promote formation of higher molecular weight structures. While the highly charged nature of the polymers limited their ability to participate in sequence-specific recognition of DNA, SPR experiments verified that they were nonetheless able to bind DNA via charge–charge interactions. Furthermore, carrying out the polymerization reaction in the presence of a polyanionic template (for example, poly-aspartic acid) resulted in a dramatic increase in average molecular weight, an outcome not observed with a non-ionic polymer template (polyethylene glycol).

The ability of DNA to serve as a template for the reversible formation of a PNA oligomer was reported by the Ghadiri group in 2009 [44]. The overall framework here relied on the use of a transthioesterification reaction as the exchange mechanism, operating between derivatives of DNA bases (for example, the adenine

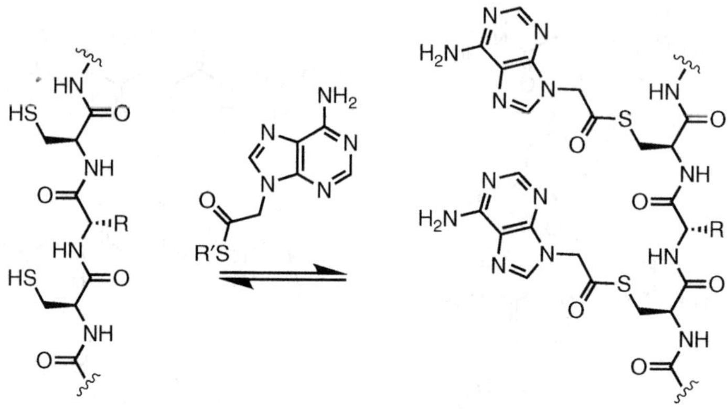

Fig. 18 Transthioesterification reaction used by Ghadiri and coworkers in the dynamic production of DNA-templated PNAs

derivative shown in Fig. 18) and peptides with an alternating XXX-cysteine sequence (where "XXX" was glutamic acid, aspartic acid, arginine, or glycine). After demonstrating that the transthioesterification reaction readily allowed formation of nucleobase-functionalized peptides, the authors examined the formation of these sequences in the presence of different single-stranded DNAs. PNA makeup was found to depend strongly on the sequence of the template DNA, consistent with a recognition-dependent process. Furthermore, introduction of different DNAs to oligomer solutions at equilibrium resulted in adaptive changes to the constitution of the oligomer.

7.2 Dynamic Decoration of Nucleic Acids

The Rayner group has reported several examples of the use of dynamic nucleic acid "decoration" for stabilizing structures of interest. In the first of these, described in 2004 [45], a mixture consisting of a self-complementary DNA hexamer bearing a 2'-amino-2'-deoxyuridine at the 3'-terminus with a pool of three aldehydes (**51–53**, Fig. 19) and sodium cyanohydride was generated. Analysis of aliquots of the

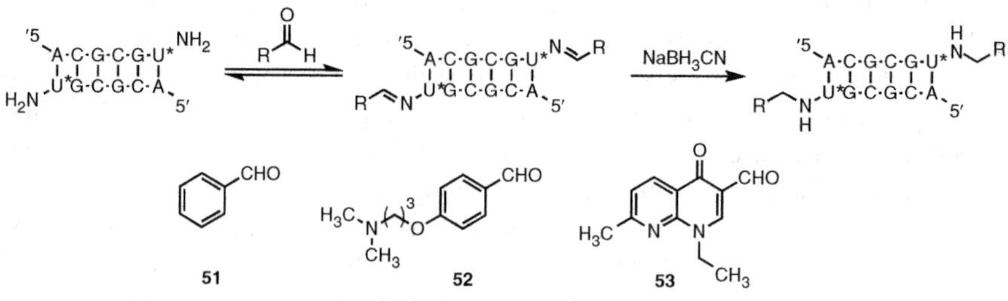

Fig. 19 Dynamic decoration of nucleic acids via imine formation

mixture taken at 2-, 4-, and 6-h time points indicated that the library composition was essentially fixed by 2 h, with the oligonucleotide derived from reductive amination of **53** being the most strongly amplified. The authors hypothesized that this selection was due to stabilization of the self-complementary duplex by conjugation with **53**, a hypothesis confirmed via thermal melting experiments. A further DCL study demonstrated proof-of-concept for identification of moieties able to stabilize RNA secondary structure as well. Continued elaboration of this concept allowed for DCC-mediated nucleic acid stabilization and in vitro aptamer selection via SELEX to operate in tandem [46]. To the extent that the DCC-SELEX strategy can be made general, it potentially dramatically expands the combinatorial diversity accessible in DCLs (numerical space), while expanding the types of chemical functionality that can be explored with SELEX (chemical or structure space).

Further work by the Rayner group has centered on the use of DCC for identifying nucleic acid modifications able to stabilize triple helices [47]. Building on work by several laboratories demonstrating that spermine was able to enhance the stability of oligonucleotide triple helices either as an additive [48, 49] or when tethered to the triplex-forming sequence [50–53], the authors hypothesized that DCC would be an efficient method for identifying other $2'$-amino modifications capable of favoring triplexes. To that end, a triplex-forming oligonucleotide incorporating a $2'$-aldehyde modified uridine (**54**) was allowed to complex with its duplex target (Fig. 20).

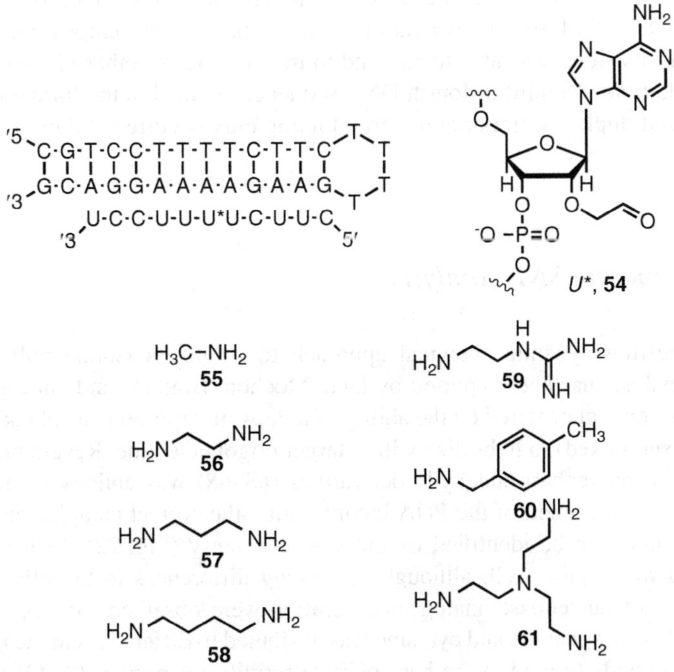

Fig. 20 Stabilization of triple helices via dynamic modification of uridine

Subsequently, an imine exchange DCL was created by the addition of amines **55–61** and sodium cyanoborohydride. Analysis of the library following equilibration showed a 9% amplification of the oligonucleotide incorporating guanidine-bearing amine **59**, and a 29% amplification of that incorporating tris(2-aminoethyl) amine **61**. Confirmation that library results reflected improved thermodynamic stabilities of the triplex structures was obtained by independent synthesis of each modified oligonucleotide followed by measurement of thermal melting profiles. Consistent with results of the DCL, the oligonucleotides incorporating **59** and **61** showed a 10.6 and 10.9 °C increase in the melting transition, respectively.

7.3 Nucleic Acid Cross-Linking

Citing as inspiration ecteinascidin 743 (**62**), a natural product known to carry out reversible alkylation of DNA [54], Wang and Rokita recently reported a dynamic crosslinking system for DNA [55]. A particularly intriguing aspect of this work involves its use of a reversible *alkylation* reaction as the exchange process. As shown in Fig. 21, the acridine-tagged quinone methide precursor **63** is treated with a fluoride source to prime the molecule for loss of acetate; this subsequently reacts with a first DNA strand ("DNA1") forming an adduct. Loss of the second acetate followed by reaction of the intermediate quinone methide with a proximate DNA strand ("DNA2") forms a covalent cross-link. The authors demonstrated that this crosslinked species was able to respond to the presence of other oligonucleotides. Thus, addition of a third or fourth DNA sequence resulted in the formation of new crosslinked duplexes. Importantly, crosslinking only occurred at duplex regions.

7.4 Dynamic SNP Analysis

An interesting dynamic chemical approach to single nucleotide polymorphism (SNP) analysis has been reported by Diaz-Mochon, Bradley, and colleagues [56]. Here, the concept centered on the ability of a PNA incorporating a "blank" site (**64**; "blank" site boxed) to hybridize with a target oligonucleotide. Reversible reaction of **64** with nucleobase aldehyde derivatives (**65–68**) was anticipated to provide selective enhancement of the PNA incorporating the correct complementary base, which could then be identified by mass spectrometry (Fig. 22). Indeed this was found to work quite well, although interesting differences in the efficiency and selectivity of nucleobase analog incorporation were observed. An approximately fivefold bias for guanine and cytosine was attributed to differences in the number of hydrogen bonds formed in the base pair; selectivity for purines (A, G) relative to pyrimidines (C, T) was also reported. Incorporation of the pyrene derivative **69** also

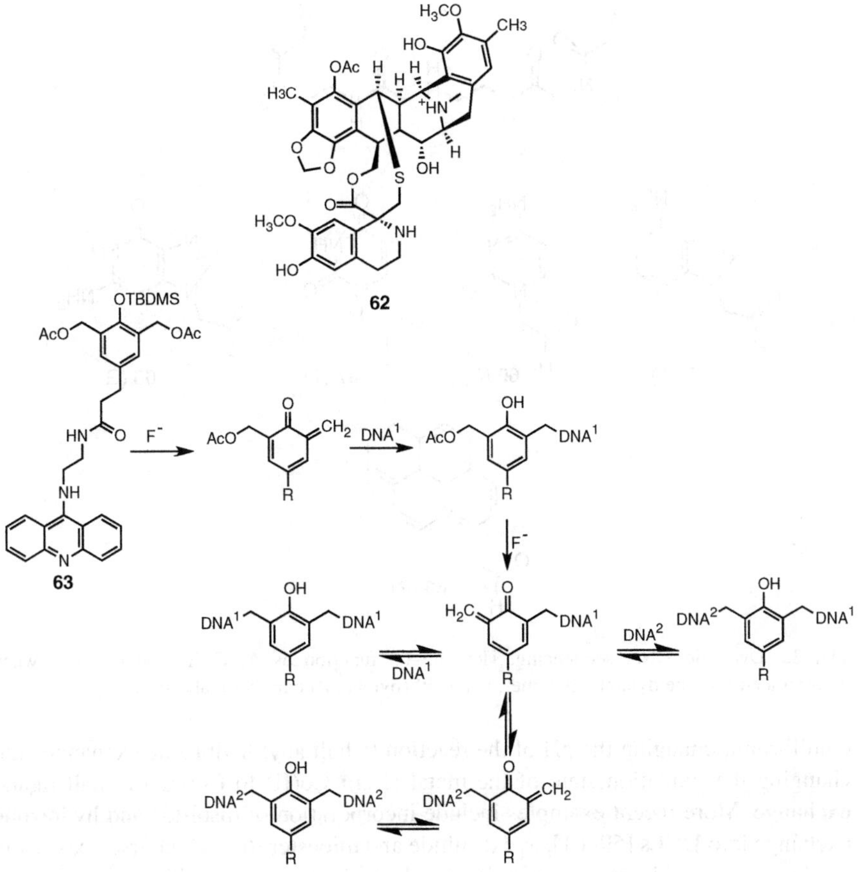

Fig. 21 Dynamic DNA crosslinking

allowed discrimination of abasic sites, by analogy to observations by Kool and Matray [57].

8 Conclusions

Future efforts to incorporate more complex reaction networks into nucleic acid-targeted DCC will undoubtedly rely on multiple compatible (and ideally orthogonal) exchange chemistries. While aspects of our earliest efforts in nucleic acid recognition involved this in an uncontrolled way (imine exchange plus metal coordination), Lehn and coworkers demonstrated the idea in a controlled system, combining acyl hydrazone exchange with the exchange of ligands around a Co(II) coordination center [58]. This also provided an orthogonal method for halting

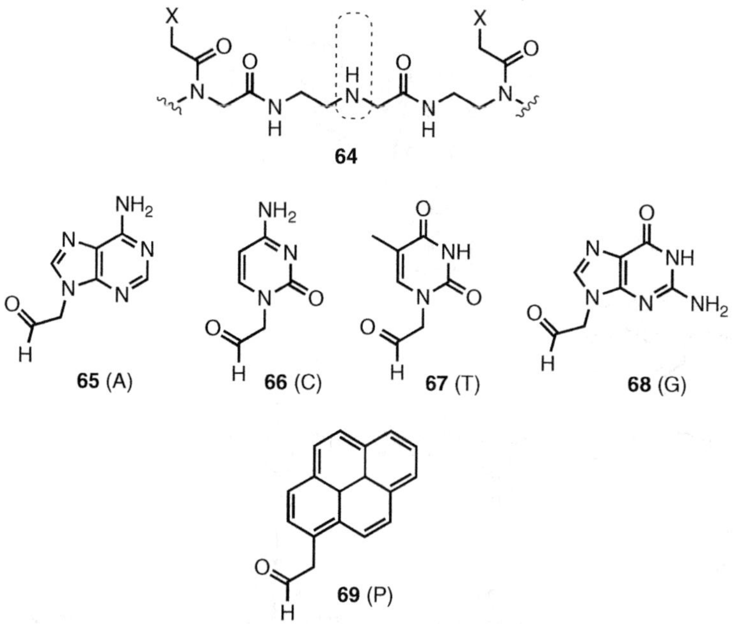

Fig. 22 Dynamic DNA sequencing. Here **65–68** function as A, C, T, and G bases when incorporated into the dynamic polymer, while **69** (pyrene, P) pairs with abasic sites

equilibrium; changing the pH of the reaction to halt acyl hydrazone exchange, and changing the oxidation state of the metal (from Co(II) to Co(III)) to halt ligand exchange. More recent examples include incorporation of disulfide and hydrazone exchange into DCLs [59–61], and disulfide and thioester [62]. Of course, expansion of the set of exchange chemistries used in these studies will require careful consideration of nucleic acid compatibility, since reactions need to operate in an aqueous milieu, close to neutral pH, and in such a way that the nucleic acid itself is not modified (except in those instances where modification of the target nucleic acid is the desired outcome). One can anticipate that such methods will pave the way for the production and analysis of DCLs of ever-increasing complexity, thus providing more versatile and valuable ways of solving the nucleic acid recognition problem – and of identifying novel methods for modifying and using nucleic acids.

References

1. Wu T, Orgel LE (1992) Nonenzymatic template-directed synthesis on oligodeoxycytidylate sequences in hairpin oligonucleotides. J Am Chem Soc 114:317–322
2. Goodwin JT, Lynn DG (1992) Template-directed synthesis: use of a reversible reaction. J Am Chem Soc 114:9197–9198

3. Zhan ZYJ, Lynn DG (1997) Chemical amplification through template-directed synthesis. J Am Chem Soc 119:12420–12421
4. Schepartz A, McDevitt JP (1989) Self-assembling ionophores. J Am Chem Soc 111:5976–5977
5. Klekota B, Hammond MH, Miller BL (1997) Generation of novel DNA-binding compounds by selection and amplification from self-assembled combinatorial libraries. Tetrahedron Lett 38:8639–8643
6. Klekota B, Miller BL (1999) Selection of DNA-binding compounds via multistage molecular evolution. Tetrahedron 55:11687–11697
7. Parkinson GN, Lee MP, Neidle S (2002) Crystal structure of parallel quadruplexes from human telomeric DNA. Nature 417:876–880
8. Bugaut A, Jantos K, Wietor J-L, Rodriguez R, Sanders JKM, Balasubramanian S (2008) Exploring the differential recognition of DNA G-quadruplex targets by small molecules using dynamic combinatorial chemistry. Angew Chem Int Ed 47:2677–2680
9. Fernando H, Reszka AP, Huppert J, Ladame S, Rankin S, Venkitaraman AR, Neidle S, Balasubramanian S (2006) A conserved quadruplex motif located in a transcription activation site of the human c-kit oncogene. Biochemistry 45:7854–7860
10. Ambrus A, Chen D, Dai J, Jones RA, Yang D (2005) Solution structure of the biologically relevant G-quadruplex element in the human c-MYC promoter. Implications for G-quadruplex stabilization. Biochemistry 44:2048–2058
11. Wong HM, Payet L, Huppert JL (2009) Function and targeting of G-quadruplexes. Curr Opin Mol Ther 11:146–155
12. Whitney AM, Ladame S, Balasubramanian S (2004) Templated ligand assembly by using G-quadruplex DNA and dynamic covalent chemistry. Angew Chem Int Ed 43:1143–1146
13. Szajewski RP, Whitesides GM (1980) Rate constants and equilibrium constants for thiol-disulfide interchange reactions involving oxidized glutathione. J Am Chem Soc 102:2011–2026
14. Ladame S, Whitney AM, Balasubramanian S (2005) Targeting nucleic acid secondary structures with polyamides using an optimized dynamic combinatorial approach. Angew Chem Int Ed 44:5736–5739
15. Jantos K, Rodriguez R, Ladame S, Shirude PS, Balasubramanian S (2006) Oxazole-based peptide macrocycles: a new class of G-quadruplex binding ligands. J Am Chem Soc 128:13662–13663
16. Nielsen MC, Ulven T (2008) Selective extraction of G-quadruplex ligands from a rationally designed scaffold-based dynamic combinatorial library. Chem Eur J 14:9487–9490
17. Morris MJ, Negishi Y, Pazsint C, Schonhoft JD, Basu S (2010) An RNA G-quadruplex is essential for cap-independent translation initiation in human VEGF IRES. J Am Chem Soc 132:17831–17839
18. Ludlow RF, Otto SJ (2008) Two-vial, LC-MS identification of ephedrine receptors from a solution-phase dynamic combinatorial library of over 9000 components. J Am Chem Soc 130:12218–12219
19. Poulsen S-A (2006) Direct screening of a dynamic combinatorial library using mass spectrometry. J Am Soc Mass Spectrom 17:1074–1080
20. Poulsen S-A, Davis RA, Keys TG (2006) Screening a natural product-based combinatorial library using FTICR mass spectrometry. Bioorg Med Chem 14:510–515
21. McNaughton BR, Miller BL (2003) A mild and efficient one-step synthesis of quinolines. Org Lett 5:4257–4259
22. Gupta RA, Shah N, Wang KC, Kim J, Horlings HM, Wong DJ, Tsai M-C, Hung T, Argani P, Rinn JL, Wang Y, Brzoska P, Kong B, Li R, West RB, van de Vijver MJ, Sukumar S, Chang HY (2010) Long non-coding RNA HOTAIR reprograms chromatin state to promote cancer metastasis. Nature 464:1071–1076
23. Roshan R, Ghosh T, Scaria V, Billai B (2009) MicroRNAs: novel therapeutic targets in neurodegenerative diseases. Drug Discov Today 14:1123–1129

24. Baird SD, Turcotte M, Korneluk RG, Holcik M (2006) Searching for IRES. RNA 12:1755–1785
25. Thomas J, Hergenrother P (2008) Targeting RNA with small molecules. Chem Rev 108:1171–1224
26. Xavier KA, Eder PA, Giordano T (2010) RNA as a drug target: methods for biophysical characterization and screening. Trends Biotechnol 18:349–356
27. Tor Y (2003) Targeting RNA with small molecules. Chembiochem 4:998–1007
28. Chenoweth DM, Dervan PB (2010) Structural basis for cyclic Py-Im polyamide allosteric inhibition of nuclear receptor binding. J Am Chem Soc 132:14521–14529
29. Houghton JL, Green KD, Chen W, Garneu-Tsodikova S (2010) The future of aminoglycosides: the end or renaissance? Chembiochem 11:880–902
30. Karan C, Miller BL (2001) RNA-selective coordination complexes identified via dynamic combinatorial chemistry. J Am Chem Soc 123:7455–7456
31. Liu Y, Tidwell RR, Leibowitz MJ (1994) Inhibition of in vitro splicing of a group I intron of Pneumocystis carinii. J Eukaryot Microbiol 41:31–38
32. Baril M, Dulude D, Gendron K, Lemay G, Brakier-Gingras L (2003) Efficiency of a programmed −1 ribosomal frameshift in the different subtypes of the human immunodeficiency virus type 1 group M. RNA 9:1246–1253
33. Wheeler TM, Thornton CA (2007) Myotonic dystrophy: RNA-mediated muscle disease. Curr Opin Neurol 20:572–576
34. Palde PB, Ofori LO, Gareiss PC, Lerea J, Miller BL (2010) Strategies for recognition of stem-loop RNA structures by synthetic ligands: application to the HIV-1 frameshift stimulatory sequence. J Med Chem 53:6018–6027
35. Scott D, Dawes G, Ando M, Abell C, Ciulli A (2009) A fragment-based approach to probing adenosine recognition sites by using dynamic combinatorial chemistry. Chembiochem 10:2772–2779
36. Voshell SM, Lee SJ, Gagné MR (2006) The discovery of an enantioselective receptor for (−)-adenosine from a racemic dynamic combinatorial library. J Am Chem Soc 128:12422–12423
37. Chung M-K, Hebling CM, Jorgenson JW, Severin K, Lee SJ, Gagné MR (2008) Deracemization of a dynamic combinatorial library induced by (−)-cytidine and (−)-2-thiocytidine. J Am Chem Soc 130:11819–11827
38. Vial L, Ludlow RF, Leclaire J, Pérez-Fernandez R, Otto S (2006) Controlling the biological effects of spermine using a synthetic receptor. J Am Chem Soc 128:10253–10257
39. Sreenivasachary N, Lehn J-M (2005) Gelation-driven component selection in the generation of constitutional dynamic hydrogels based on guanine-quartet formation. Proc Natl Acad Sci USA 102:5938–5943
40. Guschlbauer W, Chantot JF, Thiele D (1990) Four-stranded nucleic acid structures 25 years later: from guanosine gels to telomer DNA. J Biomol Struct Dyn 8:491–511
41. Spada GP, Carcuro A, Colonna FP, Garbesi A, Gottarelli G (1988) Lyomesophases formed by the dinucleoside phosphate D(GPG). Liq Cryst 3:651–654
42. Buhler E, Sreenivasachary N, Candau S-J, Lehn J-M (2007) Modulation of the supramolecular structure of G-quartet assemblies by dynamic covalent decoration. J Am Chem Soc 129:10058–10059
43. Sreenivasachary N, Hickman DT, Sarazin D, Lehn JM (2006) DyNAs: constitutional dynamic nucleic acid analogues. Chem Eur J 12:8581–8588
44. Ura Y, Beirle JM, Leman LJ, Orgel LE, Ghadiri MR (2009) Self-assembling sequence-adaptive peptide nucleic acids. Science 325:73–77
45. Bugaut A, Toulmé J-J, Rayner B (2004) Use of dynamic combinatorial chemistry for the identification of covalently appended residues that stabilize oligonucleotide complexes. Angew Chem Int Ed 43:3144–3147
46. Bugaut A, Toulmé J-J, Rayner B (2006) SELEX and dynamic combinatorial chemistry interplay for the selection of conjugated RNA aptamers. Org Biomol Chem 4:4082–4088

47. Azéma L, Bathany K, Rayner B (2010) 2′-O-Appended polyamines that increase triple-helix-forming oligonucleotide affinity are selected by dynamic combinatorial chemistry. Chembiochem 11:2513–2516
48. Hanvey JC, Williams EM, Besterman JM (1991) DNA triple-helix formation at physiologic pH and temperature. Antisense Res Dev 1:307–317
49. Thomas T, Thomas TJ (1993) Selectivity of polyamines in triplex DNA stabilization. Biochemistry 32:14068–14074
50. Tung CH, Breslauer KJ, Stein S (1993) Polyamine-linked oligonucleotides for DNA triple-helix formation. Nucleic Acids Res 21:5489–5494
51. Nara H, Ono A, Matsuda A (1995) Nucleosides and nucleotides. 135. DNA duplex and triplex formation and resistance to nucleolytic degradation of oligodeoxynucleotides containing syn-norspermidine at the 5-position of 2′-deoxyuridine. Bioconjug Chem 6:54–61
52. Barawkar DA, Rajeef KG, Kumar VA, Ganesh KN (1996) Triplex formation at physiological pH by 5-Me-dC-N-4-(spermine) [X] oligodeoxynucleotides: non protonation of N3 in X of X*G:C triad and effect of base mismatch ionic strength on triplex stabilities. Nucleic Acids Res 24:1229–1237
53. Cuenoud B, Casset F, Hüsken D, Natt F, Wolf RM, Altmann KH, Martin P, Moser HE (1998) Dual recognition of double-stranded DNA by 2′-aminoethoxy-modified oligonucleotides. Angew Chem Int Ed 37:1288–1291
54. Zewail-Foote M, Hurley LH (2001) Differential rates of reversibility of ecteinascidin 743-DNA covalent adducts from different sequences lead to migration to favored bonding sites. J Am Chem Soc 123:6485–6495
55. Wang H, Rokita SE (2010) Dynamic cross-linking is retained in duplex DNA after multiple exchange of strands. Angew Chem Int Ed 49:5957–5960
56. Bowler FR, Diaz-Mochon JJ, Swift MD, Bradley M (2010) DNA analysis by dynamic diversity. Angew Chem Int Ed 49:1809–1812
57. Matray TJ, Kool ET (1999) A specific partner for abasic damage in DNA. Nature 399:704–708
58. Goral V, Nelen MI, Eliseev AV, Lehn JM (2001) Double-level "orthodonal" dynamic combinatorial libraries on transition metal template. Proc Natl Acad Sci USA 98:1347–1352
59. Rodriguez-Docampo Z, Otto S (2008) Orthogonal or simultaneous use of disulfide and hydrazone exchange in dynamic covalent chemistry in aqueous solution. Chem Commun 2008:5301–5303
60. Orrillo AG, Escalante AM, Furlan RLE (2008) Covalent double level dynamic combinatorial libraries: selectively addressable exchange processes. Chem Commun 2008:5298–5300
61. Escalante AM, Orrillo AG, Furlan RLE (2010) Simultaneous and orthogonal covalent exchange processes in dynamic combinatorial libraries. J Comb Chem 12:410–413
62. Leclaire J, Vial L, Otto S, Sanders JKM (2005) Expanding diversity in dynamic combinatorial libraries: simultaneous exchange of disulfide and thioester linkages. Chem Commun 2005:1959–1961

Top Curr Chem (2012) 322: 139–164
DOI: 10.1007/128_2011_199
© Springer-Verlag Berlin Heidelberg 2011
Published online: 19 July 2011

Dynamic Nanoplatforms in Biosensor and Membrane Constitutional Systems

Eugene Mahon, Teodor Aastrup, and Mihail Barboiu

Abstract Molecular recognition in biological systems occurs mainly at interfacial environments such as membrane surfaces, enzyme active sites, or the interior of the DNA double helix. At the cell membrane surface, carbohydrate–protein recognition principles apply to a range of specific non-covalent interactions including immune response, cell proliferation, adhesion and death, cell–cell interaction and communication. Protein–protein recognition meanwhile accounts for signalling processes and ion channel structure. In this chapter we aim to describe such constitutional dynamic interfaces for biosensing and membrane transport applications. Constitutionally adaptive interfaces may mimic the recognition capabilities intrinsic to natural recognition processes. We present some recent examples of 2D and 3D constructed sensors and membranes of this type and describe their sensing and transport capabilities.

Keywords Bilayers · Biosensor · Constitutional dynamic chemistry · Dynamic interfaces · Nanoparticles · Quartz crystal microgravimetry · Surface plasmon resonance

Contents

1 Introduction 140
2 Biosensing Methods/Techniques 143
  2.1 Quartz Crystal Microgravimetry 143
  2.2 Surface Plasmon Resonance 146
  2.3 Electrochemical Sensing Platforms 147

E. Mahon and M. Barboiu (✉)
Institut Européen des Membranes – ENSCM-UMII-CNRS 5635, Place Eugène Bataillon, CC 047, 34095 Montpellier, Cedex 5, France
e-mail: mihai.barboiu@iemm.univ-montp2.fr

T. Aastrup
Attana AB, Björnnäsvägen 21, 11347 Stockholm, Sweden

3 Nanoplatforms for Biosensing and Membranes	147
3.1 Nanoparticles	148
3.2 Dynamic Interfaces	149
4 Conclusions	155
References	156

Abbreviations

Con A	Concanavalin A
Gb3	Globotriaosylceramide
GM1	Monosialotetrahexosylganglioside
HBM	Hybrid bilayer membrane
MIP	Molecularly imprinted polymer
NP	Nanoparticle
QCM	Quartz crystal microbalance
SAM	Self assembled monolayer
SERS	Surface Enhanced Raman Spectroscopy
SPR	Surface plasmon resonance

1 Introduction

Constitutional dynamic chemistry (CDC) and its application dynamic combinatorial chemistry (DCC) are new evolutional approaches to produce *chemical diversity* [1]. Basically, CDC implements a dynamic reversible interface between interacting components. It might mediate the structural self-correlation of different domains of the system by virtue of their basic constitutional behaviours. On the other hand, the self-assembly of the components controlled by mastering molecular/supramolecular interactions may embody the flow of structural information from molecular level toward nanoscale dimensions. Understanding and controlling such up-scale propagation of structural information might offer the potential to impose further precise order at the mesoscale and new routes to obtain highly ordered ultradense arrays over macroscopic distances.

Herein we describe forays to date in the field of adaptive interfaces to investigate biointeractions as well as discussing some emerging possibilities. In recent times, ideas from the fields of synthetic chemistry, nanochemistry and biochemistry have confluenced around the biosensor field (Fig. 1). Nanoplatforms possessing the intrinsic optical, electronic and magnetic properties of nanomaterials when combined with synthetic surface chemistry and an understanding of biomolecules can create opportunities for the transduction of biological interactions.

A biosensor can be described as a complex device for the detection of analytes that combines a biological component with a physicochemical detector component [2, 3]. This device, depending on how it functions, may take up a variety of forms from 2D surface based approaches to 3D micro and nanoplatforms. Examples of 3D

Dynamic Nanoplatforms in Biosensor and Membrane Constitutional Systems

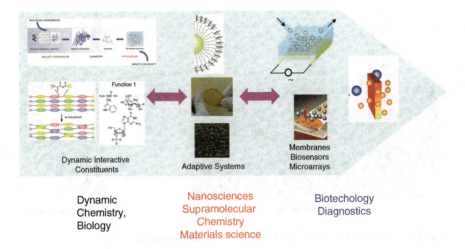

Fig. 1 Dynamic chemistry and biology toward dynamic interactive nanosystems for biotechnological and diagnostics applications

platforms include plasmonic silver or gold nanoparticles which can be applied based on their optical response (SPR); these structures have also been used to augment 2D platform sensitivity by increasing surface area by assembly or taking advantage of nanospecific amplifications (e.g. SERS).The transduction element, converting the biological interaction to a quantifiable signal, can be formed from a variety of materials such as quartz, gold, silicon, etc.

In order to interact with biological components in a specific and readable manner, an interface structure containing the biological recognition element should be bound to the transduction material. Ideally this interface will inhibit non-specific interaction maximizing response to biospecific interaction. A somewhat "bioinspired" feature to introduce to the interface architecture is that of dynamic constitutionality (Fig. 2). This essentially represents a layer in which the components may themselves reorganize within the layer under the influence of a variety of changing parameters such as molecular recognition event, temperature, pH, etc.

For an example of such a layer we only have to look to the biological model of the cell membrane, a structure which is inherently dynamic composed as it is of fluid hydrophobic material, throughout which are distributed the biological transduction elements. The cell membrane interface at which *biological signal transduction* occurs is intrinsically dynamic where a combination of *fluidity* or *mosaicism* contributes to the complexity of the recognition processes [4]. If we are then aiming to understand the nature and energy dynamics of such interactions better through biophysical study, incorporating the thermodynamically controlled dynamic behaviours of the in vivo interface in sensor design represents an important advance.

Understanding biological processes at a molecular level bridges biology and CDC [5, 6]. As "dynamic chemistry" it is underpinned by the labile, reversible yet

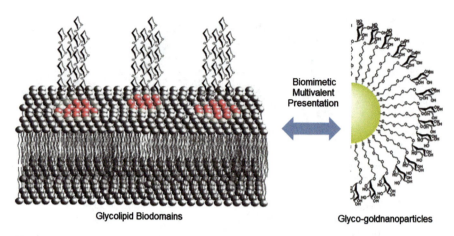

Fig. 2 Glyco-gold nanoparticles as biomimetic multivalent systems

highly specific interactions which exist throughout nature defining structures, mediating information transfer, etc. As an area of research it has been developed over the past 10 years incorporating interactions observed in nature such as hydrophobic interaction, hydrogen bonding, metal ion coordination, electrostatic interactions and π-stacking. Complementary to the novel science based around supramolecular interactions, synthetic bioinspired supramolecular designs can in turn provide a greater depth of understanding of biological mechanisms. Through the principles of self-assembly and molecular recognition allied to synthetic chemistry, we can access a range of self-organized materials which can interact on a biological scale and in a biological manner with the present and growing capability to incorporate built-in nanomaterial signal transmitters from the nanoscale. Molecular recognition events such as carbohydrate–protein or protein–protein interactions occurring at the cell surface can be understood within this concept of chemical collectivism [7–9]; indeed this collective self-assembly can be considered as a basis for the structural evolution of the biosphere [10]. Within this context, during the last decade the CDC is expressing more interest for dynamic constitutional systems, undergoing continual change in constitution, through dissociation/reconstitution of different components during complex processes. This concept can be related to a sort of *"Chemical collectivism"* or of *"Chemical Darwinism"* The reversibility of interactions between components of a system is a crucial factor and, accordingly, the dynamic interfaces might render the emergence system states self-adaptive, which mutually (synergistically) may adapt their spatial/temporal distribution based on their own structural constitution during the simultaneous formation of self-organized domains. The extension of the constitutional chemistry approach to nanoplatforms would be able to compete at multiple length scales within nanoscopic networks and to display variations in their sizes and functionality. Furthermore, we can relate this behaviour to purely synthetic compositions such as the *"Dynamic interactive systems"* [11–13] characterized by their aptitude to organize (self-control) macroscopically their distribution in response to external stimuli in coupled equilibria.

This growing multidisciplinary field of nanobiotechnology could have important contributions to make to the future of medicine [14], part of which, the consideration of molecular recognition events at the cell membrane surface as information transfer, could have important implications when applied to pharmaceutical development [15, 16]. Dynamic interfaces may provide a means to reach greater sensitivities and novel unpredicted behaviours in terms of biological interaction both for in vitro measurement and in vivo applications.

2 Biosensing Methods/Techniques

2.1 Quartz Crystal Microgravimetry

Quartz Crystal Microgravimetry (QCM) sensors work on the phenomenon of piezoelectricity. This mechanical-electrical effect was first reported in 1880 by the Curie brothers describing the generation of electrical charges on the surface of solids caused by pulling, pushing or torsion [17]. Initially it was demonstrated that there exists a linear relationship between mass adsorbed to crystal surfaces and the crystals resonant frequency in air or a vacuum [18]. Extension of this observation to study biological interactions was realized with the design of solution based systems and combination of these with microfluidics and controlled surface chemistry. Consequently, QCM has become a highly relevant analytical technique due to it sensitive solution-surface interface measurement capability. It possesses a wide detection range which at the low end can detect monolayer coverage by small molecules.

Sauerbrey provided the first treatment of the effect of mass loading on quartz resonators (Eq. (1) in [18]). He showed that an ideal layer of foreign mass results in a frequency decrease that is proportional to the deposited mass where the resonator was operated in air or vacuum. Assuming the density of the crystal and the adsorbed layer were equal, then the above-mentioned equation held valid. This equation is valid for a thin, uniform, rigidly attached mass. Application of QCM to biological samples became possible when suitable oscillator circuits for operation in liquids were developed [19]. As the Sauerbrey relationship was formulated for thin rigid films in air or vacuum, questions arose as to the validity of the Sauerbrey relationship in liquid media. For a long time it was postulated that a direct quantification of protein adsorption at functionalized surfaces was possible based on this relationship. However, frequency shifts larger than those observed in air were frequently observed in aqueous media for equivalent proteins. When used in liquid media other important frequency determining factors have to be taken into account. When in contact with a liquid, the frequency depends on the liquid density and also on its viscosity. This modified resonant frequency shift was treated theoretically by

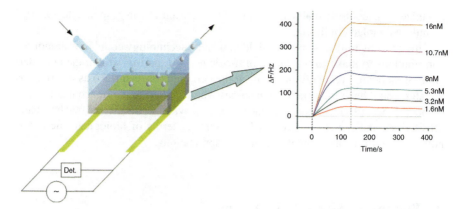

Fig. 3 Representation of QCM experimental setup and concentration assay for gold sugar nanoparticles binding on immobilized lectin layer

Kanazawa et al. [20] and their calculations are valid for rigid films immersed in liquid. However, this work is not valid for non-rigid "soft" materials. For soft material layers it was even noticed that in applying Sauerbrey's equation the mass of the viscoelastic layer is underestimated and the result is a "missing mass" which was elucidated in calculations by Voinova et al. [21].

Alongside viscoelastic properties, electrolyte contributions, surface roughness and surface energy changes may have to be considered. For example, changes in hydrophilicity can cause very large response changes in QCM as a function of surface roughness [22, 23]. Rough and hydrophilic surfaces entrap liquids in small cavities contributing to the overall mass detected. Hydrophobic surface cavities may be unwetted and so contain air or vacuum cavities resulting in an artificially large frequency change on hydrophilization of the surface. These problems can be kept to a minimum by optimizing surface smoothness. Although on analysing soft material adsorption in liquid environments the frequency shift cannot be translated directly to mass load according to the Sauerbrey relationship, the quartz crystal microbalance can be used for label free analysis of binding events. Concentration dependant measurements of the frequency shift together with the assumption of a linear relationship between ΔF and Δm allow thermodynamic and kinetic parameters of binding events to be determined as has been demonstrated in numerous examples described in the following section (Fig. 3).

2.1.1 Quartz Crystal Microgravimetry in Biosensing

QCM has been used extensively in the areas of DNA hybridization, protein adsorption studies, immunological systems and also in the areas of protein–protein and protein–carbohydrate interaction [24]. Examples prepared using carbohydrate SAMs to function as lectin biosensors are known since initially it was shown through investigation by quartz crystal microbalance that self-assembled monolayers (SAMs) of Gb3 mimics having different lengths of alkyl chains prepared on gold surfaces could interact with galactose-specific lectin *ricinus communis agglutin* (RCA$_{120}$) and Shiga toxins (Stxs) [25]. Later it was shown that the α-Gal carbohydrate antigen interacted in a specific manner with tri-Gal presenting SAMs [26]. "Click chemistry" was introduced in later work for the preparation of carbohydrate functionalized crystal surfaces which then showed selective lectin recognition [27, 28]. There are far more examples if one looks to oligonucleotide interactions with the first direct DNA detection using QCM back in 1988 [29]. Following this there were many more studies carried out using immobilized oligonucleotide surfaces. These immobilizations were generally performed using biotin–avidin [30], or gold–thiol interactions [31, 32] with varying surface performances ensuing.

Polymer adsorption has also been adapted to QCM sensing whereby biofunctional thin films are adsorbed on the crystal surface with non-specific binding controlled by tuning of polymer composition. This approach proved successful as applied to carbohydrate–protein interaction by Matsuura et al. through adsorption of lactose bearing amphiphilic polymers on hydrophobic surfaces which then showed RCA$_{120}$ and peanut lectin (PNA) affinity [33]. Carbohydrate surfaces prepared by photo insertion into an adsorbed polymer were tested by QCM and showed the predicted affinities [34] while in another example a covalently bound glycopolymer demonstrated Concanavalin A detection ability [35].

A general limitation on QCM sensitivity is the mass of the analyte which is to be adsorbed from solution. At present QCM does not have single small molecule sensitivity. The most common procedure for QCM detection of protein–carbohydrate interaction is immobilization of the small molecule as the surface bound receptor in SAMs [26–28] or polymer films [34, 36] followed by monitoring of the binding of the relatively large protein, antibody, etc. Amplification methods may thus be employed in order to study certain biorecognition processes. Two previously demonstrated ways in which to achieve amplification are mass increase, i.e. the introduction of nanoplatforms such as nanoparticles (NPs), vesicles [37] and micelles to carry the recognizing small molecule element, and also by increasing recognizing surface area, e.g. porous films, MIP multilayers, etc. [38]. Nanoparticles as frequency change enhancement probes have already been demonstrated for the biotin–streptavidin interaction [39] and also for enhanced DNA detection by QCM. DNA-conjugated nanoparticles can be used to enhance the signal produced upon hybridization to a surface bound single-stranded template [40]. The carbohydrate–lectin interaction was also recently investigated by Barboiu et al. using this NPs–QCM approach (Fig. 4) [41] It can be concluded that NPs act as efficient QCM signal enhancers. The amplification is considerable, which is to be expected given an estimated increase in the region of ~6,000 mass%. NPs and QCM can go hand in hand in order to expand the applicability of the QCM technique to investigate small molecule interactions in real-time. Frequency changes of 1,600 Hz were observed for complete NP saturation while typical protein layers would show changes in the region of 100 Hz. GlycoNPs have been described as being biomimetic in their multivalent sugar presentation and the results presented concur. Large avidity increases with

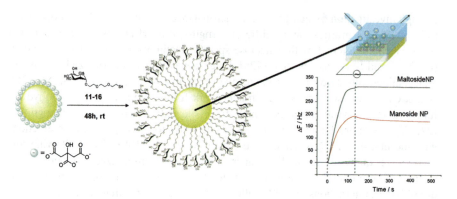

Fig. 4 Glyconanoparticles recognition by ConA functionalized QCM electrode surfaces

enhancement factors (β) of 2,000 and 25,000 for Mannoside and Maltoside nanoparticles respectively were shown, relative to the corresponding monomer [41].

2.2 Surface Plasmon Resonance

SPR is one of the most commonly employed surface sensing techniques today [42]. Like QCM it is a surface analytical technique capable of in situ monitoring of interfacial processes. Whereas QCM is an acoustic wave device where the QCM oscillation frequency and quality are related to the mass loading and the viscoelastic properties of the adsorbed materials, Surface Plasmon Resonance (SPR), another label-free detection technique, is based on the resonance coupling between incident light and a gold surface plasmon wave. In a classical SPR instrument, this occurs at a typical incident angle, causing a minimum in the reflectivity of the p-polarized light beam. The value of this critical angle depends on the thickness and refraction index of any layer adsorbed on the gold surface. The measured signals are thus proportional to the molecular weight of the adsorbed materials, and can be used to quantify the number density of different types of adsorption [43]. SPR measures "dry mass" meaning that the signal measured is not sensitive to water associated with macromolecules. This is not the case for QCM where the response may also contain an associated water contribution. Adsorbed molecular mass may thus be more directly discerned from SPR data.

It appears that in most cases the techniques are comparable in terms of resulting sensitivity. Whereas the SPR has the advantage in terms of real sensitivity, the fact the QCM technique also measures entrapped water amplifies the gravimetric response and may render it sensitivity comparable in macromolecular binding experiments [44, 45]. An added advantage of QCM over SPR is the availability of the QCM-D technique which is a measurement of the dissipation energy. A film that is viscoelastic or "soft" will not fully couple with the quartz crystal's oscillation and in doing so will dampen it. This damping thus contains information as to the film's viscoelastic properties [46].

2.2.1 Surface Plasmon Resonance in Biosensing

SPR has been applied in the study of a range of biological interactions including carbohydrate–protein, protein–protein, DNA hybridization and DNA–protein interaction [47]. Recent developments in interface engineering have allowed the technique to approach small molecule sensitivity further [48]. Binding and cross reactivity studies are becoming possible by using glyconanoparticles and lectins layers. Ramstrom et al. [49] showed that glyconanoparticles exhibit affinity ranking consistent with that of free ligands in solution. Like QCM, a major advantage of this technique is that it allows label free analysis, thus avoiding problems such as structural interference caused by fluorescent labelling techniques. It is also amenable to amplification techniques to increase sensitivity such as those using plasmonic particles [50–54].

2.3 Electrochemical Sensing Platforms

The electrochemical based biosensors are the most widely used format in biosensing. Typically the reaction under investigation would either generate a measurable current (amperometric), a measurable potential or charge accumulation (potentiometric) or measurably alter the conductive properties of a medium (conductometric) between electrodes [55].

The coupling of biomimetic dynamic interfaces with electrochemical impedance spectroscopy has shown value in the study of ion transport studies across tethered bilayers [56]. Electrochemical impedance spectroscopy may prove valuable used in conjunction with the previously described techniques [57, 58]. Again electrode construction with readily chemically adaptable surface materials such as gold, silver and glass/silica amongst others make this a promising approach for introducing functional interfaces.

3 Nanoplatforms for Biosensing and Membranes

Multivalent interactions represent multiple copies of a specific recognition element connected in some way to a central scaffold. Resultant affinity can be highly dependent on the nature of the scaffold as multivalent interactions have been shown to be influenced by such factors as shape, valency, orientation and flexibility. In this context, adaptive scaffolds can exhibit a large range of *dynamic valencies*. Lower valency categories are supramolecules such as crown ethers, cryptands, calixarenes and porphyrins and higher valency scaffolds include dendritic structures, nanoparticles or confined pores. The self-assembly of the system components controlled by mastering molecular/supramolecular interactions may embody the flow of structural information from molecular level toward nanoscale dimensions. Resultant spatio/temporal affinity is highly dependent on the nature of the scaffold as multivalent interactions have been shown to be influenced by such factors as shape, valency, orientation and flexibility. The possibility of designing and engineering nanometric multivalent platforms is at the forefront of cross-disciplinary oriented research. Natural systems have evolved for billions of years to accept complex evolutive structures. Using this model, the synthetic systems could be extended to the vast field of scientific challenges, for example, resulting in the property (function)-driven generation of new adjustable (adaptive) system structures. Various 2D platform types for such sensing have be applied, adapted to SPR, QCM and electrical methods, as well as 3D surface approaches utilizing the intrinsic physical properties. Dynamic self assembled *monolayers, supported bilayers* and *dynameric films* can represent platforms for studying multivalent interactions. Such evolutive dynamic devices at the nanoscale may possess novel properties not present in the bulk material. The integration of dynamic systems might give great potential in applications.

3.1 Nanoparticles

Nanoparticles display large surface to volume ratios and can report on biological interaction directly from nanodispersions in biological media [59]. The intrinsic nanoscale property of the material can enable highly sensitive transduction of biointeractions.

This is the case for gold and silver nanoparticles which have sensing capability based on the sensitivity of their surface plasmon absorbance band to the surrounding environment. This absorption band shows an interparticle distance dependency which is very useful in terms of selective particle aggregation studies as a means of sensing [60]. Aggregation causes a coupling of the gold nanoparticle's plasma modes, which results in a red shift and broadening of the longitudinal plasma resonance in the optical spectrum [61]. It has been observed in theoretical and experimental studies that when the individual spherical gold particles come into close proximity to one another, electromagnetic coupling of clusters becomes effective for cluster–cluster distances smaller than five times the cluster radius ($d < 5R$, where d is the centre-to-centre distance and R is the radius of the particles) and may lead to complicated extinction depending on the size and shape of the formed cluster aggregate. This effect is negligible if $d > 5R$ but becomes increasingly important at smaller distances [62]. The wavelength at which absorption due to dipole–dipole interactions occurs may vary from 520 nm (effectively isolated particles) through 750 nm (particles that are separated by only 0.5 nm), and the resulting spectra are a composite of the conventional plasmon resonance due to single spherical particles and the new peaks resulting from particle–particle interactions [63]. This interparticle plasmon coupling during biointeraction induced aggregation has been exploited for DNA and antibody detection and colorimetric detection of lectin-carbohydrate interactions using functionalized particles [64–66] among others. Non-specific adsorption of biomacromolecules can be minimized by using ethylene glycol based linkers to protect the gold core [67].

Other nanomaterials can also be employed in the fabrication of nanoparticles. The intrinsic photoluminescence of quantum dots has been utilized for specific protein sensing [68, 69]. Composites of the aforementioned gold NP, silver NP and quantum dots usually with silica or polymeric materials have also been employed as described in a recent review [70] The multivalent presentation of carbohydrate ligands significantly enhances the binding affinity toward lectins by several orders of magnitude in comparison to the free ligands in solution and strongly depends on the type and length of the spacer linkage and ligand density [71].

The final goal is subsequently to apply the nanodevices directly in vivo, using the powerful detection abilities of NPs to localize and image static and tumour cells efficiently. Interestingly, the strong binding of gold glyconanoparticles on bacterial cell surfaces gives rise to the high potential of labelling proteins on in vivo biological surfaces using carbohydrate nanoplatforms [72].

3.2 Dynamic Interfaces

Different interfaces are known to show constitutionally dynamic behaviours:

1. Biomimetic interfaces including lipid bilayers, hybrid bilayers, tethered bilayers as well as other interdigitated layers are composed of naturally existing constituents based mainly on the hydrophobic interaction.
2. Non-biomimetic interfaces which are constructed as they are from supramolecular associations or reversible covalent linkages. An approach to such an interface can be envisaged through the application of supramolecular chemistry and dynamic constitutional chemistry which are both conducive to adaptive structures.

3.2.1 Supported Bilayers

Supported bilayers represent biomimetic layers which can be supported on a range of materials and adapted for the study of biointeractions (protein–protein, lipid–lipid) including molecular recognition, ion-channel transport and intramembrane interactions. This interface type can be separated into the so-called SLBs (supported lipid bilayers), HBMs (hybrid bilayer membranes) and t-BLMs (tethered bilayer membranes).

The SLB is conventionally formed on silica surfaces and stabilized by interactions between the hydrophilic hydroxyl surface and the phospholipid headgroups. It is purely a phospholipid bilayer stabilized only by electrostatics and the hydrophobic effect, which, through incorporation of bioactive structures such as membrane proteins, glycolipids and glycoproteins, can give a biomimetic response with interacting species in solution.

A range of molecular recognition based interactions have been studied using this layer type. The carbohydrate–protein interaction was investigated by Liebau et al. who, using QCM and glycolipid as the dynamic recognition element, showed some specific lectin adhesion to a supported lipid bilayer [73].

SPR imaging was also employed to look at the same cholera toxin–GM1 interaction on a supported bilayer by Philips et al. in array experiments [74]. Another investigative technique applied to the model ganglioside–cholera toxin system was total internal reflection fluorescence microscopy applied to spatially addressable supported bilayer membranes [75]. The recognition by HIV-1 surface glycoprotein gp120 of glycosphingolipids partitioned within an SBM has also been looked at by total internal reflection fluorescence [76].

Other types of cell membrane interaction have also been examined. For example Wang et al. have used an electrochemical SPR sensor to monitor peroxidase enzyme activity within the plasma membrane [77]. As a means of improving stability and creating a closer approximation of a true cell membrane, interlayer structures between the lower phospholipid leaf and the solid supports have been tested. Cushion layers based on PEG to enable transmembrane protein insertion were tested by Munro et al. [78], while Schuster et al. have reviewed the use of S-layer proteins as a construction element [79, 80].

Solid supported bilayers of other composition have also been reported such as lipopolysaccharide bilayers which can mimic the outer bacterial membrane with this example used to monitor bacteriophage interaction by AFM [81].

3.2.2 Hybrid Bilayers

The HBM consists of two differing leaves normally generated on a gold or silica surface. The lower leaf is a fixed long chain alkyl self-assembled monolayer covalently bound to a solid support while the upper leaf is a phospholipid mono-layer [82, 83]. This type of bilayer may show increased stability due to the covalent nature of the lower leaf fixation; however it would present a system further removed in structure from the biological condition given this fixed nature and more limited fluidity [84].

In terms of biosensing applications using such layers, again cholera toxin detection on a porous silicon substrate [85] has been reported. Also biotin–avidin interaction by QCM [86], glutamate detection [87], as well as protein membrane interactions [88, 89] have been studied.

3.2.3 Tethered Bilayers

The tethered bilayer offer improvements in both stability and in terms of its approximation of the true cell membrane through incorporation of tethering links

Dynamic Nanoplatforms in Biosensor and Membrane Constitutional Systems 151

between the bilayer and solid support which allows both inner and outer phospholipid leaf to interface with an aqueous domain [90–92]. The result is a self-assembled bilayer which exhibits a closer approximation of true cell membrane fluidity [93]. Such a construction also allows for the accommodation of proteins [94], incorporated synthetic ion-channels [95] and sugar lectin systems [96]. Taylor et al. has reported on SPR detection of cholera toxin using a tethered bilayer [97]. Cornell et al. investigated ion channel based sensors with tethered bilayers [98].

3.2.4 Vesicular Systems

Vesicles have demonstrated their dynamic behaviour in studies showing evidence of rapid domain formation in response to external stimuli [9]. The binding of glycosidic domains to cholera toxin has been demonstrated using vesicles as biomimetic sensors [99]. Protein–carbohydrate interaction has also been studied using synthetic glycolipids recently by Barboiu et al. [100] (Fig. 5).

They demonstrated an affinity enhancement of mannoside vesicles when compared with the Me-α-glucose monosaccharide in solution. The 40-fold increase can be attributed to multivalent glycoside presentation, allied to their fluidity within bilayers. The dynamic hydrophobic interface between carbohydrate molecules and bilayers may mediate the structural self-correlation of supramolecular sugar clusters by virtue of their basic constitutional behaviours. The resultant dynamic system can undergo continual change in its constitution through dissociation/reconstitution of different mesophases during the vesicle recognition process. Artificial glycocalix bilayer vesicles decorated with sugars reversibly agglutinate, depending on the

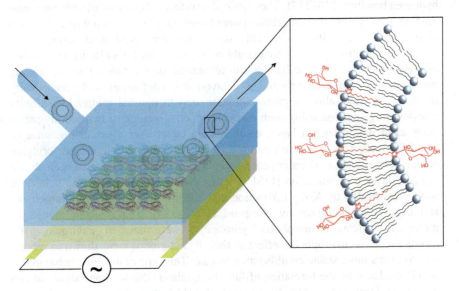

Fig. 5 Con A layers exposed to functionalized vesicle solutions within a QCM system

surface coverage of sugars. These dynamic constitutional vesicles operating via multivalent orthogonal interactions operating simultaneously are nice examples for further understanding molecular recognition at cell surfaces [101].

3.2.5 Supported Self-Assembled Monolayers and Bilayers on Surfaces

As an intermediate between solid supported layers and the inherent dynamic and nanostructured properties of phospholipid vesicle supports, silica and especially mesoporous silica nanoparticles may provide interesting platforms for dynamic bilayers. Previous studies have shown that stable bilayers can form on both amorphous [102] or functional silica [103, 104] and mesoporous nanoparticles [105] or membranes [106]. This type of biomimetic carrier has great potential as a type of trackable stabilized membrane capable of displaying cellular targeting elements in a close to natural configuration.

Besides the bilayer type dynamic interfaces there also exist dynamic interfaces involving SAMs. The cooperative dynamic nature of SAMs on gold has been confirmed previously [107]. On 2D surfaces the level of mobility in gold–thiol SAMs has been looked at and is considered quite slow ranging from 10^{-18} to 10^{-14} cm^2 s^{-1} (~1 nm/h) [109, 108] when compared to phospholipid and membrane protein, lateral mobilities of around 10^{-8} cm^2 s^{-1}.

A few different studies have shown evidence for thiol mobility in the presence of supramolecular driving forces. Mixed monolayer mobility has been studied by Rotello and co-workers and has been shown to be influenced by intramonolayer hydrogen bonding [110–112]. They noticed a surface adaptation towards increased binding affinity over time for their system based on a mixed monolayer protected cluster interacting with flavin. This shows that functional gold nanoparticles possess an adaptive quality which could be exploited for high affinity recognition of biomacromolecules. The ligands can reorganize on the particle surface during interactions to the most energetically favourable configuration. Hypothetically speaking this may allow a preconditioning step to optimal multivalent affinity monomer organization using such particles which, given the slow mobility previously cited within such a layer, could remain intact over a useful timescale or could be locked in place using, for example, a monomer–monomer bridging photoreaction. Supramolecular peptide–peptide interactions have also been shown to cause layer reorganization [113]. Another example of this dynamic behaviour was demonstrated by X-ray diffraction and reflectivity on particles immobilized at the air–water interface by Nørgaard et al. [114]. These mixed monolayer protected particles (hydrophobic/hydrophilic) were deemed environmentally responsive to the hydrophobic effect as their ligand groups were shown to reorganize to give a more stable equilibrium interface. This type of dynamic behaviour is also in evidence by the formation of functional islands due to phase separation on particle surfaces [115, 116]. In the case of gold NPs perhaps a more dynamic interface layer would result by employing a more labile anchoring group than thiol, such as an amino, phosphine or carboxylate.

3.2.6 Supramolecular Networks and Constitutionally Dynamic Polymers

Materials based on reversibly covalently bound constituents could also be applied in the construction of constitutionally dynamic systems [117]. For example, the "glycodynamers" could be exploited as adaptive layers for selective protein sensing [118] or adaptive nucleic acid [119, 120] displays could be exploited for DNA sensing applications. These sensing interfaces could be "pre-conditioned" and fixed for optimal sensing of the choice analyte using the principles already established in DCC [1].

Supramolecular networks film deposited on 2D surfaces provides an interesting route to adaptive functional surfaces [121, 122]. Such films can take the form of metallo-organic frameworks [123], hydrogen bonded networks, Van der Waals etc. or combinations of the above. There are very few examples in the literature of the employment of such a film although it has been demonstrated for hydrogen peroxide sensing [124]. An elegant use of reversible covalently structured architecture for a condition dependent detachable shell nanodelivery vehicle was demonstrated by Oishi et al. Through incorporation of disulfide bridges as a structuring element in their transmembrane delivery studies they showed that exposure to disulfide reducing cytosolic glutathione increased the efficiency of load delivery [125]. This type of layer can be imagined as a constitutionally dynamic framework which, through the incorporation of biorecognition elements, can open new avenues in adaptive biosensing [126].

Fullerene–sugar conjugates with a specific multivalent presentation have recently been used for achieving high affinity interactions with lectins and enzymes [127–129]. These nanoplatforms have been used as inhibitors for *Escherichia coli FimH*, showing interesting antiadhesive bacterial properties [128] or as strong inhibitors of glycosidase bearing an important multivalent effect [129].

3.2.7 Systems Membranes

Besides the bilayer type dynamic interfaces there also exists dynamic interfaces which can be constructed based on the principles of supramolecular and dynamic covalent chemistry and which can be applied toward sensing and targeting at the molecular level. New routes toward self-preparation of highly ordered ultradense nanoarrays over macroscopic distances led to membrane systems where the concept of self-organization and a specific function (i.e. generation of specific translocation pathways in a hybrid solid) might in principle be associated [130–134]. The resulting hybrid materials are composed from functional domains self-organized at the nanometric level, randomly ordered in the hybrid matrix. These oriented nanodomains result from the controlled self-assembly of simple molecular components which are then fixed by sol-gel transcription. These results provide new insights into the basic features that control the convergence of *supramolecular self-organization* with the conduction *supramolecular function* and suggest tools for developing the next generation of highly conductive materials. This is reminiscent of the supramolecular organization of binding sites in channel-type proteins collectively contributing to the selective translocation of solutes along the hydrophilic pathways. On the other hand, the weak supramolecular interactions' (H-bonds, van der Waals, etc.) positioning of the molecular components are typically less robust than the cross-linked covalent bonds formed in a polymerization process, like the sol-gel. Accordingly, a sole solution to overcome these difficulties is to improve the binding (association) efficiency of molecular components generating supra-molecular assemblies. At least in theory, an increased number of interactions between molecular components and the right selection of the solvent might improve the stability of the templating supramolecular systems, communicating with the inorganic siloxane network. Among these systems the nucleobases and nucleosides are well known fascinating compounds with a great ability to form directionally controlled multiple H-bonding, hydrophobic and stacking interactions [135, 136]. Their remarkable self-association properties play a critical role in the stabilization of higher-order RNA hairpins loops, double or triple helix DNA and G-quartets or G-quadruplexes [137, 138].

Many fundamental biological processes appear to depend on unique properties of inner hydrophilic domains of the membrane proteins in which ions or water molecules diffuse along the directional pathways. In the membrane protein systems simple *inner functional moieties* (i.e. carbonyl, hydroxyl, amide, etc.) are pointing toward the protein core, surrounded by the *outer scaffolding protein wall* orienting the transport direction. Bilayer or nanotube artificial membranes were developed with the hope of mimicking the natural ion-channels, which can directly benefit the fields of separations, sensors or storage-delivery devices.

Supramolecular columnar ion-channel architectures confined within scaffolding hydrophobic silica mesopores can be structurally determined and morphologically tuned by alkali salts templating [130, 131]. The dynamic character allied to reversible interactions between the continually interchanging components make them respond to external ionic stimuli and adjust to form the most efficient transporting superstructure, in the presence of the fittest cation, selected from a set of diverse less-selective possible architectures which can form by their self-assembly. Evidence has been presented that such a membrane adapts and evolves its internal structure so as to improve its ion-transport properties: the dynamic non-covalent bonded macrocyclic ion-channel-type architectures can be morphologically tuned by alkali salts templating during the transport experiments or the conditioning steps [95].

From the conceptual point of view these membranes express a synergistic adaptive behaviour: the addition of the fittest alkali ion drives a constitutional evolution of the membrane toward the selection and amplification of the specific transporting superstructures within the membrane in the presence of the cation that promoted its generation in the first place. This is a nice example of dynamic self-instructed ("trained") membranes where a solute induces the upregulation of its

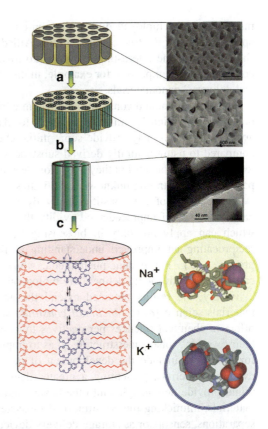

Fig. 6 Schematic representation of the synthetic route to obtain constitutional silica mesoporous membranes is (**a**) filled with mesostructured silica-CTAB, (**b**) then calcinated, (**c**) reacted with hydrophobic ODS and finally filled with the hydrophobic carriers. Generation of directional ion-conduction pathways which can be morphologically tuned by alkali salts templating within *dynamic hybrid materials* by the hydrophobic confinement of ureido-macrocyclic receptors within silica mesopores [130]

own selective membrane. This led to the discovery of the functional supramolecular architecture evolving from a mixture of reversibly insidepore exchanging devices via ionic stimuli so as to improve membrane ion transport properties. These phenomena might be considered as an upregulation of the most adapted 3D "*insidepore*" superstructure, enhancing the membrane efficiency and the selectivity by the binding of the ion-effectors (Fig. 6).

Finally these results extend the application of *CDC* from materials science to *functional Dynamic Interactive Membranes- Systems-Membranes*, using *dynamic intrapore resolution* towards *dynamic hybrid materials and membranes*. Such systems evolve to form the fittest insidepore architecture, demonstrating flexible functionality and adaptation in confined conditions.

4 Conclusions

Looking through the literature to date it is apparent that the use of constitutionally dynamic interfaces for biorecognition and membrane transport represents an emerging field. There are quite a few examples where biomimetic type supported membranes are employed due to the continual improvements in their structural approximation of the true cell membrane. Allied to in vitro cell work these studies should yield some interesting insights into membrane recognition processes and could readily be employed, for example, in the characterization of novel "targeting and delivery" nanomaterials.

Supramolecular and constitutional dynamic interfaces and layers have evidently not been consciously much employed to date in biosensing and transport applications. They may provide some initial chemical synthetic difficulties when compared to using naturally derived substances such as phospholipid and cholesterol based components but they may provide a route to nanostructured surfaces and particles demonstrating unknown specificities and behaviours.

The benefits of the constitutionally dynamic approach are many, including nanomedicines of unprecedented avidity through optimal multivalent adaptation which can apply not only in biosensing but also in nanotargetting therapies, reciprocating an improved understanding at the molecular level of biological processes, e.g. multivalent phenomena, while at the same time enhancing our capability in influencing deliberately biological processes.

Finally, the results obtained extend the application of *CDC* [1, 139] from materials science [8] to *functional constitutional devices* [95, 140]. This feature offers to chemical sciences perspectives towards self-designed systems evolving their own functional superstructure so as to improve their performances. Prospects for the future include the development of these original methodologies towards dynamic materials, presenting a greater degree of structural complexity. They might provide new insights into the basic features that control the design of new materials, mimicking the constitutional biosystems with applications in chemical separations, sensors or as storage-delivery devices.

Acknowledgments This work was financed as part of the Marie Curie Research Training Network- "DYNAMIC" (MRTN-CT-2005-019561) and of a EURYI scheme award. See www.esf.org/euryi.

References

1. Lehn JM (2007) From supramolecular chemistry towards constitutional dynamic chemistry and adaptive chemistry. Chem Soc Rev 36:151–160
2. Thevenot DR, Toth K, Durst RA, Wilson GS (2001) Electrochemical biosensors: recommended definitions and classification. Anal Lett 34:635–659
3. Rapp BE, Gruhl FJ, Lange K (2010) Biosensors with label-free detection designed for diagnostic applications. Anal Bioanal Chem 398:2403–2412
4. Vereb G, Szollosi J, Matko J, Nagy P, Farkas T, Vigh L, Matyus L, Waldmann TA, Damjanovich S (2003) Dynamic, yet structured: the cell membrane three decades after the Singer-Nicolson model. Proc Natl Acad Sci U S A 100:8053–8058
5. Lehn JM (1990) Perspectives in supramolecular chemistry – from molecular recognition towards molecular information-processing and self-organization. Angew Chem Int Ed 29:1304–1319

Dynamic Nanoplatforms in Biosensor and Membrane Constitutional Systems

6. Lehn J-M (1999) Dynamic combinatorial chemistry and virtual combinatorial libraries. Chem Eur J 5:2455–2463
7. Menger FM (1991) Groups of organic-molecules that operate collectively. Angew Chem Int Ed Engl 30:1086–1099
8. Barboiu M (2010) Dynamic interactive systems: dynamic selection in hybrid organic-inorganic constitutional networks. Chem Commun (Camb) 46:7466–7476
9. Christian DA, Tian AW, Ellenbroek WG, Levental I, Rajagopal K, Janmey PA, Liu AJ, Baumgart T, Discher DE (2009) Spotted vesicles, striped micelles and Janus assemblies induced by ligand binding. Nat Mater 8:843–849
10. Walde P (2006) Surfactant assemblies and their various possible roles for the origin(s) of life. Orig Life Evol Biosph 36:109–150
11. Nguyen R, Allouche L, Buhler E, Giuseppone N (2009) Dynamic combinatorial evolution within self-replicating supramolecular assemblies. Angew Chem Int Ed Engl 48:1093–1096
12. Giusepponne N, Lehn J-M (2006) Protonic and temperature modulation of constituent expression by component selection in a dynamic combinatorial library of imines. Chem Eur J 12:1715–1722
13. Mihai S, Cazacu A, Arnal-Herault C, Nasr G, Meffre A, van der Lee A, Barboiu M (2009) Supramolecular self-organization in constitutional hybrid materials. New J Chem 33: 2335–2343
14. Ariga K, Ji QM, Hill JP, Kawazoe N, Chen GP (2009) Supramolecular approaches to biological therapy. Expert Opin Biol Ther 9:307–320
15. Luca C, Barboiu M, Supuran CT (1991) Stability constant of complex inhibitors and their mechanism of action. Rev Roum Chim 36(9–10):1169–1173
16. Barboiu M, Supuran CT, Menabuoni L, Scozzafava A, Mincione F, Briganti F, Mincione G (1999) Carbonic anhydrase inhibitors, synthesis of topically effective intraocular pressure lowering agents derived from 5-(aminoalkyl-carboxamido)-1,3,4-thiadiazole-2-sulfon-amide. J Enz Inhib 15:23–46
17. Curie J, Curie P (1880) An oscillating quartz crystal mass detector. Comp Rend 91:294–297
18. Sauerbrey G (1959) Verwendung Von Schwingquarzen Zur Wagung Dunner Schichten Und Zur Mikrowagung. Zeitschrift Fur Physik 155:206–222
19. Nomura T, Okuhara M (1982) Frequency-shifts of piezoelectric quartz crystals immersed in organic liquids. Anal Chim Acta 142:281–284
20. Kanazawa KK, Gordon JG (1985) Frequency of a quartz microbalance in contact with liquid. Anal Chem 57:1770–1771
21. Voinova MV, Jonson M, Kasemo B (2002) 'Missing mass' effect in biosensor's QCM applications. Biosens Bioelectron 17:835–841
22. Beck R, Pittermann U, Weil KG (1992) Influence of the surface microstructure on the coupling between a quartz oscillator. J Electrochem Soc 139:453–461
23. Martin SJ, Frye GC, Ricco AJ, Senturia SD (1993) Effect of surface-roughness on the response of thickness-shear mode resonators. Anal Chem 65:2910–2922
24. Cooper MA, Singleton VT (2007) A survey of the 2001 to 2005 quartz crystal microbalance biosensor literature: applications of acoustic physics to the analysis of biomolecular interactions. J Mol Recognit 20:154–184
25. Miura Y, Sasao Y, Dohi H, Nishida Y, Kobayashi K (2002) Self-assembled monolayers of globotriaosylceramide (Gb3) mimics: surface-specific affinity with shiga toxins. Anal Biochem 310:27–35
26. Zhang Y, Telyatnikov V, Sathe M, Zeng XQ, Wang PG (2003) Studying the interaction of alpha-Gal carbohydrate antigen and proteins by quartz-crystal microbalance. J Am Chem Soc 125:9292–9293
27. Zhang Y, Luo SZ, Tang YJ, Yu L, Hou KY, Cheng JP, Zeng XQ, Wang PG (2006) Carbohydrate-protein interactions by "clicked" carbohydrate self-assembled monolayers. Anal Chem 78:2001–2008

28. Shen Z (2007) Nonlabeled quartz crystal microbalance biosensor for bacterial detection using carbohydrate and lectin recognitions. Anal Chem 79:2312–2319
29. Fawcett NC, Evans JA, Chien LC, Flowers N (1988) Nucleic-acid hybridization detected by piezoelectric resonance. Anal Lett 21:1099–1114
30. Zhou XC, Huang LQ, Li SFY (2001) Microgravimetric DNA sensor based on quartz crystal microbalance: comparison of oligonucleotide immobilization methods and the application in genetic diagnosis. Biosens Bioelectron 16:85–95
31. Mannelli F, Minunni A, Tombelli S, Wang RH, Spiriti MM, Mascini M (2005) Direct immobilisation of DNA probes for the development of affinity biosensors. Bioelectrochemistry 66:129–138
32. Hook F, Ray A, Norden B, Kasemo B (2001) Characterization of PNA and DNA immobilization and subsequent hybridization with DNA using acoustic-shear-wave attenuation measurements. Langmuir 17:8305–8312
33. Matsuura K, Tsuchida A, Okahata Y, Akaike T, Kobayashi K (1998) A quartz-crystal microbalance study of adsorption behaviors of artificial glycoconjugate polymers onto chemically modified gold surfaces and their interactions with lectins. Bull Chem Soc Jpn 71:2973–2977
34. Pei YX, Yu H, Pei ZC, Theurer M, Ammer C, Andre S, Gabius HJ, Yan MD, Ramstrom O (2007) Photoderivatized polymer thin films at quartz crystal microbalance surfaces: sensors for carbohydrate-protein interactions. Anal Chem 79:6897–6902
35. Huang MC, Shen ZH, Zhang YL, Zeng XQ, Wang PG (2007) Alkanethiol containing glycopolymers: a tool for the detection of lectin binding. Bioorg Med Chem Lett 17:5379–5383
36. Cui RJ, Huang HP, Yin ZZ, Gao D, Zhu JJ (2008) Horseradish peroxidase-functionalized gold nanoparticle label for amplified immunoanalysis based on gold nanoparticles/carbon nanotubes hybrids modified biosensor. Biosens Bioelectron 23:1666–1673
37. Hildebrand A, Schaedlich A, Rothe U, Neubert RHH (2002) Sensing specific adhesion of liposomal and micellar systems with attached carbohydrate recognition structures at lectin surfaces. J Colloid Interface Sci 249:274–281
38. Avila M, Zougagh M, Escarpa A, Rios A (2008) Molecularly imprinted polymers for selective piezoelectric sensing of small molecules. Trac-Trends Anal Chem 27:54–65
39. Kim NH, Baek TJ, Park HG, Seong GH (2007) Highly sensitive biomolecule detection on a quartz crystal microbalance using gold nanoparticles as signal amplification probes. Anal Sci 23:177–181
40. Nie LB, Yang Y, Li S, He NY (2007) Enhanced DNA detection based on the amplification of gold nanoparticles using quartz crystal microbalance. Nanotechnology 18:305501–305505
41. Mahon E, Aastrup T, Barboiu M (2010) Multivalent recognition of lectins by glyconanoparticle systems. Chem Commun (Camb) 46:5491–5493
42. Paul S, Vadgama P, Ray AK (2009) Surface plasmon resonance imaging for biosensing. IET Nanobiotechnol 3:71–80
43. Stenberg E, Persson B, Roos H, Urbaniczky C (1991) Quantitative-determination of surface concentration of protein with surface-plasmon resonance using radiolabeled proteins. J Colloid Interface Sci 143:513–526
44. Su X, Wu Y-J, Knoll W (2005) Comparison of surface plasmon resonance spectroscopy and quartz crystal microbalance techniques for studying DNA assembly and hybridization. Biosens Bioelectron 21:719–726
45. Boujday S, Méthivier C, Beccard B, Pradier C-M (2009) Innovative surface characterization techniques applied to immunosensor elaboration and test: comparing the efficiency of Fourier transform-surface plasmon resonance, quartz crystal microbalance with dissipation measurements, and polarization modulation-reflection absorption infrared spectroscopy. Anal Biochem 387:194–201
46. Zhang GZ, Wu C (2009) Quartz crystal microbalance studies on conformational change of polymer chains at interface. Macromol Rapid Commun 30:328–335

47. Scarano S, Mascini M, Turner APF, Minunni M (2010) Surface plasmon resonance imaging for affinity-based biosensors. Biosens Bioelectron 25:957–966
48. Mitchell J (2010) Small molecule immunosensing using surface plasmon resonance. Sensors 10:7323–7346
49. Manera MG, Spadavecchia J, Taurino A, Rella R (2010) Colloidal Au-enhanced surface plasmon resonance imaging: application in a DNA hybridization process. J Opt 12:8
50. Wang JL, Munir A, Zhou HS (2009) Au NPs-aptamer conjugates as a powerful competitive reagent for ultrasensitive detection of small molecules by surface plasmon resonance spectroscopy. Talanta 79:72–76
51. Lyon LA, Musick MD, Natan MJ (1998) Colloidal Au-enhanced surface plasmon resonance immunosensing. Anal Chem 70:5177–5183
52. Fang SP, Lee HJ, Wark AW, Corn RM (2006) Attomole microarray detection of MicroRNAs by nanoparticle-amplified SPR imaging measurements of surface polyadenylation reactions. J Am Chem Soc 128:14044–14046
53. Sim HR, Wark AW, Lee HJ (2010) Attomolar detection of protein biomarkers using biofunctionalized gold nanorods with surface plasmon resonance. Analyst 135:2528–2532
54. Kim S, Lee J, Lee SJ, Lee HJ (2010) Ultra-sensitive detection of IgE using biofunctionalized nanoparticle-enhanced SPR. Talanta 81:1755–1759
55. Grieshaber D, MacKenzie R, Voros J, Reimhult E (2008) Electrochemical biosensors – sensor principles and architectures. Sensors 8:1400–1458
56. Knoll W, Koper I, Naumann R, Sinner EK (2008) Tethered bimolecular lipid membranes – a novel model membrane platform. Electrochim Acta 53:6680–6689
57. Briand E, Zach M, Svedhem S, Kasemo B, Petronis S (2010) Combined QCM-D and EIS study of supported lipid bilayer formation and interaction with pore-forming peptides. Analyst 135:343–350
58. Lindholm-Sethson B, Nystrom J, Malmsten M, Ringstad L, Nelson A, Geladi P (2010) Electrochemical impedance spectroscopy in label-free biosensor applications: multivariate data analysis for an objective interpretation. Anal Bioanal Chem 398:2341–2349
59. Alivisatos P (2004) The use of nanocrystals in biological detection. Nat Biotechnol 22:47–52
60. Elghanian R, Storhoff JJ, Mucic RC, Letsinger RL, Mirkin CA (1997) Selective colorimetric detection of polynucleotides based on the distance-dependent optical properties of gold nanoparticles. Science 277:1078–1081
61. Shipway AN, Lahav M, Gabai R, Willner I (2000) Investigations into the electrostatically induced aggregation of Au nanoparticles. Langmuir 16:8789–8795
62. Gerardy JM, Ausloos M (1983) Absorption-spectrum of clusters of spheres from the general-solution of Maxwells equations.4. Proximity, bulk, surface, and shadow effects (in binary clusters). Phys Rev B 27:6446–6463
63. Ghosh SK, Pal T (2007) Interparticle coupling effect on the surface plasmon resonance of gold nanoparticles: from theory to applications. Chem Rev 107:4797–4862
64. Schofield CL, Haines AH, Field RA, Russell DA (2006) Silver and gold glyconanoparticles for colorimetric bioassays. Langmuir 22:6707–6711
65. Otsuka H, Akiyama Y, Nagasaki Y, Kataoka K (2001) Quantitative and reversible lectin-induced association of gold nanoparticles modified with alpha-lactosyl-omega-mercapto-poly(ethylene glycol). J Am Chem Soc 123:8226–8230
66. Schofield CL, Field RA, Russell DA (2007) Glyconanoparticles for the colorimetric detection of cholera toxin. Anal Chem 79:1356–1361
67. Zheng M, Davidson F, Huang XY (2003) Ethylene glycol monolayer protected nanoparticles for eliminating nonspecific binding with biological molecules. J Am Chem Soc 125:7790–7791
68. Li HA, Cao ZJ, Zhang YH, Lau CW, Lu JZ (2010) Combination of quantum dot fluorescence with enzyme chemiluminescence for multiplexed detection of lung cancer biomarkers. Anal Meth 2:1236–1242

69. Medintz IL, Uyeda HT, Goldman ER, Mattoussi H (2005) Quantum dot bioconjugates for imaging, labelling and sensing. Nat Mater 4:435–446
70. Asefa T, Duncan CT, Sharma KK (2009) Recent advances in nanostructured chemosensors and biosensors. Analyst 134:1980–1990
71. Wang X, Ramstrom O, Yan M (2009) A photochemically initiated chemistry for coupling underivatized carbohydrates to gold nanoparticles. J Mater Chem 19:8944–8949
72. Lin C-C, Yeh Y-C, Yang C-Y, Chen C-L, Chen G-F, Chen C-C, Wu Y-C (2002) Selective binding of mannose-encapsulated gold nanoparticles to type 1 pili in Escherichia coli. J Am Chem Soc 124:3508–3509
73. Liebau M, Hildebrand A, Neubert RHH (2001) Bioadhesion of supramolecular structures at supported planar bilayers as studied by the quartz crystal microbalance. Eur Biophys J Biophys Lett 30:42–52
74. Phillips KS, Wilkop T, Wu JJ, Al-Kaysi RO, Cheng Q (2006) Surface plasmon resonance imaging analysis of protein-receptor binding in supported membrane arrays on gold substrates with calcinated silicate films. J Am Chem Soc 128:9590–9591
75. Shi J, Yang T, Cremer PS (2008) Multiplexing ligand-receptor binding measurements by chemically patterning microfluidic channels. Anal Chem 80:6078–6084
76. Conboy JC, McReynolds KD, Gervay-Hague J, Saavedra SS (2002) Quantitative measurements of recombinant HIV surface glycoprotein 120 binding to several glycosphingolipids expressed in planar supported lipid bilayers. J Am Chem Soc 124:968–977
77. Wang JL, Wang F, Chen HJ, Liu XH, Dong SJ (2008) Electrochemical surface plasmon resonance detection of enzymatic reaction in bilayer lipid. Talanta 75:666–670
78. Munro JC, Frank CW (2004) Adsorption of lipid-functionalized poly(ethylene glycol) to gold surfaces as a cushion for polymer-supported lipid bilayers. Langmuir 20:3339–3349
79. Schuster B, Sleytr UB (2009) Composite S-layer lipid structures. J Struct Biol 168:207–216
80. Knoll W, Naumann R, Friedrich M, Robertson JWF, Losche M, Heinrich F, McGillivray DJ, Schuster B, Gufler PC, Pum D, Sleytr UB (2008) Solid supported lipid membranes: new concepts for the biomimetic functionalization of solid surfaces. Biointerphases 3: FA125–FA135
81. Handa H, Gurczynski S, Jackson MP, Mao GZ (2010) Immobilization and molecular : interactions between bacteriophage and lipopolysaccharide bilayers. Langmuir 26: 12095–12103
82. Plant AL (1993) Self-assembled phospholipid alkanethiol biomimetic bilayers on gold. Langmuir 9:2764–2767
83. Plant AL, Brigham-Burke M, Petrella EC, Oshannessy DJ (1995) Phospholipid/alkanethiol bilayers for cell-surface receptor studies by surface plasmon resonance. Anal Biochem 226:342–348
84. Hubbard JB, Silin V, Plant AL (1999) Self assembly driven by hydrophobic interactions at alkanethiol monolayers: mechanism of formation of hybrid bilayer membranes. Biophys J 76:A431
85. Kilian KA, Bocking T, Gaus K, King-Lacroix J, Gal M, Gooding JJ (2007) Hybrid lipid bilayers in nanostructured silicon: a biomimetic mesoporous scaffold for optical detection of cholera toxin. Chem Commun (Camb) 1936–1938
86. Mun S, Choi SJ (2009) Optimization of the hybrid bilayer membrane method for immobilization of avidin on quartz crystal microbalance. Biosens Bioelectron 24:2522–2527
87. Favero G, Campanella L, Cavallo S, D'Annibale A, Perrella M, Mattei E, Ferri T (2005) Glutamate receptor incorporated in a mixed hybrid bilayer lipid membrane array, as a sensing element of a biosensor working under flowing conditions. J Am Chem Soc 127:8103–8111
88. Mozsolits H, Wirth HJ, Werkmeister J, Aguilar MI (2001) Analysis of antimicrobial peptide interactions with hybrid bilayer membrane systems using surface plasmon resonance. Biochimica Et Biophysica Acta-Biomembranes 1512:64–76

89. Lee TH, Mozsolits H, Aguilar MI (2001) Measurement of the affinity of melittin for zwitterionic and anionic membranes using immobilized lipid biosensors. J Pept Res 58:464–476

90. Ye Q, Konradi R, Textor M, Reimhult E (2009) Liposomes tethered to omega-functional PEG brushes and induced formation of PEG brush supported planar lipid bilayers. Langmuir 25:13534–13539

91. Sinner E-K, Knoll W (2001) Functional tethered membranes. Curr Opin Chem Biol 5:705–711

92. Naumann R, Schiller SM, Giess F, Grohe B, Hartman KB, Kärcher I, Köper I, Lübben J, Vasilev K, Knoll W (2003) Tethered lipid bilayers on ultraflat gold surfaces. Langmuir 19:5435–5443

93. Shenoy S, Moldovan R, Fitzpatrick J, Vanderah DJ, Deserno M, Losche M (2010) In-plane homogeneity and lipid dynamics in tethered bilayer lipid membranes (tBLMs). Soft Matter 6:1263–1274

94. Wagner ML, Tamm LK (2000) Tethered polymer-supported planar lipid bilayers for reconstitution of integral membrane proteins: Silane-polyethyleneglycol-lipid as a cushion and covalent linker. Biophys J 79:1400–1414

95. Cazacu A, Legrand YM, Pasc A, Nasr G, Van der Lee A, Mahon E, Barboiu M (2009) Dynamic hybrid materials for constitutional self-instructed membranes. Proc Natl Acad Sci U S A 106:8117–8122

96. Vandenbussche S, Diaz D, Fernandez-Alonso MC, Pan WD, Vincent SP, Cuevas G, Canada FJ, Jimenez-Barbero J, Bartik K (2008) Aromatic-carbohydrate interactions: an NMR and computational study of model systems. Chem Eur J 14:7570–7578

97. Taylor JD, Phillips KS, Cheng Q (2007) Microfluidic fabrication of addressable tethered lipid bilayer arrays and optimization using SPR with silane-derivatized nanoglassy substrates. Lab Chip 7:927–930

98. Cornell BA, BraachMaksvytis VLB, King LG, Osman PDJ, Raguse B, Wieczorek L, Pace RJ (1997) A biosensor that uses ion-channel switches. Nature 387:580–583

99. Williams TL, Jenkins ATA (2008) Measurement of the binding of cholera toxin to GM1 gangliosides on solid supported lipid bilayer vesicles and inhibition by europium (III) chloride. J Am Chem Soc 130:6438–6443

100. Mahon E, Aastrup T, Barboiu M (2010) Dynamic glycovesicle systems for amplified QCM detection of carbohydrate-lectin multivalent biorecognition. Chem Commun (Camb) 46: 2441–2443

101. Voskuhl J, Stuart MCA, Ravoo BJ (2010) Sugar-decorated sugar vesicles: lectin-carbohydrate recognition at the surface of cyclodextrin vesicles. Chem Eur J 16:2790–2796

102. Mornet S, Lambert O, Duguet E, Brisson A (2005) The formation of supported lipid bilayers on silica nanoparticles revealed by cryoelectron microscopy. Nano Lett 5:281–285

103. Savarala S, Ahmed S, Ilies MA, Wunder SL (2010) Formation and colloidal stability of DMPC supported lipid bilayers on SiO$_2$ nanobeads. Langmuir 26:12081–12088

104. Arnal-Hérault C, Barboiu M, Pasc A, Michau M, Perriat P, van der Lee A (2007) Constitutional self-organization of adenine-uracil-derived hybrid materials. Chem Eur J 13:6792–6800

105. Liu JW, Stace-Naughton A, Jiang XM, Brinker CJ (2009) Porous nanoparticle supported lipid bilayers (protocells) as delivery vehicles. J Am Chem Soc 131:1354–1355

106. Barboiu M (2004) Supramolecular polymeric macrocyclic receptors – hybrid carrier vs. channel transporters in bulk liquid membranes. J Incl Phenom Macrocyl Chem 49:133–137

107. Lin SY, Chen CH, Lin MC, Hsu HF (2005) A cooperative effect of bifunctionalized nanoparticles on recognition: sensing alkali ions by crown and carboxylate moieties in aqueous media. Anal Chem 77:4821–4828

108. Imabayashi S, Hobara D, Kakiuchi T (2001) Voltammetric detection of the surface diffusion of adsorbed thiolate molecules in artificially phase-separated binary self-assembled monolayers on a Au(111) surface. Langmuir 17:2560–2563

109. Ionita P, Volkov A, Jeschke G, Chechik V (2008) Lateral diffusion of thiol ligands on the surface of Au nanoparticles: an electron paramagnetic resonance study. Anal Chem 80:95–106
110. Boal AK, Rotello VM (2000) Intra- and intermonolayer hydrogen bonding in amide-functionalized alkanethiol self-assembled monolayers on gold nanoparticles. Langmuir 16:9527–9532
111. Boal AK, Rotello VM (2000) Fabrication and self-optimization of multivalent receptors on nanoparticle scaffolds. J Am Chem Soc 122:734–735
112. Paulini R, Frankamp BL, Rotello VM (2002) Effects of branched ligands on the structure and stability of monolayers on gold nanoparticles. Langmuir 18:2368–2373
113. Duchesne L, Wells G, Fernig DG, Harris SA, Levy R (2008) Supramolecular domains in mixed peptide self-assembled monolayers on gold nanoparticles. ChemBioChem 9:2127–2134
114. Nørgaard K, Weygand MJ, Kjaer K, Brust M, Bjørnholm T (2004) Adaptive chemistry of bifunctional gold nanoparticles at the air/water interface. A synchrotron X-ray study of giant amphiphiles. Faraday Discuss 125:221–233
115. Gentilini C, Franchi P, Mileo E, Polizzi S, Lucarini M, Pasquato L (2009) Formation of patches on 3D SAMs driven by thiols with immiscible chains observed by ESR spectroscopy. Angew Chem Int Ed Engl 48:3060–3064
116. Gentilini C, Pasquato L (2010) Morphology of mixed-monolayers protecting metal nanoparticles. J Mater Chem 20:1403–1412
117. Barboiu M, Cazacu A, Michau M, Caraballo R, Arnal-Herault C, Pasc-Banu A (2008) Functional organic inorganic hybrid membranes. Chem Eng Proc 47:1044–1052
118. Ruff Y, Lehn JM (2008) Glycodynamers: dynamic analogs of arabinofuranoside oligosaccharides. Biopolymers 89:486–496
119. Sreenivasachary N, Hickman DT, Sarazin D, Lehn JM (2006) DyNAs: constitutional dynamic nucleic acid analogues. Chem -Eur J 12:8581–8588
120. Hickman DT, Sreenivasachary N, Lehn JM (2008) Synthesis of components for the generation of constitutional dynamic analogues of nucleic acids. Helv Chim Acta 91:1–20
121. Miyashita N, Kurth DG (2008) Directing supramolecular assemblies on surfaces. J Mater Chem 18:2636–2649
122. Nasr J, Barboiu M, Ono T, Fujii S, Lehn J-M (2008) Dynamic polymer membranes displaying tunable transport properties on constitutional exchange. J Membr Sci 321:8–14
123. Shi ZL, Lin N (2010) Structural and chemical control in assembly of multicomponent metal–organic coordination networks on a surface. J Am Chem Soc 132:10756–10761
124. Camacho C, Matias JC, Cao R, Matos M, Chico B, Hernandez J, Longo MA, Sanroman MA, Villalonga R (2008) Hydrogen peroxide biosensor with a supramolecular layer-by-layer design. Langmuir 24:7654–7657
125. Oishi M, Hayama T, Akiyama Y, Takae S, Harada A, Yarnasaki Y, Nagatsugi F, Sasaki S, Nagasaki Y, Kataoka K (2005) Supramolecular assemblies for the cytoplasmic delivery of antisense oligodeoxynucleotide: polylon complex (PIC) micelles based on poly(ethylene glycol)-S,S-oligodeoxynucleotide conjugate. Biomacromolecules 6:2449–2454
126. Legrand YM, Dumitru F, van der Lee A, Barboiu M (2009) Constitutional chirality – adriving force for self-sorting homochiral single-crystals from achiral components. Chem Commun 2667–2669
127. Nierengarten JF, Iehl J, Oerthel V, Holler M, Illescas BM, Munoz A, Martin N, Rojo J, Sanchez-Navarro M, Cecioni S, Vidal S, Buffet K, Durka M, Vincent SP (2010) Fullerene sugar balls. Chem Commun 46:3860–3862
128. Durka M, Buffet K, Iehl J, Holler M, Nierengarten JF, Taganna J, Bouckaert J, Vincent SP (2011) The functional valency of dodecamannosylated fullerenes with Escherichia coli FimH –towards novel bacterial antiadesives. Chem Commun 47:1321–1323

129. Compain P, Decroocq CK, Iehl J, Holler M, Hazelard D, Barragan TM, Mellet CO, Nierengarten JF (2010) Glycosidase inhibition with fullerene iminosugar balls: a dramatic multivalent effect. Angew Chem Int Ed 49:5753–5756
130. Barboiu M, Cerneaux S, van der Lee A, Vaughan G (2004) Ion-driven ATP pump by self-organized hybrid membrane materials. J Am Chem Soc 126:3545–3550
131. Barboiu M, Vaughan G, van der Lee A (2003) Self-organized heteroditopic macrocyclic superstructures. Org Lett 5:3073–3076
132. Cazacu A, Tong C, van der Lee A, Fyles TM, Barboiu M (2006) Columnar self-assembled ureido crown ethers: an example of ion-channel organization in lipid bilayers. J Am Chem Soc 128:9541–9548
133. Michau M, Barboiu M, Caraballo R, Arnal-Herault C, Perriat P, Van Der Lee A, Pasc A (2008) Ion-conduction pathways in self-organised ureidoarene-heteropolysiloxane hybrid membranes. Chem -Eur J 14:1776–1783
134. Michau M, Caraballo R, Arnal-Herault C, Barboiu M (2008) Alkali cation-pi aromatic conduction pathways in self-organized hybrid membranes. J Membr Sci 321:22–30
135. Arnal-Herault C, Banu A, Barboiu M, Michau M, van der Lee A (2007) Amplification and transcription of the dynamic supramolecular chirality of the guanine quadruplex. Angew Chem Int Ed Engl 46:4268–4272
136. Arnal-Herault C, Pasc A, Michau M, Cot D, Petit E, Barboiu M (2007) Functional G-quartet macroscopic membrane films. Angew Chem Int Ed Engl 46:8409–8413
137. Davis JT, Spada GP (2007) Supramolecular architectures generated by self-assembly of guanosine derivatives. Chem Soc Rev 36:296–313
138. Davis JT (2004) G-quartets 40 years later: from 5 '-GMP to molecular biology and supramolecular chemistry. Angew Chem Int Ed Engl 43:668–698
139. Ramstrom O, Lehn JM (2002) Drug discovery by dynamic combinatorial libraries. Nat Rev Drug Discov 1:26–36
140. Barboiu M, Lehn JM (2002) Dynamic chemical devices: modulation of contraction/extension molecular motion by coupled-ion binding/pH change-induced structural switching. Proc Natl Acad Sci U S A 99:5201–5206

Top Curr Chem (2012) 322: 165–192
DOI: 10.1007/128_2011_258
© Springer-Verlag Berlin Heidelberg 2011
Published online: 25 October 2011

Dynamic Assembly of Block-Copolymers

D. Quémener, A. Deratani, and S. Lecommandoux

Abstract Block copolymers (BCs) are well-known building blocks for the creation of a large variety of nanostructured materials or objects through a dynamic assembly stage which can be either autonomous or guided by an external force. Today's nanotechnologies require sharp control of the overall architecture from the nanoscale to the macroscale. BCs enable this dynamic assembly through all the scales, from few aggregated polymer chains to large bulk polymer materials. Since the discovery of controlled methods to polymerize monomers with different functionalities, a broad diversity of BCs exists, giving rise to many different nanoobjects and nanostructured materials. This chapter will explore the potentialities of block copolymer chains to be assembled through dynamic interactions either in solution or in bulk.

Keywords Block copolymer · Dynamic assembly · Nanostructure

Contents

1 Introduction 166
2 Block Copolymer Self-Assembly in Solution 168
    2.1 Micelle Preparation 169
    2.2 Spherical Micelles 171
    2.3 Cylindrical Micelles 174
    2.4 Polymersomes 175
    2.5 Other Morphologies 176

D. Quémener (✉) and A. Deratani
Institut Européen des Membranes, UMR 5635 (CNRS-ENSCM-UM2), Université Montpellier II, 2 Place E. Bataillon, 34095 Montpellier, France
e-mail: damien.quemener@iemm.univ-montp2.fr

S. Lecommandoux
Laboratoire de Chimie des Polymères Organiques, UMR CNRS 5629, 16 Avenue Pey Berland, 33607 Pessac, France

3	Self-Assembly in Bulk/Thin Films	177
	3.1 Bulk Morphologies	177
	3.2 Self-Assembly in Thin Films	180
	3.3 Directing the Self-Assembly	182
4	Conclusion	186
References		186

1 Introduction

Block copolymers (BCs) are a particular class of polymers belonging to a wider family known as soft materials [1] that, independent of the method of synthesis, can simply be considered as formed by the union of two or more chemically homogeneous polymer fragments (blocks) joined together by covalent bonds. In the simplest case of two distinct monomers, conventionally termed A and B, linear diblock (AB), triblock (ABA), multiblock, or star-block copolymers can be prepared. With the enhancement and the "democratization" of the synthetic routes towards the BC preparation, a large number of research groups from all countries are now using these BCs as building blocks in order to prepare more complex assemblies. The concept of self-assembly relates to structures and patterns that put themselves together from basic building blocks. Even if the first examples of synthetic assemblies seem to be quite recent [2–6], this concept was being discussed by the French philosopher Descartes in the seventeenth century, whom envisioned an ordered universe arising out of chaos according to natural laws through the organization of small objects into larger assemblages [7]. Today's self-assembly is the basis of many nanotechnologies [8] in medical diagnostics, flexible electronics, opal optics, and battery electrodes, to mention only a few examples. The design of the building blocks is the keystone of nanofabrication, and should contain all the necessary information to direct their self-assembly. In that way, a large variety of ordered morphologies covering several length-scales can be achieved. For instance, there is no clear picture of all the self-assembled morphologies that can be reached from a selected BCs system. Many parameters intrinsic to the copolymer structure have a strong influence on the morphology, like the interaction parameter between the two blocks, the molecular weight, the block ratio, the polydispersity, and the thermomechanical properties. Moreover, the nonergodicity in BCs implies that the same polymer can give different structures that are kinetically (meta)stable depending on the preparation process [9, 10]. Finally, the targeted morphologies will be completely different if the copolymer is pure (bulk) or diluted in a solvent system. In bulk, the minority block is segregated from the majority block forming regularly shaped and uniformly spaced nanodomains [11]. The shape of the segregated domains in a diblock is governed by the volume fraction of the minority block and block incompatibility. Figure 1 shows the equilibrium morphologies documented for diblock copolymers [12]. At a volume fraction of ~20%, the minority block forms a body-centered cubic spherical phase in the matrix of the majority block. It changes to hexagonally packed cylinders at a volume fraction

Fig. 1 Schematic representations of the morphologies obtained for diblock copolymer melts. Reprinted with permission from Khandpur et al. [12]. Copyright 1995 American Chemical Society

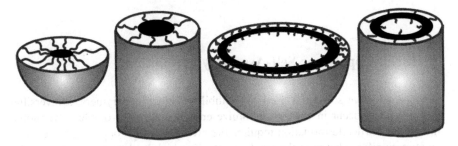

Fig. 2 Cross-sectional view of chain packing in diblock copolymer spherical, cylindrical, vesicular, and tubular micelles [18]. Copyright 2006 Wiley-VCH Verlag GmbH & Co. KGaA. Reproduced with permission

~30%. Alternating lamellae are formed at approximately equal volume fractions for the two blocks. At a volume fraction of ~38%, the minority block forms gyroid or perforated layers respectively at moderate and high incompatibility. These interesting morphological transitions have been established experimentally and can be accounted for by statistical thermodynamic theories [13]. Furthermore, the smallest dimension of a segregated domain, e.g., the diameter of a cylinder, is proportional to the two-thirds power of the molar mass of the minority block and can typically be tuned from ~5 to ~100 nm by changing the molar mass of the block [14].

In analogy to their bulk behavior, diblock copolymers also self-assemble in block-selective solvents, which solubilize one but not the other block, forming micelles of various shapes [15, 16]. If the soluble block is predominant, the insoluble block aggregates to produce spherical micelles. As the length of the soluble block is decreased relative to the insoluble block, cylindrical micelles or vesicles are formed. For a given diblock, unusual micelles with shapes differing from spheres can sometimes be induced by playing with the non-solvent quality [15, 17]. Figure 2 schematically illustrates the structure and chain packing in four types of diblock copolymer micelles where the black block is insoluble.

ABC triblock copolymers also undergo self-assembly in selective solvents or in bulk leading to more complex morphologies [19]. Over the past 10 years, approximately ten additional structures have been discovered. Block segregation pattern complexity increases further for tetra- and pentablocks, thus making the number of morphologies almost infinite.

Aside from the linear architecture, BCs can be prepared with advanced architectures such as miktoarm star structures, i.e., BCs where arms of different chemical nature are linked to the same branch point [20]. Unique segregation properties are expected of these polymers [21, 22].

In this chapter, the self-assembly of BCs is explored throughout the last examples of major contributions. The first part will be devoted to the self-assembly in solution as a way to prepare micellar nanoobjects with different shapes and behaviors. The second part will develop self-assembly in bulk and in thin films with a special paragraph dedicated to the guided assembly.

2 Block Copolymer Self-Assembly in Solution

When a selective solvent is used to "solubilize" block copolymers, a reversible assembly may occur in order to minimize energetically unfavorable solvophobic interactions. Micelle formation requires the presence of two opposing forces, i.e., an attractive force between the insoluble blocks which leads to aggregation, and a repulsive force between the soluble blocks which prevents unlimited growth of the micelle into a distinct macroscopic phase. Micelles are stabilized in the solution due to the interaction of the soluble blocks and the solvent [23].

The resulting micellar aggregates resemble, in most of their aspects, those obtained with classical low molecular weight surfactants, but the nonergodicity of BCs allows the preparation of many different kinetically frozen morphologies. From the initial basic observations of micelle formation by Merret in 1954 [24] to the last structures of living micelles obtained by Winnik and co-workers in 2007 [25], the solution self-assembly is a fast growing area leading to various nanoobjects related to the chemical nature of the polymer blocks. The dimensionless packing parameter (p), defined in (1), was introduced more than 30 years ago [26] in order to predict the micellar morphology according to the structural parameters of surfactants:

$$p = \frac{v}{a_o l_c} \tag{1}$$

where v is the volume of the hydrophobic chains, a_o is the optimal area of the head group, and l_c is the length of the hydrophobic tail. Lots of limitations have been described for the use of p [27], but it remains a nice way to explain the change of conformations in function of the molecular curvature (Fig. 3). Spherical morphology should be observed for $p \leq 1/3$ whereas cylindrical micelles are favored when $1/3 \leq p \leq 1/2$ and vesicles (or polymersomes) are usually formed when $1/2 \leq p \leq 1$.

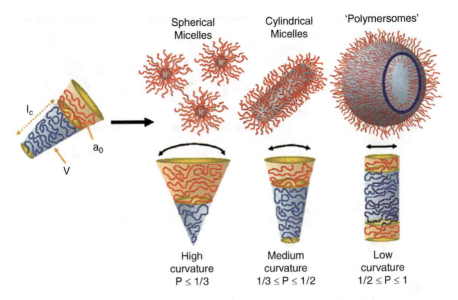

Fig. 3 Various self-assembled structures formed by amphiphilic block copolymers in a block-selective solvent. The type of structure formed is due to the inherent curvature of the molecule, which can be estimated through calculation of its dimensionless packing parameter, p [28]. Copyright Wiley-VCH Verlag GmbH & Co. KGaA. Reproduced with permission

2.1 Micelle Preparation

Micelle preparation starts from the design of the block copolymer precursors. Even if two blocks of the copolymer have to be incompatible in order to guide the self-assembly, their solubility in the dispersion solvent can vary between two extreme cases: (1) the two blocks can have a different solubility such that the use of a selective solvent will trigger the micelle formation and (2) the two blocks can have the same solubility, the solvent being non-selective. In this case, micelles will be produced by the in situ formation of multiple cross-linking points.

With amphiphilic BCs in a selective solvent, the main technique requires the copolymer to be dissolved in a good solvent for both blocks, followed by a controlled change in the solvent composition or the temperature [29]. For example, by adding gradually a solvent selective for one of the blocks, a precipitation will occur and the micelle formation will be triggered. An alternative route often used is the replacement of the good solvent by a selective solvent through a dialysis setup [30]. The solid copolymer could also be "solubilized" directly in the selective solvent but this technique involves the formation of a high proportion of large aggregates [31].

More recently, micelle formation of non-amphiphilic BCs in non-selective solvents in which both blocks are soluble has been reported [32]. In this case,

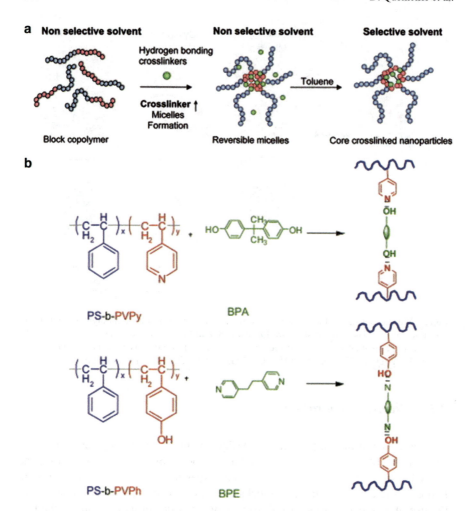

Fig. 4 (a) Micelles formation in non-selective solvent and core-crosslinked nanoparticles. (b) Chemical structures of the implied copolymers. Reprinted from de Luzuriaga et al. [37]. Copyright 2010, with permission from Elsevier

BCs can form micelles due to non-covalent cross-linking [33, 34] or chemical cross-linking of one of the blocks [35, 36]. For example, micelles could be prepared from poly(styrene-*b*-4-vinylphenol) and poly(styrene-*b*-4-vinylpyridine) micelles in non-selective solvent by addition of a low-molecular weight hydrogen-bonding crosslinker such as bis-pyridyl ethane and bisphenol A, respectively [37] (Fig. 4).

Micelles can be classified into several types with morphologies varying from spherical to vesicular or other less common structures, such as inverse micelles, bilayers, or cylinders. A recent review analyzes in more detail the parameters that afford one or another structure [29].

2.2 Spherical Micelles

The literature is abundant in describing the formation of spherical micelles from block copolymer self-assembly. From the classical diblock copolymer AB, two types of micelles could be prepared as a function of their structural parameters (Fig. 5).

Two different micellar structures can be distinguished for diblock copolymers, depending on the relative length of the blocks. If the outer block is larger than the inner one ($R_c<L$), the micelles formed consist of a small dense core and a very large and diffuse corona. Different names have been assigned to this structure, e.g., core-shell (in the sense that a shell is thicker than a corona), hairy, or star-like micelles. Based on an earlier theoretical work of de Gennes [38], and prepared for the first time by Halperin et al. [39], the second type of micelles has a large dense core and a short diffuse corona ($R_c>L$). The terms crew-cut or flower-like micelles are usually used.

From AB diblock copolymers, the two morphologies can be reached with the same polymer block system only by playing with the structural parameters. Among the most studied diblock copolymer systems, one can cite polystyrene-b-poly(ethylene oxide) (PS-b-PEO), polystyrene-b-polyisoprene (PS-b-PI), and polystyrene-b-poly(acrylic acid) (PS-b-PAA) [27]. A great number of systematic studies have been published which explore and explain the impact of the structural parameters on all the micelle characteristics, i.e., the aggregation number, the micelle core dimension, and the area per corona chain on the core surface. As an example, for PS-b-PEO it was found that the micellar aggregation number of star-like micelles increases with the copolymer molecular weight at constant composition and decreases with the PEO content for a given molecular weight [29, 40, 41].

In that case, the thermodynamic equilibrium, which is sensitive to molecular weight and chemical structure, is not fulfilled and tends to give frozen morphologies. Therefore, some empirical corrections must be applied to the existing theories.

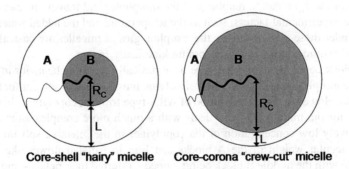

Fig. 5 Schematic representation of AB diblock copolymer micelles in a selective solvent of the A block, Rc core radius, L shell (corona) thickness. Adapted from Riess [29]. Copyright 2003, with permission from Elsevier

For example, Halperin's theory, which links the micellar hydrodynamic radius to the number of monomer units in the copolymer (2), should be adjusted by (3):

$$R_h = N_{PEO}^{0.6} N_{PS}^{0.16},$$ (2)

$$R_h = 1.77 N_{PEO}^{0.31} N_{PS}^{0.09}.$$ (3)

The major difference between the two types of micelles is probably the polymer chain dynamic during the micellization stage. The method for preparing micelles is different in relation to the block copolymer structure for practical reason. In the case of star-like micelles, the soluble block is dominant and so the block copolymer can be solubilized in the selective solvent without particular precautions. In the case of crew-cut micelles, the insoluble block becomes dominant and so the block copolymer cannot be solubilized. The solution usually followed is to solubilize the copolymer in a good solvent for both blocks and then to add slowly the selective solvent – usually water – in order to trigger the micellization. After a certain percentage of the selective solvent – typically between 5 and 35% for water – the morphology is kinetically locked or frozen. During a direct micellization in the selective solvent, the solvent is instantaneously excluded from the micellar cores thanks to hydrophobic polymer chain interactions. Therefore the chain mobility quickly falls to zero and the micelles are frozen. Actually, above the critical micellar concentration (very low in the case of copolymers), the unimers (individual soluble copolymer chains) are assembled into micelles with no further morphology evolution. In the case of crew-cut micelles, the selective solvent is slowly added so that, when the micellization process starts, the polymer chain mobility is high thanks to the presence of a large amount of good solvent in the micelle core. The chain mobility progressively decreases and the good solvent is gradually excluded as the selective solvent is added. After a certain percentage of the selective solvent, the morphology becomes kinetically frozen. The interesting point here – which is also the main difference between the two types of micelles – is that there is a small domain of water concentration (effective window) where the change of the aggregation number and the morphological transitions can be faster than the experimental factors, such as the temperature and the added water content [23]. Under these circumstances, the morphologies of micelles are basically under thermodynamic control before becoming kinetically frozen.

Triblock copolymers of ABA-type have basically the same behavior in giving a star-like morphology (Fig. 6a) when dissolving into a selective solvent for the outer A blocks. However, the micellization of ABA-type triblock copolymers in selective solvent for the middle B blocks deals with a much more complex situation [42]. A relatively low concentration of the copolymer in the selective solvent and/or a low molecular weight of the A blocks can lead to isolated flower-like micelles (Fig. 6b) with the middle B block being looped – referred to as petals – and with the two outer A blocks taking part of the same micellar core. However, if the copolymer concentration or the A block molecular weight is increased, a micelle association

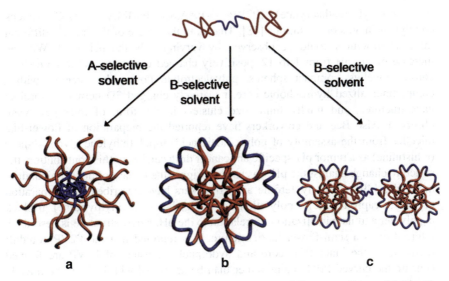

Fig. 6 Self-assembly of ABA triblock copolymers under different conditions. (a) Star-like micelle. (b) Flower-like micelle. (c) Micelle assembly

into larger aggregates can happen due to the partial conversion from loops to bridges with A blocks being located in two different micellar cores (Fig. 6c). An idea of the large diversity of micelle structures can be obtained from the theoretical study of Li and coworkers [43].

For example, Chu et al. have demonstrated the formation of bridged micelles with a poly(butylene oxide)-b-poly(ethylene oxide)-b-poly(butylene oxide) ($E_nB_mE_n$) ABA triblock copolymer dissolved in water [44]. A strong micellar association was observed with $B_{12}E_{260}B_{12}$ at a very low concentration (2 wt%) whereas a moderate association was observed with $B_5E_{91}B_5$ even at high concentration (25 wt%) [45]. Stepanek and coworkers have also observed the micelle assembly from poly[5-(N,N-diethylaminoisoprene)]-b-polystyrene-b-poly[5-(N,N-diethylaminoisoprene)] ($PAI_nPS_mPAI_n$) in dimethylformamide [42]. No micelle association could be detected with the $PAI_{11}PS_{271}PAI_{11}$ triblock copolymer even at high concentration (90 mg mL^{-1}) whereas the assembly could be evidenced in the triblock $PAI_{30}PS_{640}PAI_{30}$ from a concentration equal to 15 mg mL^{-1}.

The association of flower-like micelles through bridged conformations in an ABA triblock copolymer can be avoided by replacing the ABA triblock by a cyclic diblock copolymer. In this case, "sunflower micelles" have been observed for example by Borsali and coworkers from the assembly of cyclic and asymmetric polystyrene-b-polyisoprene diblock copolymer (PS-b-PI) and compared to the corresponding linear diblock [46, 47].

The addition of a third block to the AB diblock copolymer, referred to as ABC triblock copolymer, has recently been explored and seems to provide a higher diversity of micellar organizations. For example, Minko and coworkers have studied the self assembly of the poly(2-vinyl pyridine)-b-poly(acrylic acid)-

b-poly(*n*-butyl methacrylate) (P2VP$_{58}$-*b*-PAA$_{924}$-*b*-P*n*BMA$_{48}$) ABC triblock copolymer in aqueous solutions [48]. Thanks to the nature of the blocks, different micellar structures could be observed by varying only the pH level. With an increase of the pH from 1 to 12, positively charged centrosymmetric core-shell-corona micelles, compact spheres, polyelectrolyte flower-like micelles with a compartmentalized hydrophobic core, negatively charged 3D networks, toroidal nanostructures, and finally finite size clusters in the form of microgels were observed. Also Bae and coworkers have reported the preparation of flower-like micelles from the assembly of poly(L-lactic acid)-*b*-poly(ethylene glycol)-*b*-poly (L-histidine) as a tumor pH-specific anticancer drug carrier. In this contribution, the structure change against the pH was used to trigger the release of doxorubicin drug molecules [49]. Recently Jérôme and coworkers have described the preparation of poly(ε-caprolactone)-*b*-poly(ethylene oxide)-*b*-poly(2-vinylpyridine) triblock copolymer and its micellization. In relation to the pH, a new structure is described, referred to as a semi-flower-like structure. It is reported that the P2VP couldn't really reach the inner PCL core and hydrophobic clusters of P2VP are formed beneath the curved PEO chains rather than being mixed with PCL in the micelle core [50].

2.3 Cylindrical Micelles

In the dilute regime, the optimal shape of molecular surfactant micelles for a given temperature and cosolvent concentration can be easily predicted. However, with amphiphilic block copolymers, lots of non-equilibrium structures can exist according to the experimental conditions and independently from the polymer structure. Most BCs form spherical micelles, whereas only a narrow range of compositions is able to form cylindrical micelles [51]. The existence of cylindrical aggregates was demonstrated with a water-soluble corona block such as polybutadiene-*b*-poly(ethylene oxide) (PB-*b*-PEO) [52] and polystyrene-*b*-poly (acrylic acid) (PS-*b*-PAA) [53]. Infinitely long cylinders are energetically favorable relative to shortened cylinders with incorporated end-defects, since these structures allow uniform curvature across the entire aggregate. However, the entropic demands and molecular frustration induces the formation of defects such as end caps and branch points [28, 54]. Literature reports of giant [52] and short worms [17, 55], y-junction and end cap defects [56–61], and even worm-like micellar networks [57, 58] illustrate the complexity of block copolymer assemblies. The cylindrical shape is particularly interesting in part because of its potential applications in nanotechnology and medicine. For example, poly(ethylene oxide)-*b*-poly(ethylene-*alt*-propylene) (PEO-*b*-PEP) cylinders incorporated into epoxy resins greatly enhance the toughness of the resins [59]. Whereas spherical block copolymer micelles are able to carry dyes and other drug molecules, cylindrical micelles can orient and stretch in a flowing stream in a manner that is ideal for flow-intensive delivery applications such as phage-mimetic drug carriers and micropore delivery [60, 61]. Wang and coworkers have demonstrated the unusual propensity of polyferrocenyldimethylsilane BCs to form cylindrical micelles in solution [62–65]. Cylinders are obtained with controlled length and architecture through an epitaxially growth with a living character [25].

2.4 Polymersomes

Block copolymer vesicles, or polymersomes, are of continued interest for their ability to encapsulate aqueous compartments within relatively robust polymer bilayer shells (Fig. 7) [66, 67]. Eisenberg and coworkers were the first to report the formation of block copolymer vesicles from the self-assembly of polystyrene-*b*-poly(acrylic acid) (PS-*b*-PAA) block copolymers. They also have described the formation of a wide range of vesicle architectures in solution from the self-assembly of five different block copolymers: PS-*b*-PAA, PS-*b*-PMMA, PB-*b*-PAA, polystyrene-*b*-poly(4-vinylpyridinium methyl iodide), and polystyrene-*b*-(4-vinylpyridinium decyl iodide) [68]. Small uniform vesicles, large polydisperse vesicles, entrapped vesicles, hollow concentric vesicles, onions, and vesicles with hollow tubes in the walls have been observed and the formation mechanism discussed. Since vesicles could be prepared with low glass transition polymers such as PB [69, 70] and PPO [71], it has been established than these structures are thermodynamically stable and not trapped by the glassy nature of the hydrophobic part.

Bates and coworkers have shown the enhanced mechanical properties of polymersomes compared to liposomes evaluated from different BC systems including poly(ethylene oxide)-*b*-poly(ethylene) (PEO-*b*-PE) and poly(ethylene oxide)-*b*-polybutadiene (PEO-*b*-PB) BCs with various block compositions [72]. Moreover, the

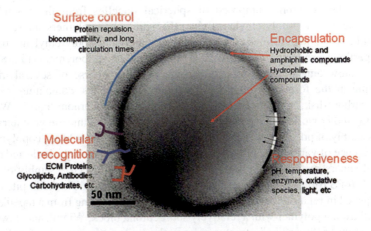

Fig. 7 Different properties that can be included into the molecular design of polymersomes [66]. Reproduced by permission of The Royal Society of Chemistry

polymersomes appeared to be inert to white cells as well as adherent cultured cells as observed with in vitro experiments.

Due to the interfacial curvatures, theoretical studies have demonstrated the preparation of stable polymersomes from the combination of AB copolymers with different molecular weight or by using an asymmetric ABC triblock copolymer [73, 74]. For example, Eisenberg and coworkers have prepared vesicles from PS-*b*-PAA diblock copolymers with artificially broaden PAA block length distributions [75]. The vesicle size was found to decrease with increasing polydispersity in agreement with the theory. This observation has been attributed to the segregation of smaller chains to the inside of the vesicle bilayer while the longer chains form the outer surface. Stable polymersomes have also been prepared by mixing two diblock copolymers (PS-*b*-PAA and PS-*b*-P4VP) with different molecular weights [76]. Due to the difference in the block length, the PAA chains were segregated into the inside of the vesicles while the outside corona consists of P4VP chains. Meier and coworkers have prepared vesicles with asymmetric membranes in aqueous media from ABC triblock copolymers of polyethylene-*b*-poly(dimethyl siloxane)-*b*-poly(methyl oxazoline), with sizes ranging between 60 and 300 nm [77]. Membrane asymmetry has also been induced in vesicles of an ABCA copolymer PEO-PS-PBD-PEO, using the high interfacial tension between PBD and water to locate the former chains towards the inner leaflet [78].

2.5 Other Morphologies

With the improved knowledge about driving the self-assembly in order to prepare targeted micelle morphology, it is now possible to increase the complexity level and functionality. For example, new morphology could be observed by mixing two micellar solutions composed of spherical micelles from the assembly of polyethylethylene-*b*-poly(ethylene oxide) (EO) diblock copolymers and segmented wormlike micelles, formed from μ-(polyethylethylene)(poly(ethylene oxide))-(polyperfluoropropylene oxide) (μ-EOF) miktoarm star terpolymers (Fig. 8) [79]. A very slow annealing process took place over the course of several months, resulting in the formation of mixed "hamburger" micelles, containing a central fluorocarbon disk, surrounded top and bottom by hydrocarbon "buns". Wooley, Pochan, and coworkers have manipulated with success the micelle geometry from the assembly of poly(acrylic acid) (PAA)-containing di- and triblock copolymers in the presence of various organic multiamines [80–87]. In this way, single and double helical superstructures have been created through solution co-assembly of PAA-*b*-PMA-*b*-PS triblock copolymers with various multiamines. The helix pitch could be adjusted in relation to the multiamine amount [86]. Starting from a tapelike seed formed from a polymer with a relatively short corona block, Winnik and coworkers have prepared "scarflike" structures from the epitaxially growth of cylindrical micelles of a block copolymer with a longer corona block [88].

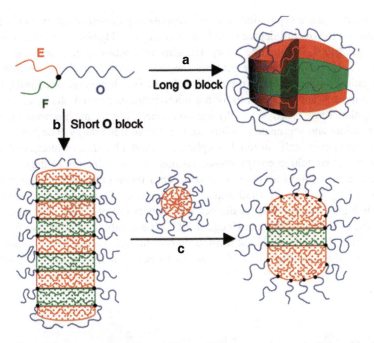

Fig. 8 Multicompartment micelle formation from μ-EOF star terpolymers and binary blends of μ-EOF/EO. (**a**) Hamburger micelle from μ-EOF with a very long PEO block. (**b**) Segmented wormlike micelle from μ-EOF with a short PEO block. (**c**) Hamburger micelle from blends of μ-EOF/EO. Reprinted with permission from Hillmyer et al. [79]. Copyright 2006 American Chemical Society

3 Self-Assembly in Bulk/Thin Films

In block copolymer melts, the solvent-polymer interaction doesn't exist anymore and other morphologies can be observed. A large amount of work deals with the self-assembly in bulk, how to predict it, and how to understand the phase change or the geometry orientation in relation to the experimental parameters. In this section we will give a quick overview of the key points to be considered when working in this field.

3.1 Bulk Morphologies

In the simplest case of linear AB diblock and ABA triblock copolymers, the phase behavior has been the subject of numerous theoretical and experimental studies over recent decades and is relatively well understood [89–96]. As mentioned before, the self-assembly process is driven by an unfavorable mixing enthalpy and small mixing entropy, while the covalent bond connecting the blocks prevents macroscopic phase separation. This microphase separation of diblock copolymers depends on the overall degree of polymerization N = $N_A + N_B$, the volume fraction f_A of the block A, and the Flory–Huggins interaction parameter χAB, which measures the incompatibility between the two blocks. Just above a value of the segregation product χAB·N predicted to be 10 (weak segregation limit), limited demixing of A and B occurs to form a micrometric-separated phase. In the strong segregation limit (χAB·N ≫ 10) the constituent blocks are thermodynamically incompatible and segregate to minimize the system free energy. The block copolymer adopts periodically ordered morphologies, and a balance of interfacial tension and entropic stretching energy considerations governs selection of the equilibrium state. Figure 9 shows a comparison between a theoretical phase diagram as first described by Leibler in 1980 and an actual phase diagrams of the equilibrium morphologies documented for diblock copolymers [12, 94]. The differences highlight the evolution of the knowledge in BC self-assembly. At a volume fraction of about 20%, the minority block forms a body-centered cubic spherical phase in the matrix of the majority block. It changes to hexagonally packed cylinders at a

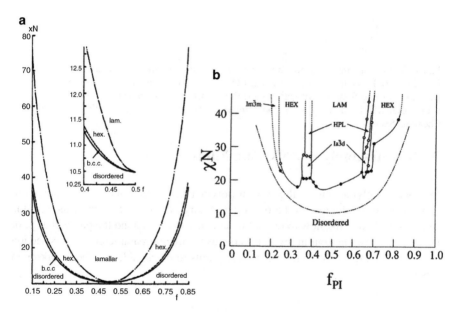

Fig. 9 (a) Phase diagram for the diblock copolymer (*solid lines*) transition line from the disordered to the bcc phase (xNt vs f), (*dashed lines*) transition line from the bcc phase to the hexagonal mesophase (xN vs f), (*dashed dotted lines*) transition line from the hexagonal to the lamellar mesophase (xN vs f). Reprinted with permission from Leibler [94]. Copyright 1980 American Chemical Society. (b) χN vs f_{PI} diagram for PI-PS diblock copolymers. *Open* and *filled circles* represent the order-order (OOT) and order-disorder (ODT) transitions, respectively. Five different ordered microstructures have been observed: body-centered cubic spherical phase (Im3m space group), hexagonally packed cylinders (HEX), alternating lamellae (LAM), gyroid (Ia3d space group), or perforated layers (HPL). Reprinted with permission from Khandpur et al. [12]. Copyright 1995 American Chemical Society

volume fraction of about 30%. Alternating lamellae are formed at approximately equal volume fractions for the two blocks. At a volume fraction of about 38%, the minority block forms gyroid or perforated layers at moderate and high incompatibility, respectively. These interesting morphological transitions have been established experimentally and could be accounted for by statistical thermodynamic theories [96]. Furthermore, the smallest dimension of a segregated domain, e.g., the diameter of a cylinder, is proportional to the two-thirds power of the molecular weight of the minority block and can typically be tuned from about 5 to 100 nm by changing the molecular weight of the block.

Several additional phases are observed experimentally, but are not thermodynamically stable [13]. Moreover, the synthetic nature of the copolymers implies some heterogeneity in the polymer structure and molecular weight distribution. An excellent review has recently been published [97], and the main conclusion is that the polydispersity index (PDI) influences all aspects of the self-assembly. For example, upon an increase of the PDI of one block, the lattice constant of an ordered structure or the size of microphase-separated domains increases, interfacial thickness increases, and phase transitions may be induced. In addition, macrophase-separation may occur as the PDI is increased at certain compositions and segregation strengths.

The self-assembly properties of ABA triblock copolymers are similar compared to the AB diblocks [98, 99], but the prepared materials show better mechanical strength [99–102], essentially because of chain configurations induced by the copolymer structure. The loop configuration places both A blocks on the same side in the same A local phase whereas in "bridge" chain configuration, the central block B is anchored into two distinct areas (Fig. 10). The bridge chain configuration brings an improved stability in the final nanostructures and can be controlled by the midblock rigidity [103].

Mapping the self-assembly of ABC triblock copolymers is much more complicated, mainly due to the increase of the parameters that determine the self-assembly behavior. Zheng and Wang reported in 1995 a theoretical prediction of the morphological phase diagram for ABC triblock copolymers in the strong segregation limit [104]. A schematic representation of all the morphologies reported is reproduced in Fig. 11.

This theoretical work has been further completed by Tang et al. [105, 106], Tyler et al. [107, 108], and more recently by Chen and Xia [109] by using self-consistent field theory simulations. Playing with the flexibility/rigidity of the polymer segments, various new morphologies could be discovered and some of them have been confirmed experimentally [110–112]. A series of papers [113–117] from Stadler and coworkers based on experimental studies have also demonstrated the high complexity of the morphology observed with such ABC triblock copolymers. They used poly(styrene-b-butadiene-b-methyl methacrylate) (PS-b-PB-b-PMMA) triblock copolymers and their hydrogenated analogs poly (styrene-b-(ethylene-co-butylene)-b-methyl methacrylate) (PS-b-PEB-b-PMMA). In this case, the morphology seems to be governed by the relatively weak

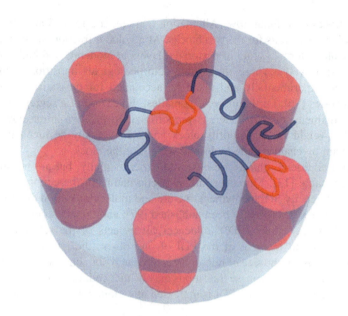

Fig. 10 Schematic depiction of "bridge" chain and "loop" chain configurations in cylindrical morphology from the ABA triblock copolymer self-assembly

incompatibility of the end blocks PS and PMMA rather than the strong incompatibility of the polybutadiene (PB) or poly(ethylene-*co*-butylene).

One of the potential applications of these ABC triblock copolymers was explored by Hillmyer and coworkers in 2005 [118]. They have prepared nanoporous membranes of polystyrene with controlled pore wall functionality from the selective degradation of ordered ABC triblock copolymers. By using a combination of controlled ring-opening and free-radical polymerizations, a triblock copolymer polylactide-*b*-poly(*N*,*N*-dimethylacrylamide)-*b*-polystyrene (PLA-*b*-PDMA-*b*-PS) has been prepared. Following the self-assembly in bulk, cylinders of PLA are dispersed into a matrix of PS and the central PDMA block localized at the PS-PLA interface. After a selective etching of the PLA cylinders, a nanoporous PS monolith is formed with pore walls coated with hydrophilic PDMA.

Other various copolymer systems, such as AB_2 graft copolymers [119–121], star-shaped molecules of the $(AB)_n$ type [122, 123], $(AB)_n$ multiblock copolymer systems [124–126], AB ring-shaped copolymers [127, 128], and ABC star-shaped terpolymers [129–131], were also widely studied.

3.2 Self-Assembly in Thin Films

If the film thickness is decreased too much (ca < 100 nm), the phase behavior of BCs in thin films can be strongly changed from the bulk behavior since it is

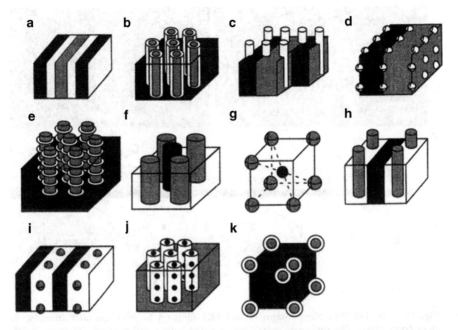

Fig. 11 Schematic representation of all the phases considered. *Dark* a, *white* b, *gray* e. (**a**) Lamellar phase. (**b**) Coaxed cylinder phase. (**c**) Lamella-cylinder phase. (**d**) Lamella-sphere phase. (**e**) Cylinder-ring phase. (**f**) Cylindrical domains in a square lattice structure. (**g**) Spherical domains in the CsCl type structure. (**h**) Lamella-cylinder-II. (**i**) Lamella-sphere-II. (**j**) Cylinder-sphere. (**k**) Concentric spherical domain in the bcc structure. Reprinted with permission from Zheng et el. [104]. Copyright 1995 American Chemical Society

influenced by interfacial interaction [132]. A block interacting preferentially with the substrate surface or free surface wets the corresponding interface and induces preferential orientation of nanodomains [133–138]. Minimization of the surface energies drives the selective wetting. These effects become more significant as the film thickness decreases [139, 140] and diverse ordered nanostructures have been achieved with a better control of orientation than bulk by tuning the film thickness and interfacial interaction [141–146].

For example, Magerle and coworkers have reported that the phase behavior in thin films of cylinder-forming BCs is dominated by surface reconstructions (Fig. 12). Comparing experiments and computer simulations based on dynamic density functional theory, a phase diagram has been established by varying the film thickness during the self-assembly of polystyrene-*b*-polybutadiene-*b*-polystyrene triblock copolymer (Fig. 13). In this diagram, the stability regions are found to be determined by the surface field and the film thickness [147].

Instead of observing the change of the morphology as a function of the film thickness, surface boundaries could also be used to control the wetting layer morphology at interfaces, the surface topographies, and the microdomain period [148]. In the case of symmetric or asymmetric wetting of the block copolymer at

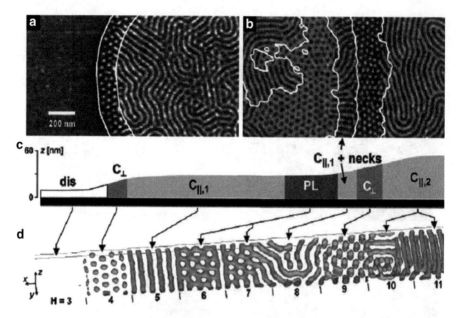

Fig. 12 (**a,b**) TM-SFM phase images of thin SBS films on Si substrates after annealing in chloroform vapor. The surface is everywhere covered with a ~10-nm-thick PB layer. *Bright* (*dark*) corresponds to PS (PB) microdomains below this top PB layer. *Contour lines* calculated from the corresponding height images are superimposed. (**c**) Schematic height profile of the phase images shown in (**a,b**). (**d**) Simulation of an $A_3B_{12}A_3$ block copolymer film in one large simulation box of $[352 \times 32 \times H(x)]$ grid points with increasing film thickness $H(x)$, $\varepsilon_{AB} = 6.5$, and $\varepsilon_M = 6.0$. The latter corresponds to a preferential attraction of B beads to the surface. Reprinted with permission from Knoll et al. [147]. Copyright 2002 by the American Physical Society

both interfaces, smooth surfaces are observed only if the film thickness (h) is respectively given by nL_0 and $(n + \frac{1}{2})L_0$, with n an integer and L_0 the microdomain period. For other thickness values, surface topographies (holes, islands) are observed in response to the frustrations born from the incommensurability between h and L_0. However, if the film is placed between two hard plates, the surface topographies could not emerge to balance these frustrations. Therefore stretching or compression of block copolymer microdomains is observed leading to a change in the microdomain period.

3.3 Directing the Self-Assembly

With recent discoveries in the self-assembly of block copolymers, a large step is now being taken towards industrial applications. However, the spontaneous process of microphase separation leads to the formation of "polycrystalline" microdomain arrays consisting of randomly oriented regions, which limits the potential

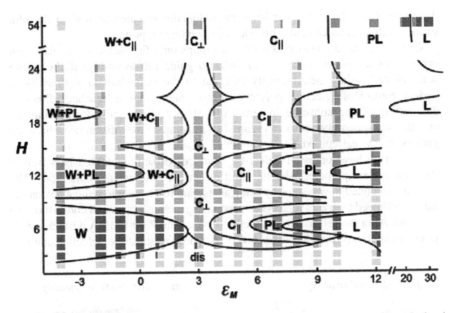

Fig. 13 Phase diagram of surface reconstructions of a $A_3B_{12}A_3$ block copolymer film calculated with MESODYN for $\varepsilon_{AB} = 6.5$. The *boxes* indicate where simulations have been done. The *boxes with two shades of gray* indicate that two phases coexist after the finite simulation time. *Smooth phase boundaries* have been drawn to guide the eye. Reprinted with permission from Knoll et al. [147]. Copyright 2002 by the American Physical Society

applications. In that way, the most used lab technique, which is spin coating, appeared to be inappropriate in treating defectless large surface areas of block copolymer films. Moreover, as stated before, the film thickness could maximize the role of interfacial energies of the substrate and the free surface. As far as industrial concerns were taking into account, researchers have tried to guide themselves through the assembly stage by external subterfuges which could be classified as active or passive guiding. Several excellent reviews have described this in detail [149, 150], the present paragraph being just a short reminder.

Among the passive guiding techniques, control of the substrate topography, in terms of surface relief structures, can modify the self-assembly of block copolymers. As an example, a topographical graphoepitaxy technique for controlling the self-assembly of block copolymer thin films was described to produce 2D periodic nanostructures with a precisely determined orientation and long-range order [151]. In this contribution, the substrate surface is patterned with 2D arrays of nanoscale posts in order to influence the self-assembly of polystyrene-*b*-polydimethylsiloxane (PS-*b*-PDMS) block copolymer.

More recently, ordered ultradense arrays in block copolymer films have been obtained over macroscopic distances using faceted surfaces of commercially available single-crystal sapphire substrates [152]. The self-assembly overrides substrate defects and uses the topography only as a guide to the orientation of the arrays.

Arrays of 3 nm diameter cylindrical microdomains oriented normal to the film surface with a center-to-center distance of 6.9 nm are reported.

Creation of heterogeneity in chemical groups onto flat surfaces has also been explored as a passive technique in order to guide self-assembly. If the length scale of the surface heterogeneity is close to the microdomain periodicity, a perfect control of selective polymer–surface interactions is reached, directing the wetting of the substrate in terms of polymer phase and thus guiding the overall self-assembly stage. Figure 14 illustrates the difference between graphoepitaxy and epitaxial self-assembly.

Ruiz et al. have prepared defect-free surfaces of self-assembled BCs from a chemical pattern prepared near the limit of current lithographic tools [154]. The pattern is obtained in two steps by using an e-beam resist layer to write a hexagonal pattern followed by use of an oxygen plasma to generate a chemical contrast on the substrate. After the self-assembly of cylinder-forming PS-b-PMMA diblock copolymer, a pattern rectification is observed compared to substrate defects leading to

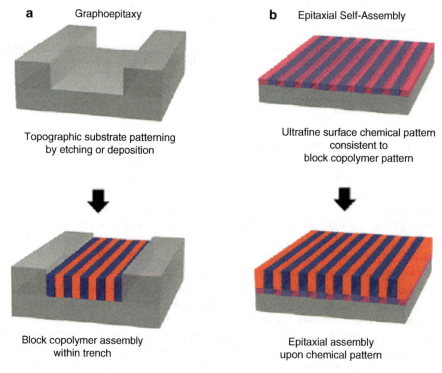

Fig. 14 Graphoepitaxy vs epitaxial self-assembly. (**a**) Graphoepitaxy utilizes topographic substrate pattern for directed block copolymer assembly. The substrate pattern remains in the finally formed nanopatterned morphology. (**b**) Epitaxial self-assembly utilizes nanoscale chemical pattern to register block copolymer assembly. Ultrafine chemical patterning requires e-beam lithography or other high-cost lithography such as EUV. Reprinted with permission from Jeong et al. [153]. Copyright 2010 American Chemical Society

Dynamic Assembly of Block-Copolymers

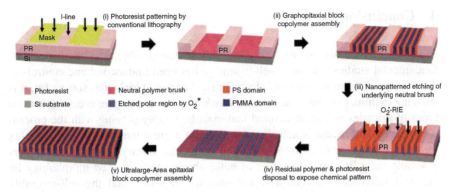

Fig. 15 Schematic illustration of ultralarge-area block copolymer lithography procedure. (*i*) Photoresist patterning by I-line lithography on a neutral polymer-brush-treated Si substrate. (*ii*) Graphoepitaxy of block copolymers within photoresist confinement. (*iii*) Polymer brush layer chemical patterning by selective etching via graphoepitaxy morphology. (*iv*) Residual photoresist and polymer disposal to expose a chemically nanopatterned substrate. (*v*) Ultralarge-area epitaxial assembly of block copolymers. Reprinted with permission from Jeong et al. [153]. Copyright 2010 American Chemical Society

perfect hexagonal assembly. It was also shown that multiplication of the density in cylinders between the substrate and the block copolymer layer was successful. More recently, highly oriented lamellar morphology was prepared by combining the two principles of graphoepitaxy and epitaxial self-assembly (Fig. 15) [153]. This combination enabled one to decrease the cost and to increase the defect free surface area covered by the nanopatterning.

Many active techniques are now used to align the BCs and an excellent review has been published recently [150]. Defects in morphology alignment or orientation can be treated by thermal or solvent annealing. In those cases, the main objective is to give enough mobility to the polymer chains to get closer to the equilibrium morphology. However, typical treatment times are greater than a few hours which could be a strong limitation in industrial applications. In this frame, flow fields could be used to control morphological orientation efficiently. For example, Mortensen et al. employed oscillatory shear in both Couette type and parallel plate cells to form ordered body-centered-cubic (bcc) lattices of micellar poly (ethylene oxide-*b*-propylene oxide-*b*-ethylene oxide) triblock copolymers [155–157]. Poly(styrene-*b*-isoprene-*b*-styrene) triblock copolymer has been transformed from bcc spheres into cylindrical microstructures with short-range order by using elongational flow deformation [158]. The reverse process is also possible by first shear-aligning cylindrical features and then heating through the cylinder-to-sphere transition [159]. Electric fields, typically of 2–20 V, can also be exploited for BCP alignment for systems in which there is a sufficient dielectric constant contrast between the polymer blocks. SCF theory has been used to probe the electric field inducement of order in bcc spherical lattices as well as the transition from spheres to cylinders [160].

4 Conclusion

Block copolymer was at the beginning only a strange object to be observed in fundamental studies, but since self-assembly has been understood and controlled, the conversion from isolated BC chains to rigid nanomaterials was successful. It is amazing to think that only one selected block copolymer can give access to various types of micelles or nanostructured materials by simply playing with the process parameters. The nature is actually able to build a great number of very complex self-assembled structures from a limited molecular library. Usually researchers assemble their BCs in solution or in bulk, observe the obtained morphology by classical microscopy techniques, and then try to understand the self-assembly pathway followed by their system. Now we are moving toward an exciting new area where it will be possible to predict and to target the final morphology. This is not a mandatory point for industrial application but it opens new avenues and could maybe lower the price of the final nanomaterials.

References

1. Hamley IW (2000) Introduction to soft matter, 2nd edn. Wiley, Chichester
2. Langmuir I, Blodgett KB (1935) Zeitschrift für wissenschaft und technische kolloidchemie. Kolloid-Zeitschrift 73:257–263
3. Bigelow WC, Pickett DL et al (1946) Oleophobic monolayers. 1 - Films adsorbed from solution in non-polar liquids. J Colloid Sci 1:513–538
4. Nuzzo RG, Allara DL (1983) Adsorption of bifunctional organic disulfides on gold surfaces. J Am Chem Soc 105:4481–4483
5. Laibinis PE, Whitesides GM et al (1991) Comparison of the structures and wetting properties of self-assembled monolayers of n-alkanethiols on the coinage metal surfaces, copper, silver, and gold. J Am Chem Soc 113:7152–7167
6. Troughton EB, Bain CD et al (1988) Monolayer films prepared by the spontaneous self-assembly of symmetrical and unsymmetrical dialkyl sulfides from solution onto gold substrates: structure, properties, and reactivity of constituent functional groups. Langmuir 4:365–385
7. Russell B (1972) A history of Western philosophy. Simon & Schuster, New York, p xi
8. Ozin GA, Hou K et al (2009) Nanofabrication by self-assembly. Mater Today 12:12–23
9. Jain S, Bates FS (2004) Consequences of nonergodicity in aqueous binary PEO-PB micellar dispersions. Macromolecules 37:1511–1523
10. Hayward RC, Pochan DJ (2010) Tailored assemblies of block copolymers in solution: it is all about the process. Macromolecules 43:3577–3584
11. Klok H-A, Lecommandoux S (2001) Supramolecular materials via block copolymer self-assembly. Adv Mater 13:1217–1229
12. Khandpur AK, Forster S et al (1995) Polyisoprene-polystyrene diblock copolymer phase diagram near the order-disorder transition. Macromolecules 28:8796–8806
13. Bates FS, Fredrickson GH (1999) Block copolymers - designer soft materials. Phys Today 52:32–38
14. Hamley IW (1998) The physics of block copolymers. Oxford University Press, New York, p 425

Dynamic Assembly of Block-Copolymers

15. Cameron NS, Corbierre MK et al (1999) E.W.R. Steacie award lecture. Asymmetric amphiphilic block copolymers in solution: a morphological wonderland. Can J Chem 77:1311–1326
16. Dingv JF, Liu GJ et al (1997) Multiple morphologies of polyisoprene-block-poly (2-cinnamoylethyl methacrylate) and polystyrene-block-poly(2-cinnamoylethyl methacrylate) micelles in organic solvents. Polymer 38:5497–5501
17. Bang J, Jain SM et al (2006) Sphere, cylinder, and vesicle nanoaggregates in poly (styrene-b-isoprene) diblock copolymer solutions. Macromolecules 39:1199–1208
18. Lazzari M, Liu G, Lecommandoux S (2006) Block copolymers in nanoscience. Wiley-VCH, Weinheim
19. Wei Z, Wang ZG (1995) Morphology of ABC triblock copolymers. Macromolecules 28:7215–7223
20. Hadjichristidis N (1999) Synthesis of miktoarm star (μ-star) polymers. J Polym Sci A Polym Chem 37:857–871
21. Hadjichristidis N, Iatrou H et al (2005) Linear and non-linear triblock terpolymers. Synthesis, self-assembly in selective solvents and in bulk. Progr Polym Sci 30:725–782
22. Gido SP, Lee C et al (1996) Synthesis, characterization, and morphology of model graft copolymers with trifunctional branch points. Macromolecules 29:7022–7028
23. Zhang L, Eisenberg A (1998) Formation of crew-cut aggregates of various morphologies from amphiphilic block copolymers in solution. Polym Adv Technol 9:677–699
24. Merrett FM (1954) Interaction of polymerizing systems with rubber and its homologs. Polymerization of MMA and styrene. Trans Faraday Soc 50:759–767
25. Wang X, Guerin G et al (2007) Cylindrical block copolymer micelles and co-micelles of controlled length and architecture. Science 317:644–647
26. Israelachvili JN, Mitchell DJ et al (1976) Theory of self- assembly of hydrocarbon amphiphiles into micelles and bilayers. J Chem Soc Faraday Trans 2(72):1525–1568
27. Kunz W, Testard F et al (2009) Correspondence between curvature, packing parameter, and hydrophilic-lipophilic deviation scales around the phase-inversion temperature. Langmuir 25:112–115
28. Blanazs A, Armes SP et al (2009) Self-assembled block copolymer aggregates: from micelles to vesicles and their biological applications. Macromol Rapid Commun 30:267–277
29. Riess G (2003) Micellization of block copolymers. Prog Polym Sci 28:1107–1170
30. Tuzar Z, Kratochvil P (1993) Micelles of block and graft copolymers in solution. In: Matijevic E (ed) Surface and colloid science. Plenum Press, New York
31. Munk P (1996) Equilibrium and nonequilibrium polymer micelles. In: Webber SE, Munk P, Tuzar Z (eds) Solvents and self organization of polymers, NATO ASI series, series E: applied sciences. Kluwer Academic Publisher, Dordrecht
32. Yoshida E, Kunugi S (2002) Micelle formation of poly(vinyl phenol)-block-polystyrene by alpha, omega-diamines. J Polym Sci A Polym Chem 40:3063–3067
33. Yoshida E, Kunugi S (2002) Micelle formation of nonamphiphilic diblock copolymers through noncovalent bond cross-linking. Macromolecules 35:6665–6669
34. Thibault RJ, Hotchkiss PJ et al (2003) Thermally reversible formation of microspheres through non-covalent polymer cross-linking. J Am Chem Soc 125:11249–11252
35. Chen DY, Peng HS et al (2003) A novel one-step approach to core-stabilized nanoparticles at high solid contents. Macromolecules 36:2576–2578
36. Ximei Y, Chen DY et al (2004) Micellization of PS-b-P4VP/formic acid in chloroform without or with the premixing of the copolymer with decanoic acid. Macromolecules 37:4211–4217
37. de Luzuriaga AR, Garcia I et al (2010) Design and stabilization of block copolymer micelles via phenol–pyridine hydrogen-bonding interactions. Polymer 51:1355–1362
38. de Gennes PG (1978) In solid state physics. Academic, New York
39. Halperin A, Tirrell M et al (1992) Tethered chains in polymer microstructures. Adv Polym Sci 100:31–71

40. Rogez D (1987) Synthesis, micellization and emulsifying properties of PS–PEO block copolymers (in French). PhD Thesis. University of Haute Alsace, France
41. Riess G, Rogez D (1982) Micellization of poly(styrene-b-ethylene oxide) block copolymers, ACS Polym Prepr (Div Polym Chem) 23:19–20
42. Giacomelli FC, Riegel IC et al (2009) Aggregation behavior of a new series of ABA triblock copolymers bearing short outer A blocks in B-selective solvent: from free chains to bridged micelles. Langmuir 25:731–738
43. Kong W, Li B et al (2010) Complex micelles from self-assembly of ABA triblock copolymers in B-selective solvents. Langmuir 26:4226–4232
44. Zhou Z, Yang Y–W et al (1996) Association of a triblock ethylene oxide (E) and butylene oxide (B) copolymer (B(12)E(260)B(12)) in aqueous solution. Macromolecules 29:8357–8361
45. Liu T, Zhou Z et al (1998) Dominant factors on the micellization of BnEmBn-Type triblock copolymers in aqueous solution. J Phys Chem B 102:2875–2882
46. Borsali R, Minatti E et al (2003) From "sunflower-like" assemblies toward giant wormlike micelles. Langmuir 19:6–9
47. Minatti E, Viville P et al (2003) Micellar morphological changes promoted by cyclization of PS-b-PI copolymer: DLS and AFM experiments. Macromolecules 36:4125–4133
48. Tsitsilianis C, Roiter Y et al (2008) Diversity of nanostructured self-assemblies from a pH-responsive ABC terpolymer in aqueous media. Macromolecules 41:925–934
49. Lee ES, Oh KT et al (2007) Tumor pH-responsive flower-like micelles of poly(L-lactic acid)-b-poly (ethylene glycol)-b-poly(L-histidine). J Control Release 123:19–26
50. van Butsele K, Cajot S et al (2009) pH-responsive flower-type micelles formed by a biotinylated poly(2-vinylpyridine)-block-poly(ethylene oxide)-block-poly(epsilon-caprolactone) triblock copolymer. Adv Funct Mater 19:1416–1425
51. Zhulina KB, Adam M (2005) Diblock copolymer micelles in a dilute solution. Macromolecules 38:5330–5351
52. Won YY, Davis HT et al (1999) Giant wormlike rubber micelles. Science 283:960–963
53. Zhang L, Eisenberg A (1995) Multiple morphologies of "crew-Cut" aggregates of polystyrene-b-poly(acrylic acid) block copolymers. Science 268:1728–1731
54. Dan N, Safran SA (2006) Junctions and end-caps in self-assembled non-ionic cylindrical micelles. Adv Colloid Interface Sci 123:323–331
55. Won Y-Y, Brannan AK, Davis HT et al (2002) Cryogenic transmission electron microscopy (cryo-TEM) of micelles and vesicles formed in water by poly(ethylene oxide)-based block copolymers. J Phys Chem B 106:3354–3364
56. Jain S, Bates FS (2003) On the origins of morphological complexity in block copolymer surfactants. Science 300:460–464
57. Jain S, Gong X et al. (2006) Disordered network state in hydrated block-copolymer surfactants. Phys Rev Lett 96:138304/1
58. Quemener D, Bonniol G (2010) Free-standing nanomaterials from block copolymer self-assembly. Macromolecules 43:5060–5065
59. Dean JM, Verghese NE et al (2003) Nanostructure toughened epoxy resins. Macromolecules 36:9267–9270
60. Kim Y, Dalhaimer P et al (2005) Polymeric worm micelles as nano-carriers for drug delivery. Nanotechnology 16:S484
61. Dalhaimer P, Bates FS et al (2003) Single molecule visualization of stable, stiffness-tunable, flow-conforming worm micelles. Macromolecules 36:6873–6877
62. Massey J, Power KN et al (1998) Self-assembly of a novel organometallic − inorganic block copolymer in solution and the solid state: nonintrusive observation of novel wormlike poly (ferrocenyldimethylsilane)-b-poly(dimethylsiloxane) micelles. J Am Chem Soc 120:9533–9540
63. Raez J, Manners I et al (2002) Nanotubes from the self-assembly of asymmetric crystalline − coil poly(ferrocenylsilane − siloxane) block copolymers. J Am Chem Soc 124:10381–10395

64. Wang XS, Arsenault A et al (2003) Shell cross-linked cylinders of polyisoprene-b-ferrocenyl-dimethylsilane: formation of magnetic ceramic replicas and microfluidic channel alignment and patterning. J Am Chem Soc 125:12686–12687
65. Massey JA, Temple K et al (2000) Self-assembly of organometallic block copolymers: the role of crystallinity of the core-forming polyferrocene block in the micellar morphologies formed by poly(ferrocenylsilane-b-dimethylsiloxane) in n-alkane solvents. J Am Chem Soc 122:11577–11584
66. LoPresti C, Lomas H et al (2009) Polymersomes: nature inspired nanometer sized compartments. J Mater Chem 19:3576–3590
67. Discher D, Eisenberg A (2002) Polymer vesicles. Science 297:967–973
68. Burke S, Shen H (2001) Multiple vesicular morphologies from block copolymers in solution. Macromol Symp 175:273–283
69. Yu Y, Zhang L et al (1997) Multiple morphologies of crew-cut aggregates of polybutadiene-b-poly(acrylic acid) diblocks with low Tg cores. Langmuir 13:2578–2581
70. Discher BM, Won YY et al (1999) Polymersomes: tough vesicles made from diblock copolymers. Science 284:1143–1146
71. Schillen K, Bryskhe K (1999) Vesicles formed from a poly(ethylene oxide) – poly(propylene oxide) – poly(ethylene oxide) triblock copolymer in dilute aqueous solution. Macromolecules 32:6885–6888
72. Lee JC-M, Bermudez H et al (2001) Preparation, stability, and in vitro performance of vesicles made with diblock copolymers. Biotechnol Bioeng 73:135–145
73. Dan N, Safran SA (1994) Self-assembly in mixtures of diblock copolymers. Macromolecules 27:5766–5772
74. Jiang Y, Chen T et al (2005) Effect of polydispersity on the formation of vesicles from amphiphilic diblock copolymers. Macromolecules 38:6710–6717
75. Terreau O, Luo L et al (2003) Effect of poly(acrylic acid) block length distribution on polystyrene-b-poly(acrylic acid) aggregates in solution. 1. Vesicles. Langmuir 19:5601–5607
76. Luo L, Eisenberg A (2002) One-step preparation of block copolymer vesicles with preferentially segregated acidic and basic corona chains. Angew Chem Int Ed Engl 41:1001–1004
77. Stoenescua R, Meier W (2002) Vesicles with asymmetric membranes from amphiphilic ABC triblocks copolymers. Chem Commun 3016–3017
78. Brannan AK, Bates FS (2004) ABCA tetrablock copolymer vesicles. Macromolecules 37:8816–8819
79. Li Z, Hillmyer MA et al (2006) Control of structure in multicompartment micelles by blending μ-ABC star terpolymers with AB diblock copolymers. Macromolecules 39:765–771
80. Pochan DJ, Chen ZY et al (2004) Toroidal triblock copolymer assemblies. Science 306:94–97
81. Li ZB, Chen ZY et al (2005) Disk morphology and disk-to-cylinder tunability of poly(acrylic acid)-b-poly(methyl acrylate)-b-polystyrene triblock copolymer solution-state assemblies. Langmuir 21:7533–7539
82. Cui HG, Chen ZY et al (2006) Controlling micellar structure of amphiphilic charged triblock copolymers in dilute solution via coassembly with organic counterions of different spacer lengths. Macromolecules 39:6599–6607
83. Cui HG, Chen ZY et al (2007) Block copolymer assembly via kinetic control. Science 317:647–650
84. Li Z, Chen Z et al (2007) Controlled stacking of charged block copolymer micelles. Langmuir 23:4689–4694
85. Hales K, Chen Z et al (2008) Nanoparticles with tunable internal structure from triblock copolymers of PAA-b-PMA-b-PS. Nano Lett 8:2023–2026
86. Zhong S, Cui H et al (2008) Helix self-assembly through the coiling of cylindrical micelles. Soft Matter 4:90–93
87. Cui H, Chen Z et al (2009) Origins of toroidal micelle formation through charged triblock copolymer self-assembly. Soft Matter 5:1269–1278

88. Gadt T, Ieong NS et al (2009) Complex and hierarchical micelle architectures from diblock copolymers using living, crystallization-driven polymerizations. Nat Mater 8:144–150
89. Inoue T, Soen T (1969) Thermodynamic interpretation of domain structure in solvent-cast films of A-B type block copolymers of styrene and isoprene. J Polym Sci B Polym Phys 7:1283–1301
90. Gervais M, Gallot B (1973) Phase diagram and structural study of polystyrene - poly(ethylene oxide) block copolymers. 1. Systems polystyrene/poly(ethylene oxide)/diethyl phthalate. Makromol Chem 171:157–178
91. Hashimoto T, Nagatoshi K (1974) Domain-boundary structure of styrene-isoprene block copolymer films cast from toluene solutions. Macromolecules 7:364–373
92. Helfand E (1975) Block copolymer theory. III. Statistical mechanics of the microdomain structure. Macromolecules 8:552–556
93. Helfand E, Wasserman ZR (1976) Block copolymer theory. 4. Narrow interphase approximation. Macromolecules 9:879–888
94. Leibler L (1980) Theory of microphase separation in block copolymers. Macromolecules 13:1602–1617
95. Bates FS (1991) Polymer-polymer phase behavior. Science 251:898–905
96. Fredrickson GH, Bates FS (1996) Dynamics of block copolymers: theory and experiment. Annu Rev Mater Sci 26:501–550
97. Lynd NA, Meuler AJ et al (2008) Polydispersity and block copolymer self-assembly. Prog Polym Sci 33:875–893
98. Matsen MW, Thompson RB (1999) Equilibrium behavior of symmetric ABA triblock copolymer melts. J Chem Phys 111:7139–7146
99. Gehlsen MD, Almdal K et al (1992) Order-disorder transition – diblock versus triblock copolymers. Macromolecules 25:939–943
100. Adams JL, Graessley WM et al (1994) Rheology and the microphase separation transition in styrene-isoprene block copolymers. Macromolecules 27:6026–6032
101. Riise BL, Fredrickson GH et al (1995) Rheology and shear-induced alignment of lamellar diblock and triblock copolymers. Macromolecules 28:7653–7659
102. Ryu CY, Lee MS et al (1997) Structure and viscoelasticity of matched asymmetric diblock and triblock copolymers in the cylinder and sphere microstructures. J Polym Sci B Polym Phys 35:2811–2823
103. Song J, Shi T et al (2008) Rigidity effect on phase behavior of symmetric ABA triblock copolymers: a Monte Carlo simulation. J Chem Phys 129:054906
104. Zheng W, Wang ZG (1995) Morphology of ABC triblock copolymers. Macromolecules 28:7215–7223
105. Tang P, Qiu F, Zhang H et al (2004) Morphology and phase diagram of complex block copolymers: ABC linear triblock copolymers. Phys Rev E Stat Nonlin Soft Matter Phys 69:031803
106. Tang P, Qiu F et al (2004) Morphology and phase diagram of complex block copolymers: ABC star triblock copolymers. J Phys Chem B 108:8434–8438
107. Tyler CA, Morse DC (2005) Orthorhombic Fddd network in triblock and diblock copolymer melts. Phys Rev Lett 94:208302
108. Tyler CA, Qin J et al (2007) SCFT study of nonfrustrated ABC triblock copolymer melts. Macromolecules 40:4654–4668
109. Xia Y, Chen J et al (2010) Self-assembly of linear ABC coil-coil-rod triblock copolymers. Polymer 51:3315–3319
110. Chatterjee J, Jain S et al (2007) Comprehensive phase behavior of poly(isoprene-b-styrene-b-ethylene oxide) triblock copolymers. Macromolecules 40:2882–2896
111. Takano A, Wada S et al (2004) Observation of cylinder-based microphase-separated structures from ABC star-shaped terpolymers investigated by electron computerized tomography. Macromolecules 37:9941–9946

Dynamic Assembly of Block-Copolymers

112. Tureau MS, Epps TH (2009) Nanoscale networks in poly[isoprene-block-styrene-block-(methyl methacrylate)] triblock copolymers. Macromol Rapid Commun 30:1751–1755
113. Auschra C, Stadler R (1993) New ordered morphologies in ABC triblock copolymers. Macromolecules 26:2171–2174
114. Stadler R, Auschra C (1995) Morphology and thermodynamics of symmetric poly(A-block-B-block-C) triblock copolymers. Macromolecules 28:3080–3097
115. Krappe U, Stadler R et al (1995) Chiral assembly in amorphous ABC triblock copolymers. Formation of a helical morphology in polystyrene-block-polybutadiene-block-poly(methyl methacrylate) block copolymers. Macromolecules 28:4558–4561
116. Breiner U, Krappe U et al (1997) Cylindrical morphologies in asymmetric ABC triblock copolymers. Macromol Chem Phys 198:1051–1083
117. Breiner U, Krappe U et al (1996) Evolution of the "knitting pattern" morphology in ABC triblock copolymers. Macromol Rapid Commun 17:567–575
118. Rzayev J, Hillmyer MA (2005) Nanoporous polystyrene containing hydrophilic pores from ABC triblock copolymer precursor. Macromolecules 38:3–5
119. Matsushita Y (2000) Studies on equilibrium structures of complex polymers in condensed systems. J Polym Sci B 38:1645–1655
120. Hadjichristidis N, Iatrou H et al (1993) Morphology and miscibility of miktoarm styrene-diene copolymers and terpolymers. Macromolecules 26:5812–5815
121. Matsushita Y, Momose H et al (1997) Lamellar domain spacing of the ABB graft copolymers. Polymer 38:149–153
122. Alward DB, Kinning DJ et al (1986) Effect of arm number and arm molecular weight on the solid-state morphology of poly(styrene-isoprene) star block copolymers. Macromolecules 19:215–224
123. Thomas EL, Anderson DM et al (1988) Periodic area-minimizing surfaces in block copolymers. Nature 334:598–601
124. Spontak RJ, Smith SD (2001) Perfectly-alternating linear (AB)n multiblock copolymers: effect of molecular design on morphology and properties. J Polym Sci B 39:947–955
125. Matsushita Y, Mogi Y et al (1994) Preparation and morphology of multiblock copolymers of the (AB)n type. Polymer 35:246–249
126. Wu L, Cochan EW et al (2004) Consequences of block number on the order – disorder transition and viscoelastic properties of linear (AB)n multiblock copolymers. Macromolecules 37:3360–3368
127. Zhu Y, Gido SP et al (2003) Microphase separation of cyclic block copolymers of styrene and butadiene and of their corresponding linear triblock copolymers. Macromolecules 36:148–152
128. Takano A, Kadoi O et al (2003) Preparation and morphology of ring-shaped polystyrene-block-polyisoprenes. Macromolecules 36:3045–3050
129. Yamauchi K, Takahashi K (2003) Microdomain morphology in an ABC 3-miktoarm star terpolymer: a study by energy-filtering TEM and 3D electron tomography. Macromolecules 36:6962–6966
130. Takano A, Kawashima W et al (2005) A mesoscopic Archimedean tiling having a new complexity in an ABC star polymer. J Polym Sci B 43:2427–2432
131. Hayashida K, Takano A et al (2006) Systematic transitions of tiling patterns formed by ABC star-shaped terpolymers. Macromolecules 39:9402–9408
132. Jung J, Park H-W et al (2010) Effect of film thickness on the phase behaviors of diblock copolymer thin film. ACS Nano 4:3109–3116
133. Matsen MW (1997) Thin films of block copolymer. J Chem Phys 106:7781–7791
134. Lee B, Park I et al (2005) Structural analysis of block copolymer thin films with grazing incidence small-angle X-ray scattering. Macromolecules 38:4311–4323
135. Park I, Lee B et al (2005) Epitaxial phase transition of polystyrene-b-polyisoprene from hexagonally perforated layer to gyroid phase in thin film. Macromolecules 38:10532–10536
136. Park HW, Im K et al (2007) Direct observation of HPL and DG structure in PS-b-PI thin film by transmission electron microscopy. Macromolecules 40:2603–2605

137. Park HW, Jung J et al (2009) New characterization methods for block copolymers and their phase behaviors. Macromol Res 17:365–377
138. Lyakhova KS, Sevink GJA et al (2004) Role of dissimilar interfaces in thin films of cylinder-forming block copolymers. J Chem Phys 120:1127–1137
139. Fasolka MJ, Mayes AM (2001) Block copolymer thin films: physics and applications. Annu Rev Mater Res 31:323–355
140. Olszowka V, Tsarkova L et al (2009) 3-Dimensional control over lamella orientation and order in thick block copolymer films. Soft Matter 5:812–819
141. Huang E, Russell TP et al (1998) Using surface active random copolymers to control the domain orientation in diblock copolymer thin films. Macromolecules 31:7641–7650
142. Xu T, Hawker CJ et al (2005) Interfacial interaction dependence of microdomain orientation in diblock copolymer thin films. Macromolecules 38:2802–2805
143. Ryu DY, Wang JY et al (2007) Surface modification with cross-linked random copolymers: minimum effective thickness. Macromolecules 40:4296–4300
144. Ham S, Shin C et al (2008) Microdomain orientation of PS-b-PMMA by controlled interfacial interactions. Macromolecules 41:6431–6437
145. Albert JNL, Baney MJ, Stafford CM et al (2009) Generation of monolayer gradients in surface energy and surface chemistry for block copolymer thin film studies. ACS Nano 3:3977–3986
146. Ramanathan M, Nettleton E et al (2009) Simple orientational control over cylindrical organic-inorganic block copolymer domains for etch mask applications. Thin Solid Films 517:4474–4478
147. Knoll A, Horvat A et al (2002) Phase behavior in thin films of cylinder-forming block copolymers. Phys Rev Lett 89:035501
148. Wang JY, Park S et al (2008) Block copolymer thin films. In: Tsui OKC, Russell TP (eds) Polymer thin films. World Scientific Publishing, Singapore
149. Kim H-C, Park S-M et al (2010) Block copolymer based nanostructures: materials, processes, and applications to electronics. Chem Rev 110:146–177
150. Darling SB (2007) Directing the self-assembly of block copolymers. Prog Polym Sci 32:1152–1204
151. Bita I, Yang JKW et al (2008) Graphoepitaxy of self-assembled block copolymers on two-dimensional periodic patterned templates. Science 321:939–943
152. Park S, Lee DH et al (2009) Macroscopic 10-terabit–per–square-inch arrays from block copolymers with lateral order. Science 323:1030–1033
153. Jeong S-J, Moon H-S et al (2010) Ultralarge-area block copolymer lithography enabled by disposable photoresist prepatterning. ACS Nano 4:5181–5186
154. Ruiz R, Kang H et al (2008) Density multiplication and improved lithography by directed block copolymer assembly. Science 321:936–939
155. Mortensen K (1992) Phase behaviour of poly(ethylene oxide)-poly(propylene oxide)-poly (ethylene oxide) triblock-copolymer dissolved in water. Europhys Lett 19:599–604
156. Mortensen K, Brown W et al (1992) Inverse melting transition and evidence of three-dimensional cubatic structure in a block-copolymer micellar system. Phys Rev Lett 68:2340–2343
157. Mortensen K, Pedersen JS (1993) Structural study on the micelle formation of poly(ethylene oxide)–poly(propylene oxide)–poly(ethylene oxide) triblock copolymer in aqueous solution. Macromolecules 26:805–812
158. Lee W-K, Kim HD et al (2006) Morphological reorientation by extensional flow deformation of a triblock copolymer styrene–isoprene–styrene. Curr Appl Phys 6:718–722
159. Koppi KA, Tirrell M et al (1994) Epitaxial growth and shearing of the body centered cubic phase in diblock copolymer melts. J Rheol 38:999–1027
160. Lin C-Y, Schick M et al (2005) Structural changes of diblock copolymer melts due to an external electric field: a self-consistent-field theory study. Macromolecules 38:5766–5773

Top Curr Chem (2012) 322: 193–216
DOI: 10.1007/128_2011_197
© Springer-Verlag Berlin Heidelberg 2011
Published online: 28 June 2011

Dynamic Chemistry of Anion Recognition

Radu Custelcean

Abstract In the past 40 years, *anion recognition* by synthetic receptors has grown into a rich and vibrant research topic, developing into a distinct branch of *Supramolecular Chemistry*. Traditional anion receptors comprise organic scaffolds functionalized with complementary binding groups that are assembled by multistep organic synthesis. Recently, a new approach to anion receptors has emerged, in which the host is dynamically self-assembled in the presence of the anionic guest, via reversible bond formation between functional building units. While coordination bonds were initially employed for the self-assembly of the anion hosts, more recent studies demonstrated that reversible covalent bonds can serve the same purpose. In both cases, due to their labile connections, the molecular constituents have the ability to assemble, dissociate, and recombine continuously, thereby creating a dynamic combinatorial library (DCL) of receptors. The anionic guests, through specific molecular recognition, may then amplify (express) the formation of a particular structure among all possible combinations (real or virtual) by shifting the equilibria involved towards the most optimal receptor. This approach is not limited to solution self-assembly, but is equally applicable to crystallization, where the fittest anion-binding crystal may be selected. Finally, the pros and cons of employing dynamic combinatorial chemistry (DCC) vs molecular design for developing anion receptors, and the implications of both approaches to selective anion separations, will be discussed.

Keywords Anions · Dynamic combinatorial chemistry · Molecular recognition · Self-assembly · Supramolecular chemistry

---

R. Custelcean (✉)
Chemical Sciences Division, Oak Ridge National Laboratory, Oak Ridge, TN 37831, USA
e-mail: custelceanr@ornl.gov

Contents

1 Introduction 194
2 Anion Recognition Via Dynamic Coordination Chemistry 195
3 Anion Recognition Via Dynamic Covalent Chemistry 201
4 Anion Recognition Via Dynamic Crystallization 205
5 Evolution Vs Intelligent Design in Self-Assembled Anion Receptors 210
6 Conclusions 213
References 214

1 Introduction

The field of anion recognition started in 1968 with the seminal paper by Park and Simmons that described a class of protonated macrobicyclic diamines, named *katapinands*, acting as halide receptors [1]. This early example of anion encapsulation has provided critical guidance for the design of the following generations of anion receptors [2]. As clearly articulated early on by Lehn [3], efficient and selective anion binding, particularly in competitive aqueous environments, requires complete sequestration of the anion from its surrounding solvent by encapsulation inside a rigid host containing a complementary cavity functionalized with binding groups geometrically constrained to prevent reorganization and effective binding of competing anions. This approach is inspired from natural anion-binding receptors, which are known to sequester the anions deep inside their structures, completely isolating them from the aqueous solvent by encapsulation inside highly organized binding cavities functionalized with hydrogen-bonding groups or metal centers. For example, the sulfate-binding protein encapsulates SO_4^{2-} through the formation of seven hydrogen bonds from aminoacid residues found in the binding cavity, 7 Å deep from the protein's surface [4]. This results in a remarkably strong and selective sulfate binding, far surpassing most synthetic receptors to date. Protein receptors have achieved such impressive levels of binding strength and selectivity over millions of years of evolution, adapting their structures in response to external environmental pressures. By comparison, a typical synthetic receptor is developed in the lab in a much shorter timeframe (from a few weeks to a couple of years), and once they are completed, their structure is fixed and unable to evolve. As a result, they are structurally and functionally much less complex than their natural counterparts. Though it has been clearly recognized that complete anion encapsulation is necessary for strong and selective binding, this has rarely been achieved, especially in aqueous environments. One major difficulty associated with most cage-type host structures is that their preparation is laborious, involving multistep syntheses and tedious separations. Furthermore, precise functionalization of their inner cavities with highly organized binding groups, as required for optimal anion recognition, is not trivial.

While traditional receptors based on relatively simple molecular scaffolds assembled by conventional organic synthesis continue to receive considerable attention, a new paradigm in anion recognition chemistry has recently begun to emerge under the broader umbrella of dynamic combinatorial chemistry (DCC) [5–10]. In this approach, the host is dynamically self-assembled in the presence of the anionic guest, via reversible bond formation between simple building units. This way, a highly complex receptor may spontaneously form in just one step. There are a couple of key elements that need to be present for such an efficient self-assembly to occur. First, the building constituents must be able to reversibly assemble and break apart in a short timescale, to allow for quick formation of the most favorable structures. The reversibility is critical for correcting the mistakes and driving the system toward a thermodynamic minimum. Along this line, coordination bonds were initially employed due to their labile nature, though more recent studies demonstrated that reversible covalent bonds can serve a similar purpose under appropriate conditions. In the absence of an anionic guest, multiple structures with comparable free energies of formation are typically accessible, thereby creating a dynamic combinatorial library (DCL) of receptors. Unlike a conventional combinatorial library, consisting of static predetermined structures, a DCL may continuously evolve via dissociation and recombination of constituents in response to external conditions. It should be mentioned here that typically not all theoretically possible combinations are observed, with most structures often remaining virtual (potential but not physically observed). This led to the term virtual combinatorial library (VCL) to be proposed for such dynamic systems [5]. When the second key element is added to the mixture, consisting of an appropriate anionic guest, the equilibria are shifted to amplify the formation of the fittest receptor offering the most favorable anion binding. The receptor is thus expressed by the anion, which plays the role of a template [11, 12]. In ideal circumstances, this leads not only to spontaneous self-assembly of a complex receptor that would have otherwise been very difficult to build by traditional synthesis, but also to optimal anion binding. Thus, in many ways, this process is reminiscent of natural evolution in protein receptors. As such, it was appropriately dubbed *Supramolecular Darwinism* [5]. Finally, the resulting receptor may be locked-in by conversion into an irreversible product or by incorporation into polymeric materials such as membranes or gels [13]. Figure 1 illustrates the conceptual difference between the traditional approach to anion receptors by multistep synthesis, and by DCC.

In the following sections, recent examples of anion recognition within the context of DCC will be discussed.

2 Anion Recognition Via Dynamic Coordination Chemistry

Coordination chemistry is particularly amenable to DCC. The labile nature of most metal–ligand coordination bonds assures facile and reversible connections among the constituents, which enables quick thermodynamic equilibration of the system. Furthermore, metals may assume a wide variety of coordination stoichiometries and geometries, thereby yielding highly diverse DCLs. It is therefore not surprising that some of the first DCC systems were based on coordination complexes,

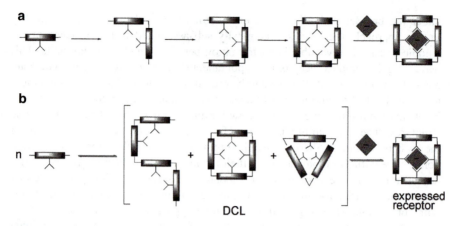

Fig. 1 Schematic comparison between anion receptors built traditionally by irreversible step-by-step synthesis (**a**) and by self-assembly via DCC (**b**)

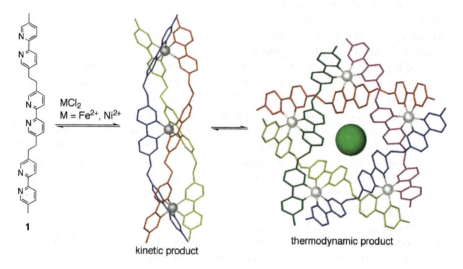

Fig. 2 Time-dependent self-assembly of linear and circular helicates

specifically double and triple helicates, as demonstrated by Lehn et al. [14]. These structures are polynuclear metal complexes of helical shape, in which two or three ligand strands self-assemble around a linear or circular array of metal ions. When mixtures of ligands and transition metal ions are employed, a VCL is dynamically generated, in which only the correctly paired metal–ligand coordination structures are expressed. As such, these systems display the main characteristics of DCC: reversible connectivity among components and selection through molecular recognition. Figure 2 illustrates the self-assembly of both linear and circular helicates from the tris(bipyridine) ligand **1** and NiCl$_2$ or FeCl$_2$ salts [15].

The linear helicate forms first as a kinetic product, upon heating for a few minutes at 170°C in ethylene glycol. Longer reaction times resulted in disassembly of the initial structure and the formation of the thermodynamic product consisting of a circular pentanuclear helicate. This structure is apparently templated by the chloride anion, which is bound strongly inside the cavity of the helical macrocycle.

This system could be considered as a VCL self-assembled by coordination bonds, in which the ligand **1** and the hexacoordinated metal cation may generate a wide variety of different-sized circular helicates through reversible disassembly/reassembly. Exactly which structure is expressed depends on the nature of the anion. The pentanuclear helicate is formed quantitatively in the presence of Cl^-. On the other hand, the corresponding hexamer is templated by the larger SO_4^{2-}, SiF_6^{2-}, or BF_4^- anions (Fig. 3). Finally, the intermediate sized Br^- generates a mixture of the pentamer and hexamer. Thus, this system represents the first clear example of anion-controlled DCC, where the anionic guest selects and amplifies a particular host from a dynamic VCL, based on size recognition [14].

A more recent example of dynamic coordination chemistry controlled by anion recognition was demonstrated with the series of ligands **2**, which formed linear binuclear triple helicates with Co^{2+} anions (Fig. 4) [16, 17]. The asymmetric ligand **2a**, functionalized at one end with an amide hydrogen-bond donor group, may form two types of helicate structures: head-to-head-to-head (HHH) and head-to-head-to-tail (HHT) isomers. With the weak hydrogen-bond acceptor ClO_4^- as counter-anion, a 3:1 mixture of HHH:HHT isomers was observed. However, with the

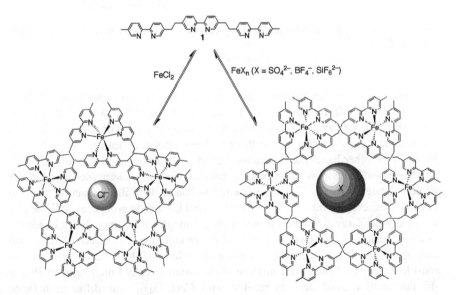

Fig. 3 Expression of two different circular helicates via anion recognition in a DCL self-assembled by metal coordination. The smaller Cl^- anion amplifies the formation of a pentanuclear helicate, whereas the larger SO_4^{2-}, SiF_6^{2-}, and BF_4^- amplify the formation of a hexanuclear helicate

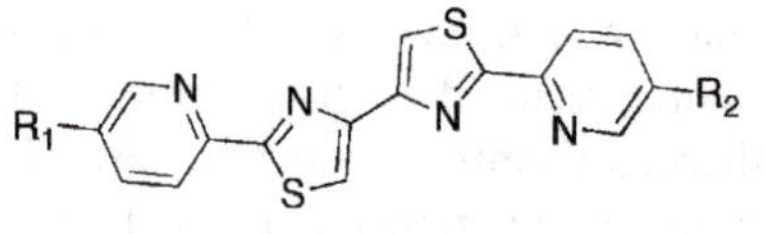

$R_1 = (S)\text{-CONHCH(CH(CH}_3)_2)CO_2Me;\ R_2 = H\ (\textbf{a})$
$R_1 = R_2 = (S)\text{-CONHCH(CH(CH}_3)_2)CO_2Me\ (\textbf{b})$
$R_1 = R_2 = H\ (\textbf{c})$

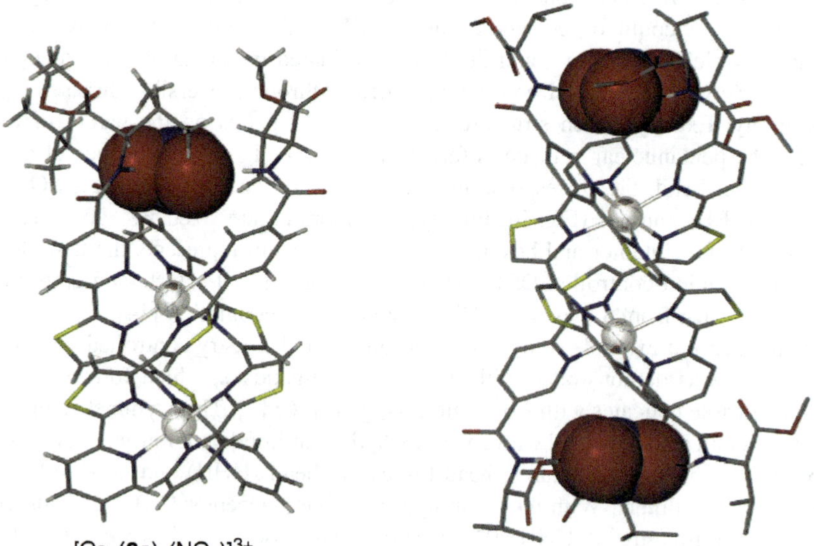

$[Co_2(\textbf{2a})_3(NO_3)]^{3+}$

$[Co_2(\textbf{2b})_3(NO_3)_2]^{2+}$

Fig. 4 Crystal structures showing nitrate recognition by the helicate receptors $[Co_2(\textbf{2a})_3]^{4+}$ (*left*) and $[Co_2(\textbf{2b})_3]^{4+}$ (*right*)

addition of NO_3^-, which is a more basic anion and thereby a stronger hydrogen-bond acceptor, the equilibrium shifted toward the formation of 95% of the HHH isomer. Thus, this helicate receptor was expressed by nitrate as a result of its better fit inside the binding cavity created by the three hydrogen-bonding amide groups (Fig. 4, left). On the other hand, the symmetric ligand **2b** functionalized at both ends with amide groups forms two anion-binding cavities when self-assembled into similar triple helicates with Co^{2+}. A nitrate anion is bound deep inside each of the two cavities by accepting three hydrogen bonds from the amide groups, as indicated by single-crystal X-ray diffraction (Fig. 4, right). Though ClO_4^- may also fit inside the binding pocket, it is bound more weakly through the formation of only two hydrogen bonds with longer contact distances, while the third amide group points outside the cavity. When a 1:1 mixture of the symmetrically functionalized **2b** and the nonfunctionalized **2c** was reacted with $Co(ClO_4)_2$, four different helicate structures were observed by NMR: $[Co_2(\textbf{2b})_3]^{4+}$, $[Co_2(\textbf{2b})_2(\textbf{2c})]^{4+}$, $[Co_2(\textbf{2b})(\textbf{2c})_2]^{4+}$, and $[Co_2(\textbf{2c})_3]^{4+}$, in a statistical 1:3:3:1 ratio. However, when NO_3^- was added to the mixture, the homoleptic helicates $[Co_2(\textbf{2b})_3]^{4+}$ and $[Co_2(\textbf{2c})_3]^{4+}$ became predominant (>95%), at the expense of the heteroleptic $[Co_2(\textbf{2b})_2(\textbf{2c})]^{4+}$

and [Co$_2$(**2b**)(**2c**)$_2$]$^{4+}$ complexes. Thus, the nitrate anions induced the self-sorting of the helicates by selectively expressing the fully functionalized [Co$_2$(**2b**)$_3$]$^{4+}$ receptor via recognition by amide hydrogen bonding [17].

Similar self-sorting through anion recognition was observed for mixtures of coordination complexes with formula [Co(bipy)$_x$(**3**)$_{3-x}$]$^{2+}$. The homoleptic complex [Co(**3**H$_2$)$_3$]$^{8+}$ was found to bind chloride relatively strongly, presumably through hydrogen bonding by the protonated amine groups of the ligand. A DCC was subsequently created by mixing **3**, 2,2′-bipyridine (bipy), and Co(NO$_3$)$_2$ in CD$_3$CN. Under these conditions, a library consisting of eight complexes, representing all possible combinations of ligands and diastereomers, could be observed by NMR and ESI-MS. Upon diprotonation of **3** and addition of Cl$^-$, the homoleptic complex [Co(**3**H$_2$)$_3$]$^{8+}$, which is complementary to chloride, and [Co(bipy)$_3$]$^{2+}$ were amplified through self-sorting [18].

Anion recognition was also demonstrated to control the self-assembly of metallamacrocycles. For example, ligand **4** is capable of forming various cyclic coordination complexes with Ni^{2+} in acetonitrile, such as triangle, square, or pentagon, as well as polymeric products. Exactly which of these structures are formed depends on the identity of the counteranions, which were found to play a templating role [19]. In the presence of BF$_4^-$ or ClO$_4^-$, exclusive formation of coordination squares was observed by NMR, ESI-MS, and single-crystal X-ray diffraction. The larger SbF$_6^-$ anion, on the other hand, templated the self-assembly of

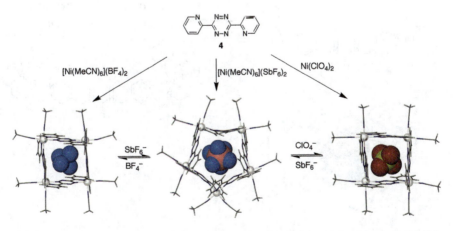

Fig. 5 Anion-templated self-assembly of two coordination squares and a pentagon, and their dynamic interconversion

a coordination pentagon (Fig. 5). In the presence of the smaller NO_3^- anion, a coordination triangle was formed, as inferred by ESI-MS.

The anion encapsulation in these coordination polygons was clearly size selective. Thus, the square discriminates BF_4^- against the larger PF_6^-, or ClO_4^- against IO_4^-. However, no shape discrimination was apparent, as the linear Br_3^- or the spherical I^- were also found to template the coordination square. Most interestingly, the coordination polygons displayed dynamic interconversion via anion exchange (Fig. 5). Thus, the pentagon was easily converted into the square when treated with excess ClO_4^- or BF_4^-. Partial conversion of the squares into the pentagon was also possible, although only with a very large excess of SbF_6^- and after refluxing for 2 days. This system may therefore be considered as a DCL, although one that is biased towards the square metallamacrocycle, which appears to be intrinsically more stable than the pentagon.

More recently, ligand **5**, functionalized with urea hydrogen-bonding groups, was employed in the construction of a small DCL [20]. This ligand was found to self-assemble with Pd cations into [2 + 2] and [3 + 3] metallacycles, as evidenced by NMR and ESI-MS. The structure of the dimer was also determined by single-

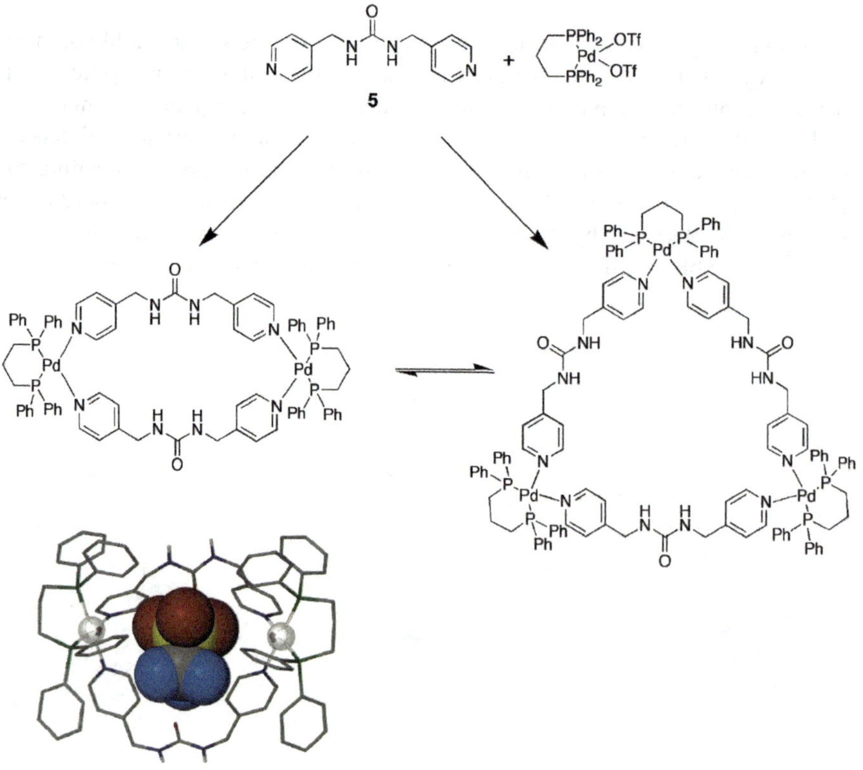

Fig. 6 Dynamic equilibrium between two metallamacrocycle anion receptors. The X-ray structure of the [2 + 2] macrocycle encapsulating $CF_3SO_3^-$ is shown in the *lower left corner*

crystal X-ray diffraction, which revealed the encapsulation of a triflate anion (Fig. 6). The two macrocycles interconvert in solution, and the equilibrium may be shifted by the change in solvent, concentration, and temperature. More importantly, the addition of H$_2$PO$_4^-$ was found to amplify the [2 + 2] isomer, which was attributed to the selective binding of this anion by the smaller macrocycle.

3 Anion Recognition Via Dynamic Covalent Chemistry

Under appropriate conditions, reversible covalent bonds may also be employed in DCC [7, 8]. Unlike conventional organic synthesis that is dominated by kinetically controlled chemistry, covalent bond formation under DCC is carried out under thermodynamic equilibration, which insures error-checking and thus the formation of the most stable product akin to traditional self-assembly with noncovalent interactions. The main advantage of DCC with covalent bonds is that the ensuing structures are significantly more robust than those assembled by coordination bonds. However, due to the stronger nature of the covalent bonds, equilibration is significantly slower, which requires the use of a catalyst to help form the product in a reasonable amount of time. Once the final structure is obtained, it can be locked-in by the removal of the catalyst or conversion into an irreversible product. A number of reversible covalent bond formations and exchanges have been employed in DCC, including imine and hydrazone formation, transesterification, transamidation, acetal exchange, boronate formation, disulfide exchange, and alkene and alkyne metathesis [7, 8]. Figure 7 illustrates some of these reactions.

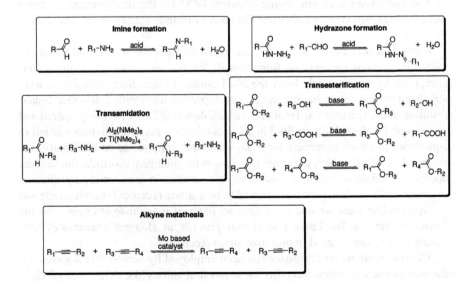

Fig. 7 Selected examples of reversible covalent bond forming and exchange reactions amenable to DCC

Imine formation and exchange are well-established reactions for DCC [21] and proceed most efficiently in mildly acidic solutions (pH around 4–5). Lewis acid catalysts like $Sc(OTf)_3$ have also been found to catalyze transiminations. Once the equilibrium is reached the imines may be irreversibly reduced to amines for stabilization. Hydrazone formation and exchange is another well-studied reaction for DCC [22]. It requires more acidic conditions (pH < 4) and higher temperatures but the resulting hydrazones are thermodynamically more stable than imines, and the reactions can be more easily driven to completion. Most applications of hydrazones in DCC involve acylhydrazines, which provide better reversibility. An attractive feature of acylhydrazones is that they contain hydrogen-bonding sites that can act as anion-binding groups in the resulting structures. Reversible disulfide exchange is one of the most frequently utilized reactions in DCC, and has been extensively employed for the development of receptors [23]. It typically involves thiols as starting materials, which are oxidized in neutral or slightly basic aqueous solutions to disulfides, followed by thiol-disulfide exchange under mild conditions. Transesterification is one of the first reactions applied to DCC, and it works best under anhydrous conditions using KOMe as catalyst [24]. Transamidation can be made reversible, and thereby amenable to DCC with metal-based catalysts such as $AlCl_3$, $Sc(OTf)_3$, $Al_2(NMe_2)_6$, or $Ti(NMe_2)_4$ [25]. However, more exploratory research needs to be carried out to optimize the catalysts and the conditions as well as to probe the full scope of the reaction before transamidation can be employed dependably in DCC. Such efforts may well be worthwhile considering that amides are exceptionally stable and provide hydrogen-bonding groups that may be exploited for anion binding. Alkynes are also significantly stable, and have recently been employed in self-assembly of macrocycles catalyzed by molybdenum-based catalysts [26].

The first example of employing covalent DCC for the development of anion receptors was provided by Kubik and Otto using disulfide exchange, as illustrated in Fig. 8 [27].

Kubik had previously shown that the cyclic hexapeptide **6** can bind sulfate or iodide in aqueous solvents by formation of 2:1 complexes in which the anions accept six hydrogen bonds from the N–H amide groups. Such complexes were stabilized by covalently linking the two cyclopeptide units with a designed linker, resulting in the receptor **7a**. To obtain a disulfide-based DCL, **7b** was prepared and mixed with the dithiol linkers **8a–f** in 2:1 acetonitrile/water. The mixture was left to equilibrate thermodynamically for 7 days, and the composition of the library was analyzed by HPLC, which indicated that **7b** was the dominant disulfide. Subsequent addition of iodide or sulfate as potassium salts resulted in substantial amplification of receptors **7c–e**. **7c** in particular proved to be a strong receptor for both iodide and sulfate, binding these anions more than an order of magnitude stronger than the reference receptor **7a**. These results thus provide an eloquent example of how covalent DCC can be used to optimize anion receptors.

The reversible imine chemistry has been employed by Sessler and Katayev for the development of anion receptors for tetrahedral anions like SO_4^{2-} and HPO_4^{2-} [28, 29]. Thus, the dipyrrole **9** functionalized with aldehyde groups, and the

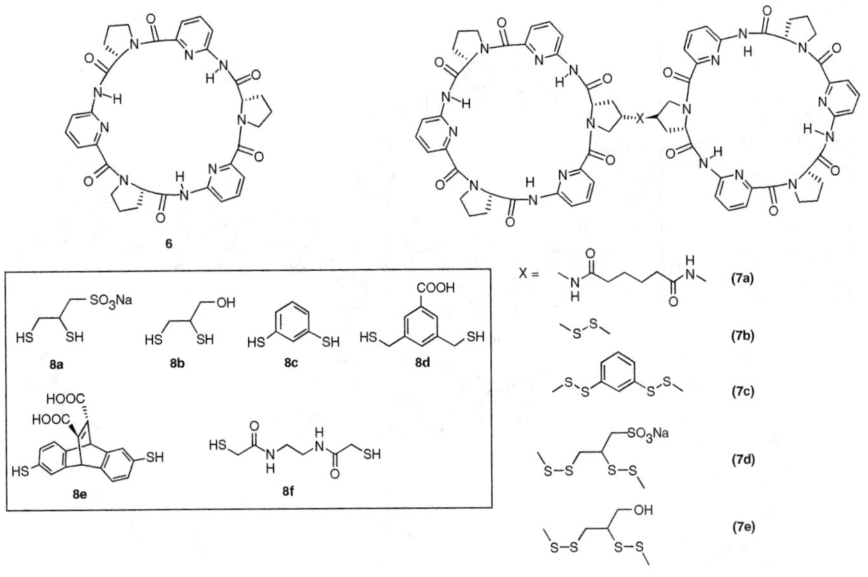

Dynamic Chemistry of Anion Recognition

Fig. 8 Anion receptor optimization by DCC based on disulfide exchange

diamines **10** and **11**, led to the formation of a DCL consisting of receptors **12–16** (Fig. 9). These receptors represent the macrocycles resulting from the [1 + 1] (**12, 13**), [2 + 2] (**14, 15**), and [3 + 3] (**16**) imine condensation reactions.

In the presence of various acids, such as HCl, HBr, CF_3COOH, HNO_3, $HClO_4$, or $HReO_4$, a mixture of all macrocycles was observed by mass spectrometry. Phosphoric and sulfuric acids, on the other hand, selectively expressed the larger receptors **14–16**, due to the recognition of the corresponding tetrahedral divalent anions HPO_4^{2-} and SO_4^{2-} through hydrogen bonding with the dipyrrole and amide groups. Specifically, starting from diamine **10**, and using phosphoric acid as a catalyst and template, led to the isolation of the **16**·H_3PO_4 salt. Sulfuric acid, on the other hand, led to the initial isolation of the **14**·$(H_2SO_4)_2$ salt. Upon neutralization with triethylamine, and addition of tetrabutylammonium hydrogen sulfate, the initial [2 + 2] product underwent ring expansion to the [3 + 3] macrocycle over a period of 5 days, and was isolated as the **16**·H_2SO_4 salt. The largest macrocyclic receptor **16** thus appears to be the thermodynamically favored product, which was amplified as a result of the good fit of sulfate or hydrogenphosphate anions inside its binding cavity. Single crystal X-ray diffraction revealed that the receptor is diprotonated (**16**H_2^{2+}), has a highly folded conformation, and binds tightly SO_4^{2-} and HPO_4^{2-}, via 11 and 12 hydrogen bonds, respectively (Fig. 10) [29]. Notably, the hydrogen bonding of HPO_4^{2-} in **16**H_2^{2+}·HPO_4^{2-} is reminiscent of binding of the same anion in the phosphate-binding protein. In direct contrast, no [3 + 3] condensation product could be obtained from the diamine **11**, with either sulfuric acid or phosphoric acid, presumably due to its reduced hydrogen-bonding ability.

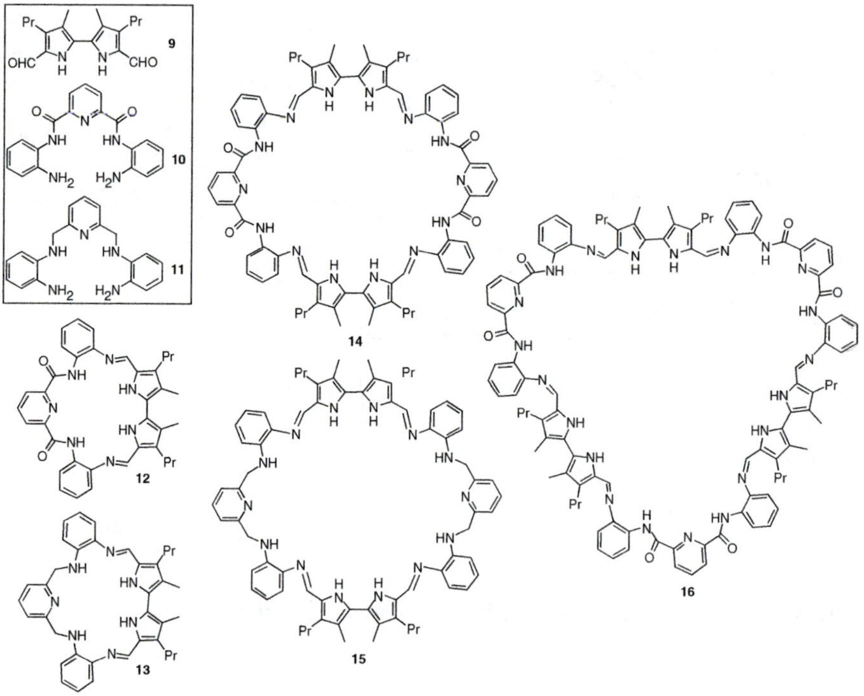

Fig. 9 DCL of macrocyclic anion receptors (**12–16**) obtained by reversible imine chemistry between the dialdehyde **9** and diamines **10, 11**

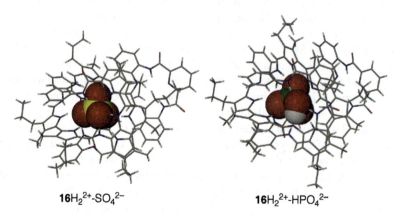

Fig. 10 Binding of SO_4^{2-} (*left*) and HPO_4^{2-} (*right*) by **16**H_2^{2+}, as revealed by single-crystal X-ray diffraction

4 Anion Recognition Via Dynamic Crystallization

In the previous sections, it has been illustrated how DCLs of anion receptors may be self-assembled in solution via either metal coordination or reversible covalent bonds. The resulting libraries were able to evolve continuously by dissociation and recombination of the constituents in response to the added anions, which drove the equilibria towards the assembly of the fittest receptors. An alternative way to express particular members of the library is through a stabilizing phase change, such as polymerization, sol–gel formation, or crystallization, where the amplification is driven not so much by the molecular recognition between the host and guest in solution, but by the relative stability of the competing self-assembled structures in the newly formed phase [30]. For the particular case of crystallization, which will be analyzed in detail here, the selection is ultimately determined by the relative solubility of the existing self-assembled structures in solution. As the least soluble combination precipitates and is removed from solution, it drives the disassembly of all competing aggregates towards the crystallization of the selected solid [31]. The crystallization itself may display dynamic features, involving dissolution of the initially formed crystals and recrystallization into different forms in response to environmental changes such as temperature, solvent, or solution composition [32]. It may therefore be considered that the collection of all possible crystal structures from a set of molecular or ionic components in a given solution constitutes a DCL. In most cases, such a library is a virtual one (VCL), as typically no more than one crystalline form is observed under a given set of conditions. However, it is sometimes possible that multiple crystal forms may concomitantly appear from solution, dynamically interconverting through dissolution/recrystallization, thereby defining a bona fide crystalline DCL.

Anions may play various roles in such dynamic crystallization processes. For example, they may template the formation of certain aggregates in solution that are particularly fitted for self-assembly into a stable crystal form through favorable spatial orientation of complementary groups. Alternatively, the anion may act as a building unit, linking the other components in the crystal by hydrogen bonding or metal coordination. In such cases, certain anions may be selected based on size or shape recognition [33]. Next, some selected examples of dynamic anion recognition through crystallization will be discussed.

The bis(pyridyl) urea **17** can form various coordination polymers from aqueous solutions with different Zn salts (Fig. 11) [34, 35]. One source of the observed structural variety is the conformational flexibility of the ligand. Gas-phase electronic structure calculations found three low-energy conformers for **17**, differing in the relative orientation of the pyridyl N atom and the C=O group [34]. Among them, the *anti–anti* was found to be the most stable conformer, followed by *syn–anti* and *syn–syn*, which are 1.6 and 3.2 kcal/mol higher in energy, respectively. However, the calculated barrier of rotation for the pyridyl-urea bond is only 2.9 kcal/mol, suggesting that in addition to these three flat-shaped isomers, there is virtually an infinite number of possible conformers with intermediate pyridyl-urea

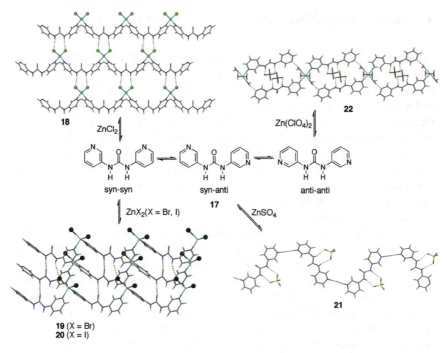

Fig. 11 DCL of coordination polymers self-assembled from the conformationally flexible ligand **17** and different Zn salts in ethanol/water solvent

dihedral angles. Upon coordination of the pyridyl groups to Zn cations, even more structural diversity may be generated, considering the variety of coordination modes the metal may assume, including the participation of solvent molecules to coordination. Finally, different anions may interact in multiple ways with the resulting structures, including hydrogen bonding to the urea groups and solvent molecules, and coordination to the metal centers. All these possible structures thus define a diverse DCL of coordination polymers, which may be either virtual or real. Crystallization may then express the most stable (least soluble) structure under the given conditions.

As Fig. 11 shows, the structure of the coordination polymer expressed by crystallization from ethanol/water mixtures depends on the nature of the anion present. Crystallization of **17** with ZnCl$_2$ resulted in a one-dimensional coordination polymer **18** with the ligand "frozen" in a twisted geometry close to the *syn–syn* conformation. Each Zn center is coordinated by two pyridyl groups from adjacent ligand molecules, and two Cl$^-$ anions. The resulting coordination chains are linked into layers by chelating urea···Cl$_2$Zn hydrogen bonds. ZnBr$_2$ and ZnI$_2$ form similar one-dimensional coordination polymers (**19**, **20**), but, compared to the chloride analog, only one of the two halides links the chains into layers by hydrogen bonding to urea groups, with additional urea···urea hydrogen bonds also crosslinking the chains. With ZnSO$_4$, on the other hand, the *syn–anti* conformer is expressed

Dynamic Chemistry of Anion Recognition

by crystallization, resulting in one-dimensional zigzag coordination chains (21) that chelate sulfate anions by urea hydrogen bonding [34]. The sulfate anions form additional hydrogen bonds to five water molecules coordinated to Zn cations (not shown), thereby crosslinking the coordination chains in an intricate three-dimensional structure. Under similar conditions, $Zn(ClO_4)_2$ forms with the *anti–anti* conformer of 17 looped coordination chains (22) connected through pyridyl–Zn coordination bonds and pyridyl···water hydrogen bonds. There is also a noncoordinating ligand 17 in the structure (not shown), found in the *syn–anti* conformation, which is associated with the coordination chains by urea···urea hydrogen bonding. The ClO_4^- anions in the structure are bound to the urea groups by chelating hydrogen bonding. Finally, under the same conditions as for crystallization of 18–22, $Zn(NO_3)_2$ did not form any coordination polymer [34].

When 17 was competitively self-assembled with Zn^{2+} in a 1:1 ratio from an ethanol/water mixture containing equivalent amounts of Cl^-, Br^-, I^-, NO_3^-, ClO_4^-, and SO_4^{2-}, some interesting dynamic crystallization phenomena occurred. First of all, no oxoanions were found in the precipitated solids, presumably due to the weaker interactions of these anions with the coordination polymers, compared to the halides. Powder X-ray diffraction and FTIR analyses confirmed that the initially precipitated solid, isolated after 30 min, was a mixture of two crystalline phases, containing mostly 18 and small amounts of the isomorphous 19 and 20. Elemental analysis indicated that all three halides were present in this mixture, in a $Cl^-/Br^-/I^-$ ratio of 9.1/4.1:1. If the initial crystals were left in the supernatant liquid for 5 days, complete dissolution of 18 and recrystallization into 19 and 20 was observed. Elemental analysis indicated that these crystals contain again all three halides, in a $Cl^-/Br^-/I^-$ ratio of 3.3:2.4:1. The relative halide content continued to change in time when the crystals were left in contact with the supernatant solution, finally reaching equilibrium after 2 weeks, when the $Cl^-/Br^-/I^-$ ratio was found to be 1.5:1.6:1. These results indicate that 18 was the kinetic product of crystallization that was eventually converted into the thermodynamic products 19 and 20, through dynamic dissolution/recrystallization. X-ray diffraction analysis of single crystals obtained from pairwise halide mixtures revealed the formation of solid solutions, with the two anions sharing the same sites in the crystals. The relative amounts of halides found in the two structures differed substantially. Thus, crystals of 18 obtained from an equimolar mixture of Cl^- and Br^- preferentially included chloride, in a Cl^-/Br^- ratio of 2.4. On the other hand, crystals of 19 isolated from the same experiment contained a Cl^-/Br^- ratio of 1.1. This disparity may be attributed to the difference in the anion-binding sites in the two crystals. In 18 there are two equivalent halide sites involving strong urea hydrogen bonding interactions that favor the more basic chloride. In 19, on the other hand, there are two different halide sites, of which only one is hydrogen-bonded to urea. Accordingly, the hydrogen-bonded site favored chloride (66% Cl^-, 34% Br^-), whereas the nonhydrogen-bonded site favored the less basic bromide (40% Cl^-, 60% Br^-). An even more pronounced segregation of halides was observed in crystals of 20 obtained from equimolar mixtures of Cl^- and I^-. In these crystals, the hydrogen-bonded site strongly preferred chloride (85% Cl^-,

15% I$^-$), whereas the opposite was observed for the nonhydrogen-bonded site (26% Cl$^-$, 74% I$^-$), in agreement with the stronger hydrogen-bond accepting ability of Cl$^-$ compared to I$^-$ [35].

Besides extending the concept of DCC to anion-binding coordination polymers, the previous example hinted at the possibility of exploiting the crystallization process for the selective separation of anions. This approach is based on the competitive crystallization of organic or metal-organic cationic structures from anionic mixtures, where anion selectivity may be achieved through specific recognition inside the growing crystal [35–42]. Similar to dynamic receptors in solution, the anion binding sites in crystal nuclei are dynamically self-assembled around the anion, which typically serves as a template. As the nuclei develop into fully-grown crystals that fall out of solution, the separation process is completed by simple filtration of the precipitate. Like with traditional anion receptors, achieving high selectivity by crystallization requires a combination of good complementarity between the anionic guest and the anion-binding site in the crystal, and organizational rigidity of the crystalline host to prevent structural reorganization and accommodation of competing anions [43]. The latter requirement, however, is at odds with the concept of DCC, which embodies structural diversity and dynamic adaptation. Accordingly, the structures generated in a DCC are organizationally flexible by default. For example, **17** formed completely different crystal structures with various anions, with the ligand conformation and the metal coordination geometry undergoing significant changes to accommodate different anions. How then can high selectivity be attained in such dynamic systems? As it turns out, this could be achieved with biased DCLs, in which, irrespective of the nature of the anion, a particular receptor structure is persistently expressed from a multitude of virtual structures. The following example of anion separation by selective crystallization demonstrates this concept.

The tripodal tris(pyridylurea) ligand **23** crystallizes with MSO$_4$ salts (M = Mg^{2+}, Zn^{2+}, Cd^{2+}, Co^{2+}, Mn^{2+}) from water/methanol solutions into three-dimensional hydrogen bonded frameworks with NaCl topology, consisting of alternating M(H$_2$O)$_6$$^{2+}$ and SO$_4$(**23**)$_2$$^{2-}$ ions (Fig. 12) [44, 45]. The sulfate anion is encapsulated by two **23** ligands that provide 12 complementary hydrogen bonds from six urea groups. Notably, despite its very flexible nature, **23** self-assembles into essentially identical hydrogen-bonded frameworks with other divalent oxoanions such as CO$_3$$^{2-}$, SO$_3$$^{2-}$, and SeO$_4$$^{2-}$, despite their different shapes, sizes, and hydrogen-bonding preferences [44]. Monovalent anions like F$^-$, Cl$^-$, Br$^-$, I$^-$, NO$_3$$^-$, and ClO$_4$$^-$ are completely excluded as a result of their charge mismatch and their inability to form alternative structures with **23**. Single-crystal X-ray diffraction analysis indicated that, like sulfate, CO$_3$$^{2-}$, SO$_3$$^{2-}$, and SeO$_4$$^{2-}$ self-assembled into virtually identical capsules with two molecules of **23**. These capsules, however, have relatively poor complementarity for carbonate and sulfite, which engage in repulsive NH$\cdots$C and NH$\cdots$S interactions. The measured thermodynamics of crystallization were consistent with the structural observations, with the tetrahedral SO$_4$$^{2-}$ and SeO$_4$$^{2-}$ that fit well inside the capsules also displaying the most negative crystallization enthalpies [46]. The high organizational rigidity

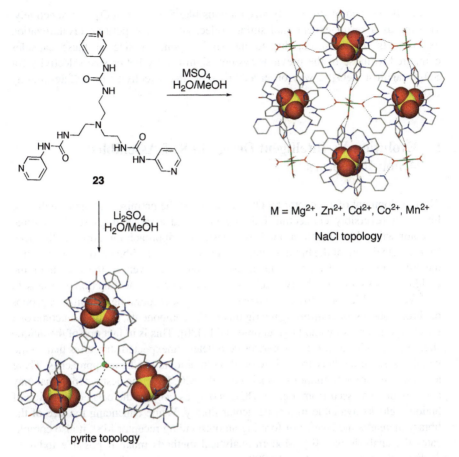

Fig. 12 Biased DCLs consisting of persistent capsule receptors self-assembled from **23** and MSO$_4$ (M = Mg^{2+}, Zn^{2+}, Cd^{2+}, Co^{2+}, Mn^{2+}) or Li$_2$SO$_4$ into crystalline frameworks with NaCl and pyrite topologies, respectively

of the capsules, manifested by the fact that they did not distort their structures to optimize the binding of the poorly fitting CO$_3^{2-}$ and SO$_3^{2-}$, appeared to be critical for the anion separation selectivity observed in competitive crystallizations from aqueous mixtures, which followed the order: SO$_4^{2-}$ > SeO$_4^{2-}$ ≫ CO$_3^{2-}$ > SO$_3^{2-}$. Thus, this system represents a case of biased DCL, where the same hydrogen bonding capsules persistently crystallize, irrespective of the nature of the divalent oxoanion present, thereby insuring exceptional anion separation selectivity.

More recent studies indicated that the capsules also persist in crystalline frameworks self-assembled from completely different cations [47]. Thus, when **23** was crystallized with Li$_2$SO$_4$, similar SO$_4$(**23**)$_2^{2-}$ capsules formed, linked this time by Li(H$_2$O)$^+$ cations into a cubic coordination framework with pyrite topology (Fig. 12). Once again, the capsules preserved their structures in the presence of differently shaped or differently sized anions like SO_3^{2-} or SeO_4^{2-}, respectively, yielding as a result exceptional sulfate selectivity in competitive crystallization experiments. Furthermore, due to the slightly smaller size of these capsules compared to those in the previous system, significantly better size selectivity for SO_4^{2-} against SeO_4^{2-} was observed, rivaling the selectivity of sulfate-binding protein.

5 Evolution Vs Intelligent Design in Self-Assembled Anion Receptors

The development of a receptor by DCC is based on the premise that, given a diverse library of reversibly connecting building units and a targeted guest, the optimal receptor will self-assemble around the guest. This approach circumvents the need for accurate design of the binding site, relying instead on the library to evolve towards the fittest receptor. Though some considerations are given to the design of the building blocks in the library, such as selecting appropriate functional groups to achieve reversible connectivities and guest binding, incorporating solubilizing groups or chromophores, or controlling the rigidity of the components, the final outcome of a DCC experiment is by and large unpredictable [30]. This is in fact one of the unique strengths of DCC, as it often generates completely unexpected structures that would be otherwise be unlikely to be developed by traditional synthetic approaches. There are, however, some limitations associated with DCC. First, the best receptor that may possibly be expressed from a given DCL is only as good as the best combination of building blocks available in that particular library. While enhancing the size of the library increases the chance of forming an outstanding receptor [48], it also complicates the analysis part [9]. Modern analytical methods make it possible today to analyze mixtures of more than 10,000 compounds. However, the interactions among components in such large systems may be rather complex, as the involved species are interconnected through multiple equilibria that may shift unpredictably. As a result, the fittest receptor is not always expressed [49]. This may be particularly problematic when the library components contain functional groups for guest binding (e.g., hydrogen-bonding groups), which may self-associate or even interfere with the reversible exchange reactions among building blocks.

At the opposite end of the spectrum from DCC is molecular modeling, which often involves theoretical calculations and computer design methods to identify the best host for a given guest. In direct contrast to DCC, the design process precisely identifies the structure of the receptor and its binding site prior to synthesis. While the design of receptors has for a long time been rather crude, mainly relying on chemical intuition or a simple ChemDraw sketch, computational chemistry has now advanced to the point where it is possible to calculate accurate three-dimensional structures and interaction energies of elaborate molecules and supramolecular complexes. One drawback associated with many molecular modeling approaches is that the number of receptors under consideration is typically small, with the selection often being biased by the researcher's prior experience or the receptor's synthetic accessibility (the commercially available molecules are always selected first!). Consequently, the result is often a less than optimal receptor. Another factor constraining the variety of structures in receptor design is the human imagination, which although vast, is no match for the stunning diversity of naturally evolved receptors that DCC aspires to emulate. A compromise solution that combines the rigor of computational modeling with the diversity characteristic of combinatorial methods is de novo structure-based design methods, initially developed for drug molecules. In this approach, the binding groups are first defined based on the structural knowledge of the binding site. Next, appropriate scaffolds for the binding groups are identified on a computer from large libraries that often contain millions of molecular fragments. More recently, similar approaches have been developed for synthetic hosts, including anion receptors [50]. For example, an optimal linker for covalently connecting two cyclopeptide molecules of **6** into a receptor was identified using de novo structure-based design methods as implemented in the HostDesigner software [51].

Although molecular design can rapidly identify optimal receptor molecules, their assembly by traditional multistep synthesis remains tedious and inefficient. It is possible, however, to design simple molecular components that have the ability to self-assemble into complex receptors. This way, the best of two worlds may be combined: combinatorial synthesis and screening performed in silico, which avoids the experimental difficulties associated with handling very large libraries, and dynamic self-assembly under thermodynamic control. These concepts have recently been embodied in the design and self-assembly of a highly efficient sulfate receptor, as detailed below.

Figure 13 illustrates the steps involved in the de novo computer-aided design of a self-assembled coordination cage receptor for SO_4^{2-} binding in water [52]. The design process started with the consideration of sulfate's strong hydrophilic nature. Effective binding of SO_4^{2-} from aqueous environments, it was reasoned, requires complete encapsulation into a receptor functionalized with strong and complementary binding groups. Previous electronic structure calculations [53], supported by X-ray crystallographic studies [44–47, 54–56], indicated that such a complementary binding could be achieved by six urea groups arranged tetrahedrally around sulfate, through the formation of 12 hydrogen-bonding interactions. Starting with this predetermined binding site, the challenge was to identify a suitable host that provides the required tetrahedral arrangement of urea groups. Since it was realized that such a host would be difficult to assemble by conventional multistep synthesis, a more efficient self-assembly strategy was sought. Specifically, an M_4L_6 tetrahedral coordination cage with each of its six edges functionalized with a urea group was targeted. As depicted in Fig. 13, such a cage receptor could be constructed starting from four $Ni(bpy)_3^{2+}$ vertices oriented tetrahedrally around the $SO_4(urea)_6^{2-}$ binding site. The resulting geometry dictates that each of the six edges would have the form bpy-link-urea-link-bpy. The final design task consisted of identifying appropriate linking fragments that bridge the bpy and urea groups while maintaining the perfect tetrahedral arrangement. This task was executed using a de novo structure-based design approach implemented in the

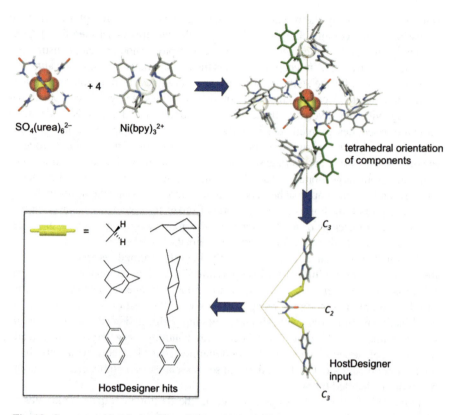

Fig. 13 Computer-aided design of a self-assembled tetrahedral receptor for sulfate encapsulation

HostDesigner software. Specifically, a library containing over 8,000 hydrocarbon fragments was screened, allowing for the Ni(bpy)$_3^{2+}$ vertices to translate and rotate about the C_3 axes of the tetrahedron, and for the urea groups to rotate about the C_2 axes. In the end, a small number of candidate links were retained, which were manually screened by constructing the whole cages and evaluating their ability to retain tetrahedral symmetry and optimal sulfate binding geometry upon optimization. Figure 13 shows the top six links identified by this process.

The simplest edge ligand **24** identified by HostDesigner was subsequently synthesized, and its ability to self-assemble into the targeted tetrahedral receptor was tested. Upon mixing **24** with NiSO$_4$ in a 6:4 ratio, the [Ni$_4$(**24**)$_6$(SO$_4$)](SO$_4$)$_3$ complex self-assembled quantitatively from water. Single-crystal X-ray diffraction studies confirmed the tetrahedral structure of the receptor and its optimal encapsulation of sulfate via 12 hydrogen bonds, as theoretically predicted (Fig. 14). The sulfate binding strength was estimated from competition experiments involving BaSO$_4$ precipitation, which established a lower limit of 6×10^6 M^{-1} for the apparent association constant for SO$_4^{2-}$ binding by the tetrahedral cage receptor [52]. This is one of the strongest sulfate bindings in water ever reported for a synthetic receptor, and rivals the

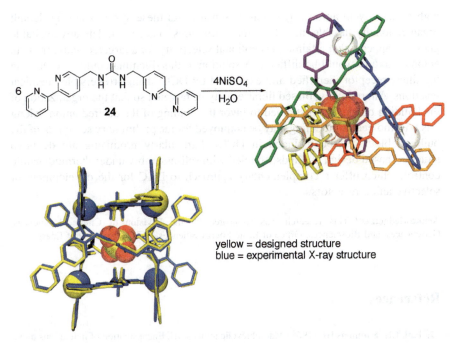

Fig. 14 Self-assembly of tetrahedral [Ni$_4$(**24**)$_6$(SO$_4$)]. The X-ray structure of the sulfate-encapsulating cage receptor is shown on the *right*, while the *picture* in the *lower left corner* depicts an overlay of the experimental (*blue*) and predicted (*yellow*) structures

sulfate-binding protein. These results attest to the power of combining computer-aided design methods with dynamic self-assembly for developing highly effective receptors.

6 Conclusions

DCC has emerged as an effective new tool for the development of anion receptors. Unlike the traditional multistep synthesis, which is laborious and inefficient, DCC allows the host to evolve from a dynamic self-assembled library whose members interconvert by reversible bonds, with the selection driven by the chemical recognition of the anionic guest. Now that the viability of this approach has been demonstrated with a number of proof-of-principle studies using small DCLs based on coordination bonds or reversible covalent bonds, the field is ready to move to the next level by involving much larger libraries that could provide truly outstanding anion receptors with exceptional binding efficiencies. The greatest difficulties remain in the analytical characterization and the control and understanding of the multiple interconnecting equilibria involved in such large libraries. Another challenge associated with the inherent adaptability of DCLs is to achieve

214   R. Custelcean

high selectivity in anion separation or sensing. As these systems are by default organizationally flexible, morphing their structures to accommodate any available guests, it appears that attaining exceptional selectivity for a targeted anion from an anionic mixture would be difficult. A solution to this dilemma could be to lock-in the fittest receptor identified and expressed by DCC, using irreversible chemical reactions. Alternatively, biased libraries may be targeted so that the organization of the building blocks is constrained to favor the binding of the desired anion. Along this line, molecular modeling may be employed for the preliminary screening of the building blocks to be included in DCLs. Particularly promising are de novo structure-based design methods, coupled with self-assembly under thermodynamic control, which offer a complementary approach to DCC for the development of selective anion receptors.

Acknowledgement This research was sponsored by the Division of Chemical Sciences, Geosciences, and Biosciences, Office of Basic Energy Sciences, U.S. Department of Energy.

References

1. Park CH, Simmons HE (1968) Macrobicyclic amines. III. Encapsulation of halide ions by in, in-1,(k + 2)-diazabicyclo[k.l.m]alkane-ammonium ions. J Am Chem Soc 90:2431–2432
2. Sessler JL, Gale PA, Cho W-S (2006) Anion receptor chemistry. RSC Publishing, Cambridge
3. Graf E, Lehn J-M (1976) Anion cryptates – highly stable and selective macrotricyclic anion inclusion complexes. J Am Chem Soc 98:6403–6405
4. Pflugrath JW, Quiocho FA (1985) Sulfate sequestered in the sulfate-binding protein of salmonella-typhimurium is bound solely by hydrogen-bonds. Nature 314:257–260
5. Lehn J-M (1999) Dynamic combinatorial chemistry and virtual combinatorial libraries. Chem Eur J 5:2455–2463
6. Lehn J-M (2007) From supramolecular chemistry towards constitutional dynamic chemistry and adaptive chemistry. Chem Soc Rev 36:151–160
7. Corbett PT, Leclaire J, Vial L et al (2006) Dynamic combinatorial chemistry. Chem Rev 106:3652–3711
8. Rowan SJ, Cantrill SJ, Cousins GRL et al (2002) Dynamic covalent chemistry. Angew Chem Int Ed 41:898–952
9. Ladame S (2008) Dynamic combinatorial chemistry: on the road to fulfilling the promise. Org Biomol Chem 6:219–226
10. Reek JNH, Otto S (2010) Dynamic combinatorial chemistry. Wiley-VCH, Weinheim
11. Vilar R (2008) Anion templates in synthesis and dynamic combinatorial libraries. Struct Bond 129:175–206
12. Vilar R (2008) Anion recognition and templation in coordination chemistry. Eur J Inorg Chem 357–367
13. Barboiu M (2010) Dynamic interactive systems: dynamic selection in hybrid organic–inorganic constitutional networks. Chem Commun 46:7466–7476
14. Hasenknopf B, Lehn J-M, Boumediene N et al (1997) Self-assembly of tetra- and hexanuclear circular helicates. J Am Chem Soc 119:10956–10962
15. Hasenknopf B, Lehn J-M, Boumediene N et al (1998) Kinetic and thermodynamic control in self-assembly: sequential formation of linear and circular helicates. Angew Chem Int Ed 37:3265–3268

Dynamic Chemistry of Anion Recognition

16. Harding LP, Jeffery JC, Riis-Johannessen T et al (2004) Anion control of the formation of geometric isomers in a triple helical array. Dalton Trans 2396–2397
17. Harding LP, Jeffery JC, Riis-Johannessen T et al (2004) Anion control of ligand self-recognition in a triple helical array. Chem Commun 654–655
18. Telfer SG, Yang X-J, Williams AF (2004) Complexes of 5,5′-aminoacido-substituted 2,2′-bipyridyl ligands: control of diastereoselectivity with a pH switch and a chloride-responsive combinatorial library. Dalton Trans 699–705
19. Campos-Fernandez C, Schottel BL, Chifotides HT et al (2005) Anion template effect on the self-assembly and interconversion of metallacyclophanes. J Am Chem Soc 127:12909–12923
20. Diaz P, Tovilla JA, Ballester P et al (2007) Synthesis, structural characterization and anion binding studies of palladium macrocycles with hydrogen-bonding ligands. Dalton Trans 3516–3525
21. Meyer CD, Joiner CS, Stoddart JF (2007) Template-directed synthesis employing reversible imine bond formation. Chem Soc Rev 36:1705–1723
22. Nguyen R, Huc I (2003) Optimizing the reversibility of hydrazone formation for dynamic combinatorial chemistry. Chem Commun 8:942–943
23. Otto S, Furlan RLE, Sanders JKM (2000) Dynamic combinatorial libraries of macrocyclic disulfides in water. J Am Chem Soc 122:12063–12064
24. Rowan SJ, Sanders JKM (1997) Building thermodynamic combinatorial libraries of quinine macrocycles. Chem Commun 1407–1408
25. Eldred SE, Stone DA, Gellman SH et al (2003) Catalytic transamidation under moderate conditions. J Am Chem Soc 125:3422–3423
26. Zhang W, Moore JS (2005) Reaction pathways leading to arylene ethynylene macrocycles via alkyne metathesis. J Am Chem Soc 127:11863–11870
27. Otto S, Kubik S (2003) Dynamic combinatorial optimization of a neutral receptor that binds inorganic anions in aqueous solution. J Am Chem Soc 125:7804–7805
28. Katayev EA, Pantos GD, Reshetova MD et al (2005) Anion-induced synthesis and combinatorial selection of polypyrrolic macrocycles. Angew Chem Int Ed 44:7386–7390
29. Katayev EA, Sessler JL, Khrustalev VN et al (2007) Synthetic model of the phosphate binding protein: solid-state structure and solution-phase anion binding properties of a large oligopyrrolic macrocycle. J Org Chem 72:7244–7252
30. Beeren SR, Sanders JKM (2010) History and principles of dynamic combinatorial chemistry. In: Reek JNH, Otto S (eds) Dynamic combinatorial chemistry. Wiley-VCH, Weinheim
31. Hutin M, Cramer CJ, Gagliardi L et al (2007) Self-sorting chiral subcomponent rearrangement during crystallization. J Am Chem Soc 129:8774–8780
32. Cui XJ, Khlobystov AN, Chen XY et al (2009) Dynamic equilibria in solvent-mediated anion, cation and ligand exchange in transition-metal coordination polymers: solid-state or recrystallization? Chem Eur J 15:8861–8873
33. Custelcean R (2010) Anions in crystal engineering. Chem Soc Rev 39:3675–3685
34. Custelcean R, Moyer BA, Bryantsev VS et al (2006) Anion coordination in metal-organic frameworks functionalized with urea hydrogen-bonding groups. Cryst Growth Des 6:555–563
35. Custelcean R, Haverlock TJ, Moyer BA (2006) Anion separation by selective crystallization of metal-organic frameworks. Inorg Chem 45:6446–6452
36. Custelcean R, Sellin V, Moyer BA (2007) Sulfate separation by selective crystallization of a urea-functionalized metal-organic framework. Chem Commun 1541–1543
37. Custelcean R (2009) Ion separation by selective crystallization of organic frameworks. Curr Opin Solid State Mater Sci 13:68–75
38. Custelcean R, Jiang D-E, Hay BP et al (2008) Hydrogen-bonded helices for anion binding and separation. Cryst Growth Des 8:1909–1915
39. Banerjee S, Adarsh NN, Dastidar P (2010) Selective separation of the sulfate anion by in situ crystallization of Cd-II coordination compounds derived from bis(pyridyl) ligands equipped with urea/amide hydrogen-bonding backbone. Eur J Inorg Chem 3770–3779

40. Adarsh NN, Tocher DA, Ribas J (2010) Metalla-macro-tricyclic cryptands: anion encapsulation and selective separation of sulfate via in situ crystallization. New J Chem 34:2458–2469
41. Adarsh NN, Dastidar P (2010) A borromean weave coordination polymer sustained by urea-sulfate hydrogen bonding and its selective anion separation properties. Cryst Growth Des 10:483–487
42. Xia Y, Wu BA, Li SG et al (2010) Anion binding of a bis(pyridylcarbamate) receptor bearing a diethylene glycol spacer. Supramol Chem 22:318–324
43. Custelcean R, Moyer BA (2007) Anion separation with metal-organic frameworks. Eur J Inorg Chem 1321–1340
44. Custelcean R, Remy P, Bonnesen PV et al (2008) Sulfate recognition by persistent crystalline capsules with rigidified hydrogen-bonding cavities. Angew Chem Int Ed 47:1866–1870
45. Wu B, Liang JJ, Yang J et al (2008) Sulfate ion encapsulation in caged supramolecular structures assembled by second-sphere coordination. Chem Commun 1762–1764
46. Custelcean R, Bock A, Moyer BA (2010) Selectivity principles in anion separation by crystallization of hydrogen-bonding capsules. J Am Chem Soc 132:7177–7185
47. Custelcean R, Remy P (2009) Selective crystallization of urea-functionalized capsules with tunable anion-binding cavities. Cryst Growth Des 9:1985–1989
48. Ludlow RF, Otto S (2010) The impact of the size of dynamic combinatorial libraries on the detectability of molecular recognition induced amplification. J Am Chem Soc 132:5984–5986
49. Corbett PT, Sanders JKM, Otto S (2005) Competition between receptors in dynamic combinatorial libraries: amplification of the fittest? J Am Chem Soc 127:9390–9392
50. Hay BP (2010) De novo structure-based design of anion receptors. Chem Soc Rev 39: 3700–3708
51. Reyheller C, Hay BP, Kubik S (2007) Influence of linker structure on the anion binding affinity of biscyclopeptides. New J Chem 31:2095–2102
52. Custelcean R, Bosano J, Bonnesen PV et al (2009) Computer-aided design of a sulfate-encapsulating receptor. Angew Chem Int Ed 48:4025–4029
53. Hay BP, Firman TK, Moyer BA (2005) Structural design criteria for anion hosts: strategies for achieving anion shape recognition through the complementary placement of urea donor groups. J Am Chem Soc 127:1810–1819
54. Custelcean R, Moyer BA, Hay BP (2005) A coordinatively saturated sulfate encapsulated in a metal-organic framework functionalized with urea hydrogen-bonding groups. Chem Commun 5971–5973
55. Ravikumar I, Lakshminarayanan PS, Arunachalam M et al (2009) Anion complexation of a pentafluorophenyl-substituted tripodal urea receptor in solution and the solid state: selectivity toward phosphate. Dalton Trans 4160–4168
56. Jia CD, Wu BA, Li SG et al (2010) A fully complementary, high-affinity receptor for phosphate and sulfate based on an acyclic tris(urea) scaffold. Chem Commun 46:5376–5378

Supramolecular Naphthalenediimide Nanotubes

Nandhini Ponnuswamy, Artur R. Stefankiewicz, Jeremy K. M. Sanders, and G. Dan Pantoş

Abstract Amino acid functionalized naphthalenediimides (NDIs) when dissolved in chloroform form a dynamic combinatorial library (DCL) in which the NDI building blocks are connected through reversible hydrogen bonds forming a versatile new supramolecular assembly in solution with intriguing host–guest properties. In chlorinated solvents the NDIs form supramolecular nanotubes which complex C_{60}, ion-pairs, and extended aromatic molecules. In the presence of C_{70} a new hexameric receptor is formed at the expense of the nanotube; the equilibrium nanotube – hexameric receptor can be influenced by acid–base reactions. Achiral NDIs are incorporated in nanotubes formed by either dichiral or monochiral NDIs experiencing the "sergeants-and-soldiers" effect.

Keywords Circular dichroism · Fullerenes · Host–guest · Ion-pairs · Sergeants-and-soldiers · Supramolecular chemistry · Supramolecular polymers

Contents

1	Introduction	218
2	Synthesis	218
3	Solid State Characterisation of α-Amino Acid Functionalised Naphthalenediimides	221
4	Nanotube Characterisation	231
	4.1 Solution State Characterisation	231
	4.2 Majority Rules Study	232
	4.3 Theoretical Studies of the CD of NDI	233
	4.4 The "Sergeants-and-Soldiers" Effect	236
5	C_{60} Encapsulation	240
	5.1 UV–Vis Spectroscopy	240

N. Ponnuswamy, A.R. Stefankiewicz, and J.K.M. Sanders
Department of Chemistry, University of Cambridge, Lensfield Road, Cambridge CB21EW, UK

G.D. Pantoş (✉)
Department of Chemistry, University of Bath, Claverton Down, Bath BA27AY, UK
e-mail: g.d.pantos@bath.ac.uk

5.2	CD Spectroscopy	241
5.3	^{13}C-NMR Experiments	242
5.4	Self-Sorting of Nanotubes	242
6	Self-Assembled C_{70} Receptors	243
7	Proton-Driven Switching Between C_{60} and C_{70} Receptors	245
8	Complexation of Polyaromatic Hydrocarbons	251
9	Complexation of Ion Pairs	254
10	Conclusions	258
References		259

1 Introduction

Dynamic combinatorial chemistry is defined as combinatorial chemistry under thermodynamic control; that is, in a dynamic combinatorial library (DCL), all constituents are in equilibrium with each other [1–6]. This requires the interconversion of library members into one another through a reversible chemical process, which can involve covalent bonds or non-covalent interactions such as hydrogen bonds. The composition of the library is determined by minimising the free energy of the whole system, which often is dominated by the thermodynamic stability of each of the library members under the particular conditions of the experiment. If a particular library member can be stabilised, either by binding to an external template, mutual interactions with other library members or due to a change in the DCL conditions, its free energy is lowered and consequently, in general, the equilibrium shifts towards its formation.

Hydrogen bonds between neutral partners in solution typically have energies between 0 kJ/mol and 20 kJ/mol [7] and have a preference for a linear arrangement of the three atoms involved (X−H⋯A angle ~180°). Due to its labile nature, equilibrium is often reached rapidly. Reinhoudt, Timmerman and co-workers were the first to describe a DCL based on hydrogen-bonded assemblies: using combinations of donor and acceptor hydrogen-bonded motifs, they built complex superstructures held together by an impressive number of hydrogen bonds [8, 9]. Rebek and co-workers have also utilised multiple hydrogen-bonding interactions to generate dynamic libraries composed of closed and spherical capsules [10, 11].

Amino acid functionalized naphthalenediimides (NDIs) when dissolved in chloroform form a DCL in which the NDI building blocks are connected through reversible hydrogen bonds forming a versatile new supramolecular assembly in solution with intriguing host–guest properties. These studies were prompted by the observation of an unusual arrangement of NDIs in a crystal structure and this chapter summarises the results and insights obtained to date.

2 Synthesis

Microwave dielectric heating is a mild and efficient method for the one-pot and stepwise synthesis of symmetrical and N-desymmetrised NDI derivatives of amines and α-amino acids. For the synthesis of symmetrical NDI derivatives, the reaction is carried out at 140 °C for 5 min in a dedicated microwave reactor in a pressure-resistant reaction vessel. Using this method, the products are obtained in high yield and purity after a simple aqueous workup; the acid-labile protecting groups (trityl, benzyl, *tert*-butyl, Boc and Pmc) are stable under the mild reaction conditions and undesired self-condensation side-products of amino acid esters, such as dipeptides, diketopiperazines and higher oligomers, are not observed. In the case of unprotected tyrosine and serine, the reaction is completely selective for the formation of the symmetrical imide without any trace of ester formation [12].

The method described above and outlined in Scheme 1 is particularly suitable for the synthesis of symmetrically substituted NDIs but originally it was of limited value for the synthesis of naphthalenemonoimide (NMI) and N-desymmetrised NDI derivatives. This is due to the difficulty of selective imide formation in a cross-conjugated dianhydride containing two equivalent electrophilic sites such as 1,4,5,8-naphthalenetetracarboxylic dianhydride (NDA). For all the aliphatic amines and amino acids tested, carrying out the reaction for 5 min at 140 °C only led to the formation of a 1:2:1 statistical mixture of dianhydride:monoimide:diimide (Table 1).

The selective and efficient synthesis of NMI and stepwise synthesis of N-desymmetrised NDI derivatives was achieved through the development of a stepwise synthetic procedure which involves heating the reaction mixture at a lower temperature (40 °C or 75 °C) for 5 min prior to the final heating stage at 140 °C (Scheme 2). Molecular modelling at ab initio level (direct SCF, 6-31G**) provided a reasonable explanation for the selectivity observed in favour of NMI over a statistical mixture on lowering the temperature. Lowering the temperature allows for the selective nucleophilic attack of the amino group on one of the anhydrides of NDA to generate NMI (and its open form precursors). The subsequent increase in temperature provides enough energy for the second nucleophilic attack on the NMI anhydride and the final dehydration to generate NDI.

For alkyl amines, a direct correlation between the steric bulk at the α-carbon and the yield of the reaction was found: amines attached to a secondary carbon gave higher yields than amines connected to a tertiary carbon, while amines connected to a quaternary carbon led only to the formation of an amide-carboxylic acid intermediate, rather than the corresponding imide. In the case of amino acids whose α-carbons are tertiary, a lower temperature was surprisingly required for high NMI selectivity in the first step (40 °C instead of 75 °C). This was explained by the presence of the COOR group, which assists in the collapse of the tetrahedral intermediate precursor to the imide formation. The amino acid derived NMIs were obtained as a mixture of open and closed forms due to the addition of triethylamine in the reaction. At high temperatures this promotes the formation of

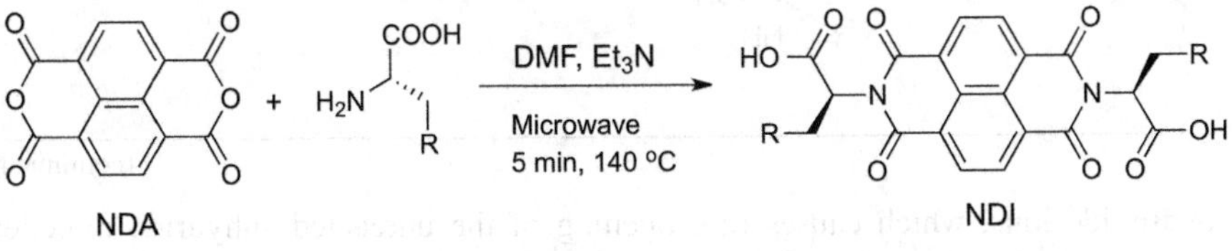

Scheme 1 Synthesis of amino acid derived symmetrical NDIs

Table 1 Selected examples of amino acid derived symmetrical NDIs. For more examples see [12]

Entry	Amino acid	NDI	Yield (%)
1		1	95
2		2	92
3		3	86
4		4	94
5		5	87
6		6	91
7		7	95
8		8	90
9		9	55

(continued)

hydroxide ions, which causes ring opening of the unreacted anhydride moieties leading to the formation of open by-products. This process however is not detrimental to the synthesis of N-desymmetrised NDIs (Tables 2 and 3).

Table 1 (continued)

Entry	Amino acid	NDI	Yield (%)
10	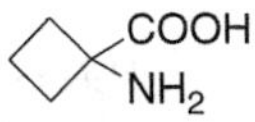 COOH, NH₂	10	85
11	 COOH, NH₂	11	59

Scheme 2 General protocol for the microwave synthesis of NMIs and desymmetrised NDIs

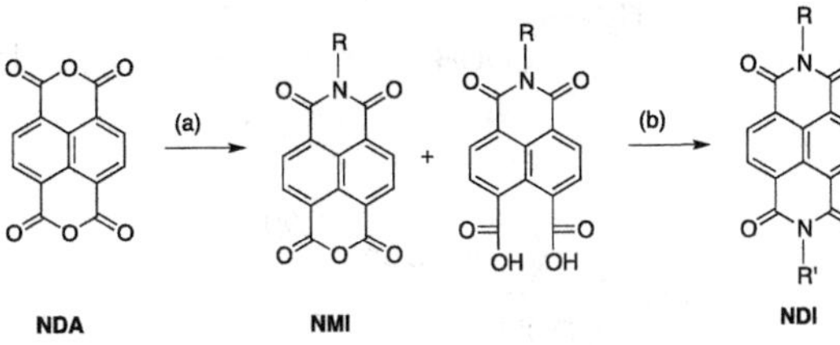

NDA   NMI   NDI

For alkyl amines: (a) R-NH₂, DMF i) 75 °C, 5 min, ii) 140 °C, 5-15 min
(b) R'-NH₂, DMF i) 75 °C, 5 min, ii) 140 °C, 5-15 min
For amino acids: (a) R-NH₂, Et₃N, DMF i) 40 °C, 5 min, ii) 140 °C, 5 min
(b) R'-NH₂, Et₃N, DMF i) 140 °C, 5 min

A higher degree of selectivity in favour of NMI was obtained when the amino acid derivatives contained aromatic side chains: trityl (100% selectivity), benzyl (70–90% selectivity) and alkyl (<60% selectivity). Also the amino acid esters gave higher selectivity towards NMI than the corresponding amino acids. This trend was rationalised by considering the solubility of the amino acid derivative and its reactivity towards NDA in DMF at room temperature. The NDA is insoluble in DMF at room temperature, but is rapidly solubilized by an amino acid containing aromatic side chains, which is itself soluble in DMF. It was proposed that the dissolution of NDA in DMF is probably due to π–π interactions between its extended aromatic core and the amino acid aromatic side chains. This enhanced solubility of NDA in DMF leads to a 1:1 mixture of NDA and the amino acid in solution and hence promotes the selective formation of the NMI. In the cases where sonication and heating were required in order to dissolve the reagents completely, a strong preference for the formation of the NDI was observed. The N-desymmetrised synthesis was primarily used to synthesise monochiral NDIs for their use in "sergeant-and-soldiers" experiments (see below).

3 Solid State Characterisation of α-Amino Acid Functionalised Naphthalenediimides

Single crystals of NDIs **1, 3–8** suitable for X-ray analysis were obtained by slow evaporation or vapour diffusion of solvents ranging from low polarity, such as benzene and dichloromethane, to the polar acetonitrile, acetone and dimethyl

Table 2 Selected examples of NMIs and N-desymmetrised NDIs. For more examples see [13]

Entry	Amine-1	NMI yield (%)	Amine-2	NDI yield (%)
1	H_2N (structure)	86	H_2N—STrt, COOH (structure)	94
2	H_2N (alkyne)	78	H_2N (branched)	69
3	H_2N—STrt, COOH (structure)	91	H_2N—$(\,)_3$NHBoc, COOH (structure)	87
4	H_2N—, OBn, COOMe (structure)	90	H_2N (branched)	91
5	H_2N—CONHTrt, COOH (structure)	87	H_2N COOMe	90

Table 3 Some N-desymmetrised NDIs derived from condensing the achiral amino acids shown with the NMI derived from S-trityl cysteine [14]

Entry	Amino acid	Monochiral-NDI	Yield (%)
1	COOH, NH_2 (cyclopropyl)	12	81
2	COOH, NH_2 (cyclobutyl)	13	73
3	$HOOC$—NH_2	14	67

sulfoxide. The molecular structures of **1**, **3–8** as well as crystallisation conditions are summarised in Table 4.

NDI **1** (entry 1 in Table 4) stands out by being the only derivative of this set that forms a supramolecular nanotube in the solid state. This compound crystallises

Table 4 X-ray structures and crystallisation conditions of molecules **1, 3–8**

X-ray structure	Amino acid	NDI	Solvent	Nanotube (Yes/No)
	H-L-Cys (Trt)-OH	**1**	CH_2Cl_2	Yes
	H-L-Ser (Bzl)-OH	**3**	CH_3CN	No
	H-L-Ile-OH	**4**	CH_2Cl_2	No
	H-L-Phe-OMe	**5**	$(CH_3)_2CO$ CH_3CN C_6H_6	No No No
	H-L-Tyr-OH	**6**	$(CH_3)_2SO$	No

(continued)

from CH_2Cl_2 in the trigonal $P3_1$ space group in which the amino acid side chains adopt a *syn* geometry with respect to the NDI plane (Fig. 1).

The *syn* geometry of **1** allows three S-trityl groups of three different molecules to interdigitate and the carboxylic groups of **1** to dimerize through two strong

X-ray structure	Amino acid	NDI	Solvent	Nanotube (Yes/No)
	H-L-Leu-NHMe	7	(CH$_3$)$_2$CO	No
	H-Cyc-OH	8	(CH$_3$)$_2$SO	No

Table 4 (continued)

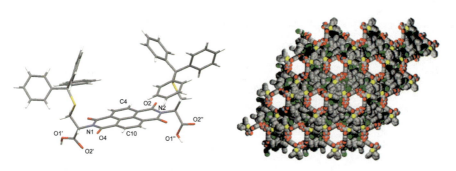

Fig. 1 Side and top views of the crystal packing of **1**

intermolecular hydrogen bonds (O1'H···O2″, 2.61 Å, 161° and O1″H···O2′, 2.63 Å, 161°). Closer inspection of the crystal structure reveals that **1** assembles in a hydrogen-bonded nanotubular supramolecular structure, in which NDI cores i and $i + 3$ are coplanar, forming the walls of the nanotube (Fig. 2a). This structure appears to be reinforced by two weak C–H···O hydrogen-bond interactions (C10H···O2, 3.16 Å, 142° and C4H···O4, 3.15 Å, 143°) per unit, formed between NDIs i and $i + 3$ (Fig. 2b).

In this arrangement, the angle between the central N1–N2 axis of the NDI core and the central axis of the nanotube is 60°, whereas the distance between two sequential aromatic cores is on average 4.8 Å. The nanotube has an average inner diameter of 12.4 Å and it contains diffuse electron density attributed to disordered water molecules.

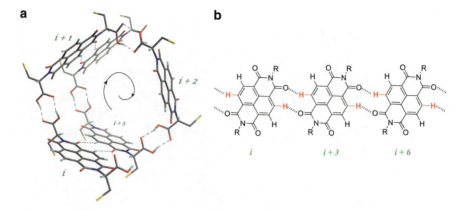

Fig. 2 (a) Top view of the molecule **1** indicating H-bonding interactions observed in the self-assembled nanotube. The *arrow* indicates the direction of the nanotube assembly. (b) Schematic representation of the proposed non-classical H-bonding interactions between the NDIs units

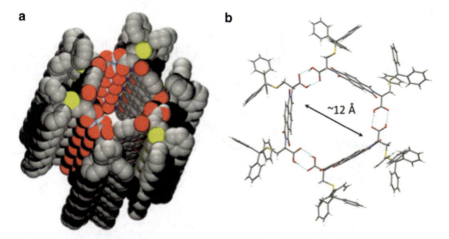

Fig. 3 Side and top views of the hydrogen-bonded nanotubular structure formed by **1**. Most of the hydrogen atoms, water molecules and CH$_2$Cl$_2$ molecules have been removed for clarity [15]

The importance of the amino acid side chain in the solid state packing can be readily observed by comparing the crystal structures of **1** and **4** (Fig. 3). Both derivatives crystallise from CH$_2$Cl$_2$, but unlike **1**, compound **4** crystallises in the *anti* geometry in the orthorhombic *C*222 space group. The intermolecular contacts between molecules of **4** are similar to those present in the crystal lattice of **1**: the dimerization between the carboxylic groups through two strong intermolecular hydrogen bonds (O2″H···O1″ distance: 2.63 Å) is also supported by two weak C–H···O hydrogen-bond interactions (C4H···O4, 3.17 Å) per unit (Fig. 4).

Despite having the same H-bonding interactions, the crystal packing view revealed significant differences between **1** (Fig. 1) and **4**. NDI **4** forms a helical polymeric

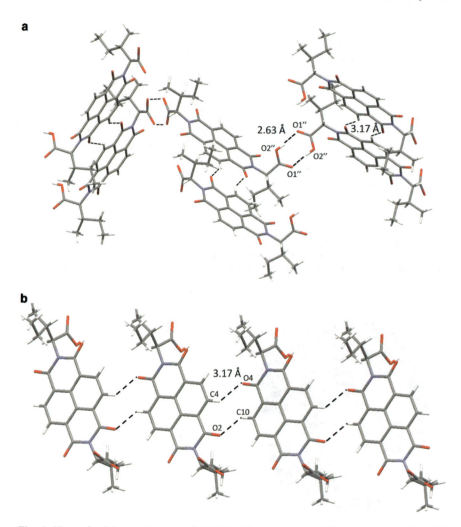

Fig. 4 View of solid state structure of 4 highlighting two types of H-bonding interactions (**a**) COOH dimerisation and (**b**) weak CH···O interactions

structure via carboxylic acid dimerization (Fig. 5a); however, in contrast to **1**, no porous architecture was observed (Fig. 5c). The difference in solid state packing behaviour between NDIs **1** and **4** as well as the opposite arrangement of the side chains (*syn* and *anti*, respectively) is directly related to the intermolecular interactions between the amino acid side chains. In the case of **4**, the flexible and relatively short alkyl chains of L-isoleucine form a tightly packed and compact structure in the solid state.

NDI **5** is an eloquent example of the delicate balance between the intermolecular forces that determine crystal packing. Analysis of the X-ray structures of **5** crystallised from three different solvents – acetone, acetonitrile and benzene – reveals that the solid state arrangement of this L-phenylalanine methyl ester derivative of NDI is

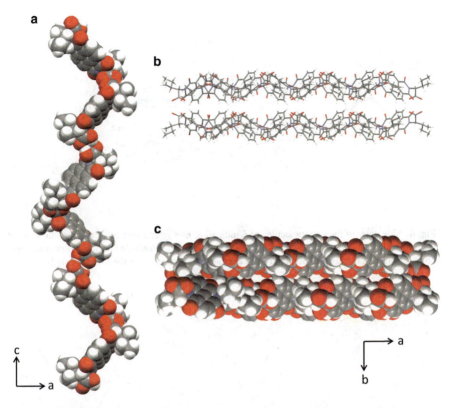

Fig. 5 (a) Side view of the helical hydrogen bonded structure of **4** in the solid state. Top views of the crystal packing of **4** capped sticks (**b**) and space filling representation (**c**)

governed by π–π interactions and complemented by a set of CH···O hydrogen bonds (Fig. 6). In this arrangement, irrespective of the solvent of crystallisation, the π–π distance between the phenyl groups and the NDI core is in the range 3.38–3.44 Å, while the 4-CH···OCH$_3$ hydrogen bond distance is 3.24–3.30 Å.

The crystallisation solvent influences the packing of these NDI stacks as can be seen in Fig. 6. In acetonitrile, separate stacks of NDI **5** are interconnected by weak C–H···O hydrogen-bond interactions (CH···O, 2.72–2.83 Å, and C···O, 3.16–3.28 Å, Fig. 7a). Slightly different behaviour was observed in the crystal lattice of **5** crystallised from benzene and acetone; these two solvents yielded similar packing arrangements. The crystal packing is less dense than that obtained from acetonitrile, and consists of alternating NDI stacks with solvent molecules trapped in between. Benzene does not disturb π–π contacts and is not directly involved in these intermolecular interactions (Fig. 7b), suggesting that in the present case the solvent has limited influence on the final molecular packing. This observation was confirmed when crystals obtained from acetone had a similar packing structure.

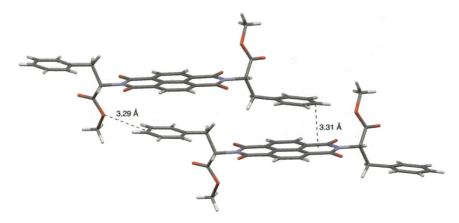

Fig. 6 Side view of the two molecules of **5** indicating π–π interaction between aromatic units of the adjacent NDIs and the CH···OCH₃ hydrogen bonds from acetonitrile (similar interactions were observed in the crystals of **5** obtained from acetone and benzene; see the text for details)

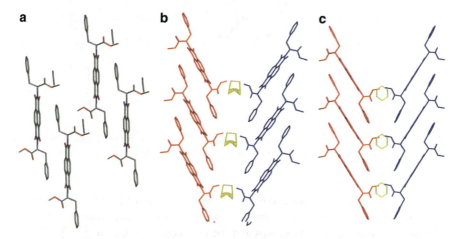

Fig. 7 The crystal packing of **5** crystallised from acetonitrile (**a**), acetone (**b**) and benzene (**c**) showing solvent molecules intercalated between two stacks of NDIs. The hydrogen atoms have been removed for clarity; the acetone molecules in (**b**) are disordered over two sites

NDI derivatives **3** and **6** also contain electron rich aromatic units in their structure, and have been crystallised from acetonitrile and dimethyl sulfoxide, respectively. In contrast to **5**, both of them possess free carboxylic acid groups, which, along with the aromatic units, were expected to influence the solid-state packing. As depicted in Fig. 8, both of these structures crystallise in a regular network of π-stacked architectures. The distances between aromatic units involved in π–π interactions are similar for both molecules and range from 3.53 Å to 3.55 Å. In the case of **3** this arrangement is further reinforced by intermolecular hydrogen bond interactions between carboxylic acid groups (OH···O distance: 2.51 Å). Molecule **6** presents a more complex structure with three separate intermolecular interactions taking place:

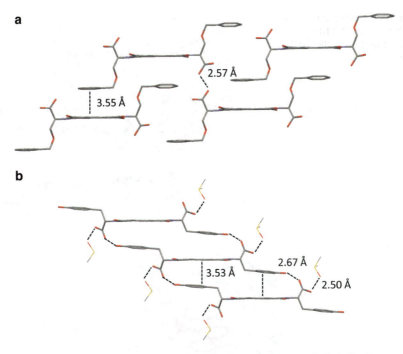

Fig. 8 (a) Side view of the crystal packing of **3** crystallised from acetonitrile indicating π–π and hydrogen bonding interactions between adjacent molecules in the solid state; (**b**) Side view of the crystal packing of **6** crystallised from DMSO indicating π–π interactions between aromatic units of the adjacent NDIs and hydrogen bonding interactions between carboxylic acids and solvent molecules

π–π interactions between aromatic groups (NDI and 4-hydroxyphenyl, 3.53 Å), and two distinct hydrogen bonding interactions (Fig. 8b). Both of them involve carboxylic acid groups, which interact on one site with a solvent molecule (DMSO, OH···O distance: 2.57 Å) and on the other site with an OH group of a tyrosine moiety from a neighbouring NDI (OH···O distance: 2.67 Å).

Crystals of achiral amino acid derived NDI **8** suitable for X-ray diffraction study were obtained from slow diffusion of water in DMSO. In this hydrogen bond acceptor solvent, the NDIs are arranged in a π-stacked structure, with the NDI cores n and $n + 1$ tilted at ca. 30° with respect to each other, while n and $n + 2$ are parallel to each other (dihedral angle 0.6°, Fig. 9).

A derivative that stands apart in this analysis is NDI **7**, in which the carboxylic acids (or esters) have been replaced by amide groups, thus allowing the possibility of a new type of hydrogen bonding interaction. The crystal structure of molecule **7** was obtained by slow evaporation of an acetone solution and the analysis showed a complex network of molecules connected via a large number of hydrogen bonds.

Each NDI **7** is connected with four distinct molecules via strong hydrogen bonding interactions between the amide units (Fig. 10, NH···O distance: 2.85 Å). Even though there are no pendant aromatic units and solvent molecules are not

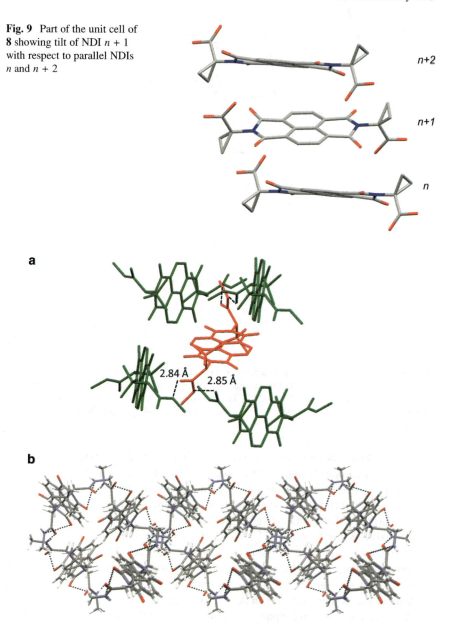

Fig. 9 Part of the unit cell of **8** showing tilt of NDI $n + 1$ with respect to parallel NDIs n and $n + 2$

Fig. 10 (a) Crystal structure of molecule **7** and its hydrogen bonding interactions with four adjacent NDI molecules in the solid state. (b) Top view of the crystal packing of **7** indicating a complex network of hydrogen bonding interactions between amide units of **7**

directly involved in the interactions between the molecules, no tubular object could be observed. This might be because the amide moieties of **7** have the *s-trans* geometry and they adopt *anti* geometry with respect to the NDI plane, thus preventing formation of the nanotube. The latter factor seems to be critical for the formation of the supramolecular nanotube in the solid state as only derivative **1** with *syn* geometry of the amino acid side chains was found to assemble into the tubular architecture. This also suggests that there is a very narrow line between formation of a nanotube and other architectures in the solid state whereby modifications of the chemical structure of the amino acid, crystallisation solvent and the introduction of new hydrogen bonding units do not result in formation of porous structure like that observed for derivative **1**.

4 Nanotube Characterisation

The helical supramolecular nanotubes of NDI **1,** observed in the solid state, were identified in $CHCl_3$ and 1,1,2,2-tetrachloroethane (TCE) solution by means of circular dichroism (CD) and NMR spectroscopies, and were further studied using molecular modelling.

4.1 Solution State Characterisation

The first evidence for the presence in solution of NDI nanotubes of **1** came from CD spectroscopy, which measures the optical rotation of circular polarised light by chiral molecules and therefore is an indicator of chirality. A molecule with a chiral centre usually produces a small intrinsic CD signal; however, if the molecule can aggregate to form a chiral supramolecular species, then a much stronger CD signal can be generated, as the entire structure expresses chirality. In chloroform, **1** had an intense CD signal at 383 nm, corresponding to the absorbance of the naphthalene core. By contrast, the corresponding methyl ester derivative of **1** (L-**1** ester) was CD silent at this wavelength. The L,L-enantiomer of **1** (L-**1**) derived from the corresponding L-amino acid forms P-helices with a positive CD signal, while the D,D-enantiomer (D-**1**) forms M-helices with a negative CD signal [15].

CD spectroscopy provided the compelling evidence that the nanotubes are held together by hydrogen bonding: addition of a competing hydrogen bonding solvent such as methanol reduced the CD signal dramatically to what would be expected for the intrinsic chirality of a non-aggregated NDI monomer (Fig. 11). Similar CD results were also obtained with other amino acid derivatives of NDI (**2–5**) and mono-chiral NDI derivatives (**12–15**) [14].

In ^{1}H-NMR spectroscopy, the NDI aromatic protons appear as a broad singlet at 8.65 ppm in $CDCl_3$. The presence of this broad signal, rather than two doublets as would be expected from the structure depicted in Fig. 2, was rationalised by the dynamic nature of the nanotube: rapid exchange of NDI units out of and into the nanotube averages the two non-equivalent sites (Fig. 2b). When the ^{1}H NMR spectra were recorded at 263 K the expected lack of symmetry is clearly visible (Fig. 12) [16].

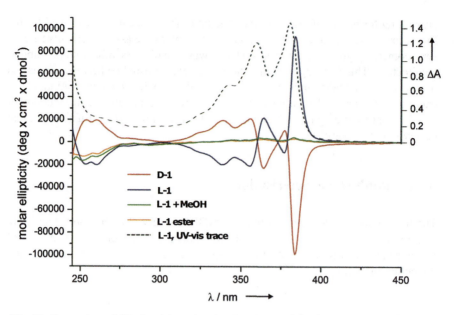

Fig. 11 Comparison of CD signal for D-1, L-1, L-1 + methanol and L-1-ester [15]

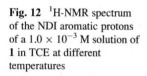

Fig. 12 ^{1}H-NMR spectrum of the NDI aromatic protons of a 1.0×10^{-3} M solution of **1** in TCE at different temperatures

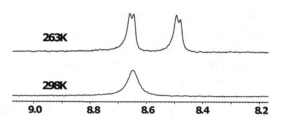

A similar splitting pattern of the NDI ^{1}H peaks could also be observed at room temperature for an N-desymmetrised monochiral NDI molecule, **12**. Figure 13 compares the naphthyl region of the ^{1}H-NMR spectra of **12** in 5% MeOD in CDCl$_3$ (in which nanotubes cannot form) and in pure CDCl$_3$ (in which they can). In the methanolic solution, molecule **12** has a C$_2$ axis of symmetry intersecting the two nitrogen atoms and therefore the naphthyl protons are observed as two doublets coupling to one another. In pure CDCl$_3$ solution, these protons split into four doublets, indicative of the asymmetric environment of the NDI.

4.2 Majority Rules Study

Some chiral self-assembling systems are able to incorporate enantiomers of opposite chirality into the supramolecular structure because the chiral centre is relatively remote from the moiety that is involved in self-assembly [18, 19]. By contrast, NDI

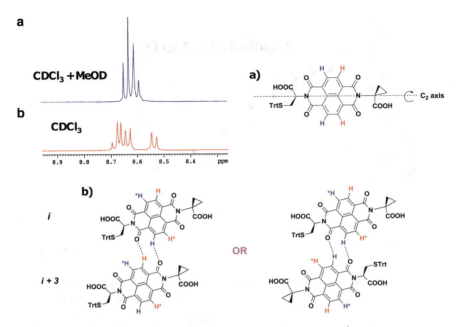

Fig. 13 Comparison of the naphthyl proton signals observed in the ¹H-NMR spectrum of molecule **12** in (**a**) CDCl₃ + MeOD: NDI core contains two distinct sets of protons (*blue and red*) and (**b**) CDCl₃: nanotube formation further differentiates the *blue and red* NDI protons as non hydrogen-bonded (H*) and hydrogen-bonded (H) [17]

monomers are highly discriminating when it comes to the formation of chiral nanotubes: one enantiomer cannot be incorporated into a helix made of the other enantiomer because the directionality created at the chiral centre is critical to self-assembly. Therefore, when mixed, the two enantiomers self-sort, as shown by a "majority rules" study [20–22]. Equimolar TCE solutions of chiral NDIs L-**1** and D-**1** were titrated with small quantities of the opposite enantiomer solution and the CD trace recorded for each mixture. The resulting graph (Fig. 14) of the CD_{max} points plotted against percentage of the starting NDI present was approximately linear, suggesting simple dilution with no interaction between the two enantiomers.

4.3 Theoretical Studies of the CD of NDI

In order to correlate the solid state and solution phase structures, molecular modelling using the exciton matrix method was used to predict the CD spectrum of **1** from its crystal structure and was compared to the CD spectrum obtained in CHCl₃ solutions [23]. The matrix parameters for NDI were created using the Franck–Condon data derived from complete-active space self-consistent fields (CASSCF) calculations, combined with multi-configurational second-order perturbation theory (CASPT2).

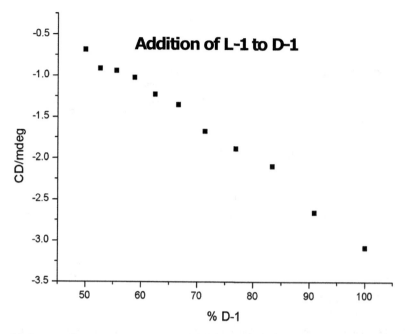

Fig. 14 Datasets from majority rules studies involving mixtures of L-1 and D-1 [17]

The minimal active space needed to describe the electronic structure of the NDI moiety includes the five occupied and five unoccupied π-orbitals of the naphthalene core and four lone pair orbitals of the carbonyl groups. The 5π[4n]5π active space contains 14 electrons and electronic transitions arise from seven states. Only two of the seven states, $1\ ^1B_{2u}$ and $1\ ^1B_{3u}$, show transitions in the region of interest between 320 and 420 nm. Other transitions have no effect on the bands in this region and hence were not considered. The main features in the experimental absorbance spectrum were reproduced using the most intense Frank–Condon transitions (Fig. 15). The calculated spectrum (dashed lines) showed a red shift of 9 nm relative to the experiment, which may be due to the representation of each transition by only two charges, and also due to the neglect of other transitions.

The dependence of the calculated CD spectrum on the angle between two adjacent NDI planes was also studied (Fig. 16). The intensity of the CD signal decreases uniformly for greater angles, until no interaction is observed for coplanar chromophores. If the angle is increased further, a change in the sense of the helix occurred which inverted the signs of the bands, similar to a reversal of chirality at the α-carbon.

In order to analyse the effect of the nanotube length on the CD spectrum, the modelling was applied to several oligomers of molecule **1** (Fig. 17). In general, the intensity of the spectra increases with the number of monomers added to the structure. The distance between the chromophore centres across the tube core is about 12 Å and the interaction between such monomers (i and $i + 2$) is responsible

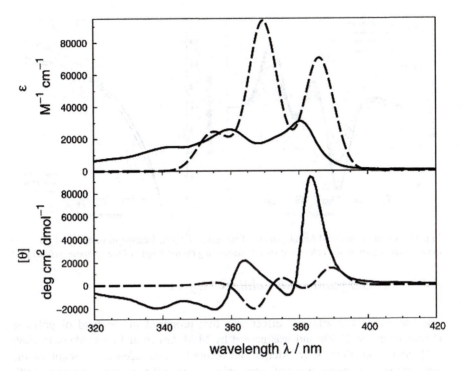

Fig. 15 Experimental spectrum from solid state (*dashed line*) and calculated spectrum of a heptamer of molecule **1** (*solid line*). *Upper panel*: Absorbance spectrum, *lower panel*: CD spectrum [23]

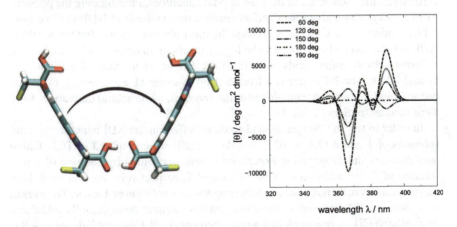

Fig. 16 Dependence of the calculated CD spectrum on the angle between the NDI planes of two monomers [23]

for the increase in going from a dimer to a trimer. As further monomers are added, the increase in CD intensity gradually becomes less pronounced, as expected (Fig. 17).

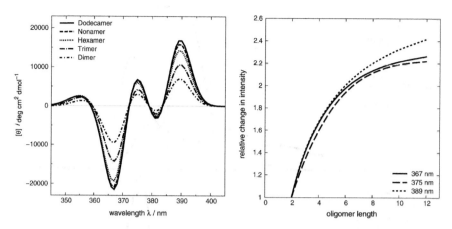

Fig. 17 Spectra calculated for oligomers of molecule **1** (*left*). Change in intensity of the three most intense bands in the calculated spectra depending on the length of the oligomer (*right*) [23]

4.4 The "Sergeants-and-Soldiers" Effect

The "sergeants-and-soldiers" effect was first proposed in the field of polymer chemistry in the 1960s and was named by M.M. Green and co-workers in 1989 [24]. Since that time, E.W. Meijer and others have broadened the scope of the concept to take in many facets of supramolecular as well as polymer chemistry [18]. In a system displaying "sergeants-and-soldiers" behaviour a chiral derivative, the "sergeant", imposes its chirality on a structure formed mainly out of achiral derivatives, the "soldiers". In the case of NDI nanotubes, investigating the possibility of a "sergeants-and-soldiers" effect requires the synthesis of NDIs derived from achiral amino acids. Glycine is perhaps the most obvious choice, but the resulting NDI had previously been found to be highly insoluble in organic solvents. Several different achiral amino acids were therefore employed to synthesise a range of achiral NDIs. The NDIs derived from *S*-trityl cysteine (**1** and **1**-ester), or *N*-Boc lysine (**2** and **2**-ester) were employed as the sergeant, while achiral derivatives **8–11** were used as soldiers (Table 1).

In order to test the "sergeants-and-soldiers" effect in the NDI nanotube system, solutions of **1** and **8** (2.1×10^{-4} mol dm^{-3} each) were prepared in TCE. Chiral amplification and propagation experiments were performed by addition of **1** to a solution of **8** and addition of **8** to solution of **1**, respectively. As a control, both experiments were then repeated substituting **8** with a solution of **1**-ester. The overall concentration of NDI moieties therefore remains constant throughout the additions. The collated CD$_{max}$ points plotted against percentage of **1** present indicate that **8** is incorporated into nanotubes formed of NDI **1**, thus producing a more intense CD signal when compared to the control experiment (Fig. 18). Maximum amplification of the CD signal (2.5-fold) is observed at around 70% of **8**.

A further experiment with the same protocol was used to test that the "sergeants-and-soldiers" behaviour was not unique to **1**. Separate solutions of both

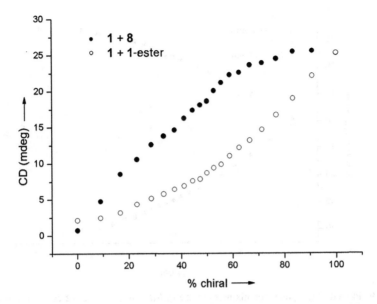

Fig. 18 Plotted CD_{max} (383.5 nm) for all experiments. Concentration of all NDIs = 2.1×10^{-4} mol dm^{-3} [17]

enantiomers of **2** (L-**2** and D-**2**) were each diluted with **8** and with their methyl ester derivatives (L-**2**-ester and D-**2**-ester) as control, thus leading to the formation of mixed nanotubes composed of "sergeants" **2** and "soldiers" **8** (Fig. 19).

Of the four achiral NDIs studied, only derivatives **8** and **10** acted as "soldiers", while derivatives **9** and **11** were not incorporated into supramolecular nanotubes. This suggests that the superstructure of the heterogeneous nanotubes is subtly different to that of the homogenous **1** nanotube. The difference is likely to be due to replacement of the α-proton with a significantly bulkier group. The steric repulsion of two alkyl groups on **8** is balanced by the cyclopropyl ring holding them tightly together, meaning the N–C–C(O) angle in **8** is within the range of those of **1** and **2**, both of which readily form nanotubes (Fig. 20). However, the added rigidity imposed by the cyclopropyl moiety forces **8** to take up conformations that are slightly different to the minimum energetic conformation for the chiral amino acid-derived NDIs (which present the α-proton in the most sterically crowded position).

The importance of the cyclopropyl strain in **8** is demonstrated by the fact that achiral NDIs **9** and **11** derived from achiral amino acids with reduced steric strain were observed to have no significant "sergeants-and-soldiers" activity. In achiral NDI **10**, the ring strain is lower than in **8** but higher than in both **9** and **11**, and as expected it acts as a "soldier", albeit inferior to **8** (Fig. 21). This supports the idea that bond angle must be key to the molecule's suitability as a soldier: it is the only significant molecular distinction between NDIs **8**, **9** and **10**, yet accounts for the variation in their properties.

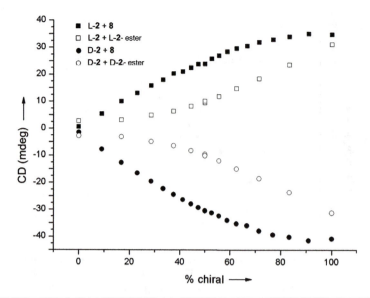

Fig. 19 Plotted CD$_{max}$ points for mixtures of L-**2**/**8** (*filled squares*) or L-**2**/L-**2**-ester (*open squares*); and D-**2**/**8** mixtures (*filled circles*) or D-**2**/D-**2**-ester (*open circles*). Concentration of all NDIs = 2.1 × 10^{-4} mol dm^{-3} [17]

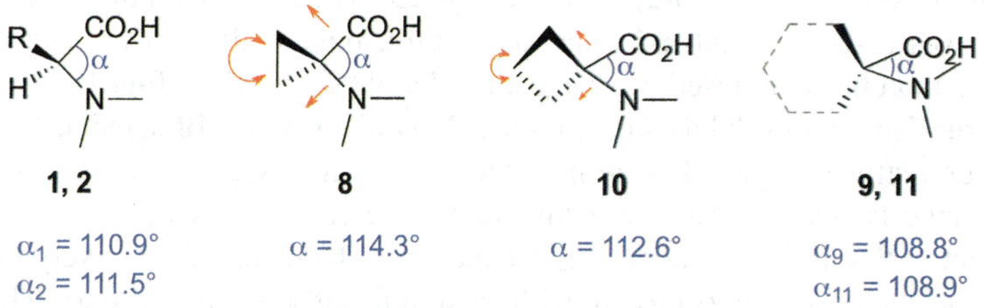

1, 2	**8**	**10**	**9, 11**
α_1 = 110.9°	α = 114.3°	α = 112.6°	α_9 = 108.8°
α_2 = 111.5°			α_{11} = 108.9°

Fig. 20 Ring strain in NDI **8** counteracts the repulsion of the alkyl carbons to bring the N–C–C(O) angle back within the range of an NDI derived from achiral amino acid with only one R group (**1**, **2**). The lesser ring strain in **10** produces a reduced version of this effect, while it is not present at all in unstrained **9** or **11**. The degree of difference in the N–C–C(O) angle is exaggerated for clarity in the diagrams. Where crystal structures of the NDIs were not available, estimates have been made by averaging the values found in crystal structures of related compounds in the CSD [25]

As can be seen in Fig. 21, solutions of chiral NDIs **1** containing small percentages of **8** tend to show stronger CD signals than 100% chiral solutions of the same total NDI concentration. The effect is small but reproducible, and must be the outcome of subtle geometrical changes in the size of the exciton coupling between adjacent chromophores, thus leading to an increased CD signal [17].

The "sergeant-and-soldier" experiment was applied in a system that contained a monochiral NDI **12–14**, and the achiral derivatives **8**, **9** and **10** [14]. This is an extreme case of the "sergeant-and-soldier" experiment in which one chiral amino

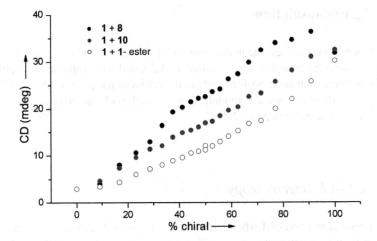

Fig. 21 Example data sets for NDIs **8** (*black filled circles*) and **10** (*grey filled circles*) acting as soldiers to **1** as sergeant. Control data using **1**-ester as inactive soldier is also plotted (*open circles*). Concentration of all NDIs = 2.1×10^{-4} mol dm^{-3}

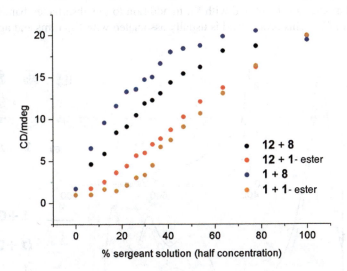

Fig. 22 "Sergeant-and-soldier" behaviour of monochiral NDI **12** [14]

acid residue controls the assembly of a mixed NDI nanotube by imposing its chirality upon at least three other achiral centres. The experiments were carried out using a modified protocol in which the concentration of the "soldier" solution was doubled compared to the "sergeant" solution (Fig. 22). These experiments showed that the monochiral **12** is a better sergeant than **13**, which is in agreement with the results obtained in the "sergeant-and-soldier" experiments using achiral and chiral NDIs.

5 C$_{60}$ Encapsulation

The nanotubular cavities (mean diameter: 12.4 Å) were found to be effective hosts for C$_{60}$ molecules (van der Waals radius: 10.3 Å), and were capable of solubilising C$_{60}$ in solvents such as chloroform, where the fullerene has poor solubility. UV–vis, CD, ^{13}C-NMR and molecular modelling were used to characterize this NDI nanotube-C$_{60}$ host–guest complex [26].

5.1 UV–Vis Spectroscopy

A chloroform solution of **1** when left to stand over solid C$_{60}$ shows a drastic colour change from pale yellow to brown within a few minutes. This is different from the expected summation of colours for a saturated solution of C$_{60}$ in chloroform (pale purple) and the initial pale yellow solution (Fig. 23, inset). The visible region of the absorption spectrum (Fig. 23, inset) is marked by the appearance of a broad band centred at 452 nm (identified with *), in addition to the absorbance bands characteristic of C$_{60}$. This broad band is usually associated with C$_{60}$ films and aggregates

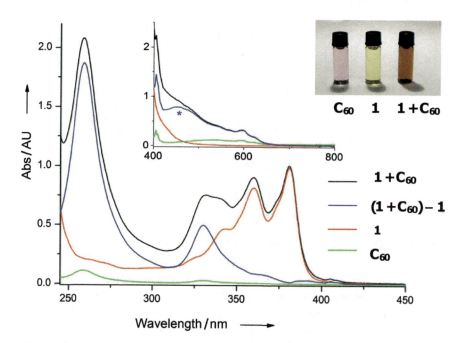

Fig. 23 UV–vis trace of C$_{60}$, **1**, **1** + C$_{60}$ and the C$_{60}$ contribution to **1** + C$_{60}$ in CHCl$_3$. The concentration of **1** for the measurement of the visible spectrum (400–800 nm) was 40 times higher than the UV region [26]

and has been attributed to interactions between fullerenes, implying that fullerenes inside the nanotube are in close proximity, forming a one-dimensional C_{60} array.

The uptake of C_{60} corresponds to the increase in absorbance at 258 and 328 nm in the absorption spectrum of **1** + C_{60} (Fig. 23). Comparison of a solution of **1** + C_{60} with a saturated solution of C_{60} in chloroform showed that the C_{60} concentration increased 16-fold in the presence of NDI nanotubes (**1**). In contrast with these results, the methyl ester of **1**, which is unable to form hydrogen-bonded supramolecular nanotubes, did not enhance the solubility of C_{60} in chloroform, supporting the thesis that the C_{60} molecules are complexed in the inner nanotubular cavity. The increase in absorbance at 258 nm also led to an estimate of [NDI]/[C_{60}] stoichiometry, revealing that an average of 3.6 NDI units were encapsulating one C_{60} molecule. Similar results were also obtained with other amino acid derivatives of NDI.

5.2 CD Spectroscopy

In the UV region, where the NDI chromophore absorbs, the CD signal of the complex is broadly unchanged from that of the nanotube alone, indicating that the structure is preserved. In the visible region, where the host is silent, a weak induced circular dichroism (ICD) signal at 595 and 663 nm is observed, suggesting that C_{60} also experiences the helicity of the environment (Fig. 24). It is not clear whether this is a direct sensing of the amino acid chirality, or whether the nanotube actually has a long-range chiral supercoiled structure.

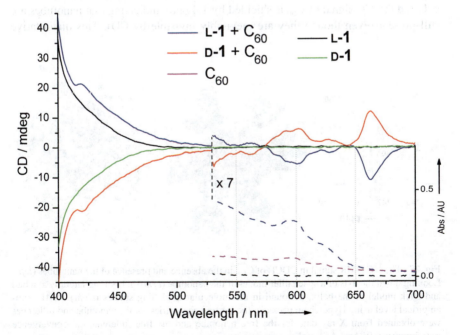

Fig. 24 Comparison of the CD spectra of **1** and **1** + C_{60} for the L and D enantiomers [26]

5.3 ^{13}C-NMR Experiments

Further evidence for the uptake of C_{60} came from ^{13}C NMR spectroscopy, in which the signal for C_{60} is shifted upfield by over 1.4 ppm upon complexation by NDI nanotubes (Fig. 25), indicative of a shielding effect due to the proximity of the NDI aromatic units (and possibly also C_{60}–C_{60} proximity). The retention of high symmetry of the fullerene upon binding within the nanotube indicates that spinning of the fullerene is fast on the ^{13}C NMR timescale.

^{13}C NMR experiments in TCE-d_2 confirmed that this functional behaviour of the nanotubes is also present in hybrid nanotubes composed of a mixture of chiral **1** and achiral **8** [14]. The C_{60} uptake is lower for the hybrid nanotubes with increasing amounts of incorporated achiral NDIs. A similar effect was observed for the monochiral NDI, in which nanotubes containing the cyclopropyl side chain **12** are better receptors for C_{60} than the analogues formed from **13**. These observations highlight the importance of the rigidity in the NDIs derived from achiral amino acids.

5.4 Self-Sorting of Nanotubes

Complexation of C_{60} also remarkably demonstrated the self-sorting of nanotubes of opposite helicity. A 1:1 mixture of L-**1** and D-**1** is capable of encapsulating C_{60} as shown by UV–vis spectroscopy: the [NDI]/[C_{60}] ratio of 3.9 matches (within experimental error) that is obtained for optically pure samples of either L-**1** or D-**1**, and the ^{13}C signal of C_{60} is shielded by 1.4 ppm, indicating that nanotubes are still present (even though they are inevitably invisible by CD). This quantitative

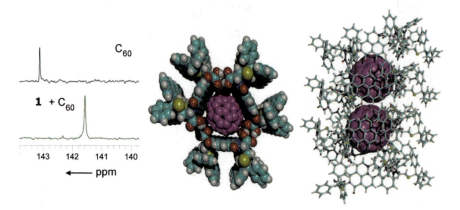

Fig. 25 The ^{13}C NMR signal in CDCl$_3$ of C_{60}, in the absence and presence of the nanotubes (*left*). Looking down the axis of a space filler model of the nanotube (*centre*). Looking laterally at a ball and stick model of the helix, as found in the molecular modelling studies (*right*). AM1 semi-empirical level using Hyperchem package; the starting geometries for the nanotube and fullerenes were obtained from X-ray data for the free nanotube and the free fullerene and convergence criterion was 0.01 kcal Å^{-1} mol^{-1}

6 Self-Assembled C₇₀ Receptors

As described above, the NDI nanotubes can complex a "string" of C_{60} molecules inside the tubular cavity. However, C_{70}, leads to complete destruction of the nanotubes by templating the formation of a new discrete receptor. The disassembly of the nanotube was immediately evident from the change in the CD spectrum of a solution of NDI on the addition of C_{70} (Fig. 26a) [27]. Equally striking differences were observed in the ^1H NMR spectra obtained from solutions prepared in $CDCl_3$ (Fig. 26b). The radically different CD and ^1H NMR spectra and the fact that derivatives **1** and **2** give similar CD and NMR signatures in the presence of C_{70} is indicative of the formation of a new supramolecular structure. ^1H NMR titration experiments (supported by CD data) led to the conclusion that six NDIs interact with one molecule of C_{70}, thus forming a hexameric receptor.

The ^1H NMR spectrum (Fig. 26b) of nanotubes of **3** (**3**$_N$) shows a high degree of symmetry experienced by the NDI units, as opposed to the signature of the C_{70} receptor, which shows four doublets for the aromatic protons: three of these have a similar chemical shift, while one is shifted downfield by 0.78 ppm. Similarly, there are two signals for the α-protons of the C_{70} receptor rather than one in the nanotube. ^{13}C NMR spectra of C_{70} in the absence and in the presence of NDI indicate clear shift differences with roughly equal effects (2.0–2.4 ppm) on all five inequivalent carbon signals. The four aromatic signals observed in the ^1H NMR spectrum of the C_{70} receptor might arise, at first sight, from four different situations (Fig. 27b–e): four inequivalent protons on a single NDI molecule (b), four equivalent protons on each of four inequivalent NDI molecules (c) or two pairs of protons on two inequivalent NDI molecules (d, e). Situations (c) and (d) can be ruled out, as they would give rise to four different signals of the α-protons rather than two. Situation (e) can also be ruled out, as it would result in four singlets for the aromatic protons. The number and multiplicity of the signals in the ^1H NMR spectrum of **3** + C_{70} clearly point to the situation pictured in Fig. 28b. Therefore, all NDI molecules are equivalent, yet none of them lie on a symmetry element of the assembly.

A detailed symmetry analysis based on spectroscopic data showed that a 6:1 stoichiometry with a D_3 point-group symmetry is the only plausible structural class of the C_{70} receptor. Thus, at the "poles" of the C_{70} receptor, three NDI molecules have to bind to each other in a C_3-symmetrical manner, requiring angles of 120° (Fig. 28). Weak CH⋯O hydrogen bonding to an imide or carboxylic acid carbonyl group could explain the chemical shift change of more than 0.7 ppm for one aromatic proton in the C_{70} receptor (Fig. 26b). Indeed, such an arrangement can give rise to a favourable trimeric interaction mode (Fig. 28), accounting for

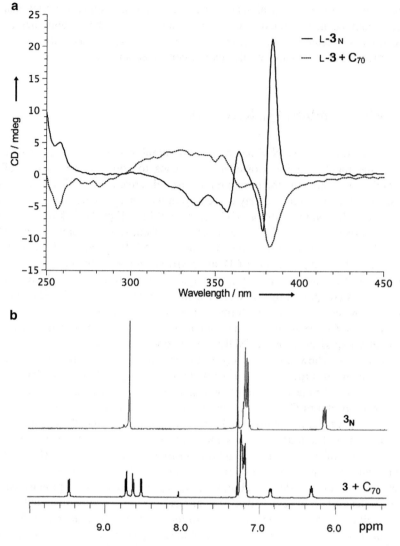

Fig. 26 (a) CD spectra of **3** in the absence and in the presence of C$_{70}$ at [**3**] = 3.86 × 10^{-4} M in CHCl$_3$; (b) ^{1}H NMR spectra of empty nanotube and C$_{70}$ receptor of **3** in CDCl$_3$

formation of one trimeric "half" of the receptor. In the proposed arrangement, consistent with D_3 point symmetry for the complex, one carboxylic residue per NDI unit would remain free for carboxylic acid dimerization with the NDI counterpart of the other hemisphere, at the equator of the C$_{70}$ receptor.

In chloroform, the monochiral NDIs were found to form the C$_{70}$ receptor in the presence of an excess of C$_{70}$. However, unlike **1**, they did not form the receptor exclusively, rather the equilibrium position between receptor and nanotube is different for each monochiral NDI, as indicated by ^{1}H NMR experiments.

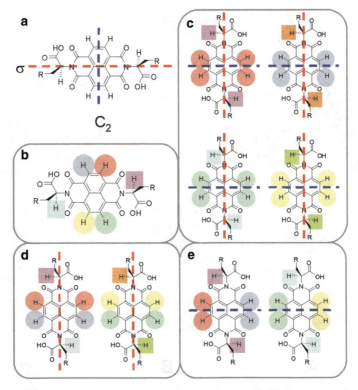

Fig. 27 (a) Symmetry elements of an NDI derivative: symmetry plane σ (*red*) and twofold axis (*blue*); (**b–e**) different combinations of symmetry elements and number of inequivalent NDI molecules that give rise to four aromatic signals [27]

As expected, **12** NDI was the most efficient at forming the hexameric capsule with 70% of the material being incorporated, while only 30% of **14** forms the receptor. Figure 28 shows a cartoon representation of the proposed geometry of the C_{70} receptor including the possible orientations for an N-desymmetrised NDI [17].

Only one α-proton signal is seen in the ^{1}H NMR spectrum of the C_{70} receptor formed from **12**, which means that all the chiral ends of the NDIs are in the same environment. It is unclear whether the single α-proton signal of the monochiral component is associated with the equatorial or axial position, but this demonstrates that the arrangement of NDIs in the receptor is ordered rather than random (Fig. 28).

7 Proton-Driven Switching Between C_{60} and C_{70} Receptors

Morphological switching between nanotube, hexameric receptor and monomers is readily achieved by simple protonation-deprotonation reactions. This system can be described as a library of dynamic, size selective fullerene receptors whose structure

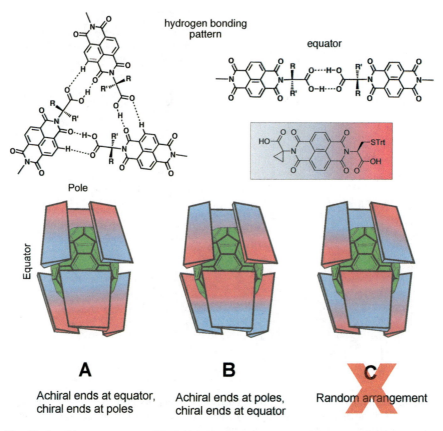

Fig. 28 Possible arrangements of NDI **12** in the C_{70} receptor. The single α-proton signal means either **A** or **B** are possible, but **C** can be eliminated

and recognition properties depend on the position of the acid/base equilibrium (Fig. 29) [28]. Unexpected differences in the sensitivity to base-induced dissociation of the nanotubes derived from different amino acids were also uncovered.

Chiro-optical studies were carried out in chloroform solution with four structurally diverse hydrogen-bonded nanotubes **1–4** using triethylamine (TEA) and methanesulfonic acid (MSA) as base and acid triggers. CD spectra of a chloroform solution of **2** (red trace) after sequential additions of base and acid are shown in Fig. 30. Addition of 1 equiv. of base caused a dramatic decrease of the CD signal intensity (green trace), attributed to the dissociation of the nanotube by the breaking of hydrogen bonds between NDI components. Subsequent addition of a stoichiometric amount of acid re-established the original spectrum (blue trace). This process is reversible, as demonstrated by the essentially complete recovery of the CD over several cycles (Fig. 30, inset). Comparison of these measurements for all four NDI derivatives revealed that the amount of base required to dissociate the nanotube architecture depends on the nature of the amino acid side chain.

Supramolecular Naphthalenediimide Nanotubes

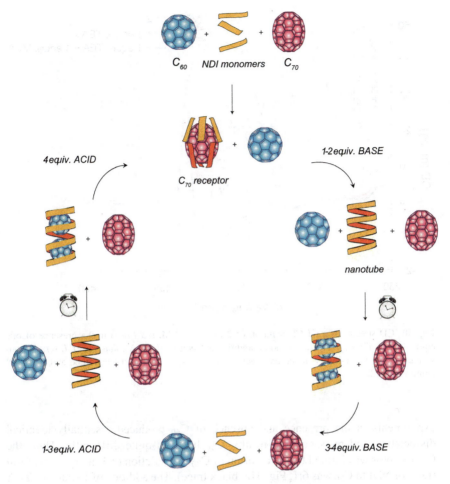

Fig. 29 Schematic representation of the proton-driven cyclic morphological switching between NDI monomers, C_{60} and C_{70} receptors (the clock indicates the slow kinetics of C_{60} uptake)

The nanotubes of **1**, **3** and **4**, all with apolar substituents, required more base (4, 2 and 2 equiv. per NDI, respectively) for complete dissociation than that needed for dissociation of **2** (1 equiv.), with a polar side chain. This may be a consequence of differences in solvation and/or creation of a more non-polar environment, which would raise the pK_a in a manner that is reminiscent of carboxylic groups in enzyme active sites [28]. It is not clear whether removal of, on average, one proton per $(COOH)_2$ link (which would still allow connection via a single, charge-assisted hydrogen bond) leads to the dissociation of the nanotubes or whether both protons need to be removed. In all cases, the nanotubes re-assembled when MSA was added to neutralise the base.

Investigations were performed to examine whether the C_{60} and C_{70} guests have any influence on the host's resistance to base-induced dissociation. While acid/base

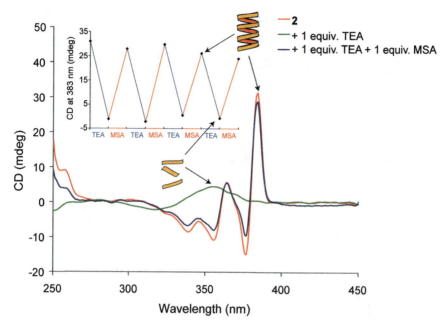

Fig. 30 CD spectra of a CHCl$_3$ solution of **2** (7×10^{-4} M, *red trace*) in the presence of one equivalent of TEA (*green trace*) and an additional 1 equiv. of MSA (*blue trace*). *Inset*: demonstration of the reversibility of the base–acid driven switch between the nanotube and free NDI components [28]

experiments in the presence and absence of C$_{60}$ produced essentially identical dissociation and re-association results, C$_{70}$ behaved quite differently. Thus, the C$_{70}$ receptor was formed by the addition of C$_{70}$ to a solution of **1** in dry chloroform (ratio of NDI to C$_{70}$ was 6:1, Fig. 31a, black trace). The addition of 1 equiv. of TEA to the **1** + C$_{70}$ complex leads to the disassembly of the C$_{70}$ receptor and the formation of a supramolecular nanotube, as indicated by the CD spectrum showing a characteristic positive signal at 383 nm (Fig. 31a, red trace). This remarkable morphological switching reveals a difference in stability of the H-bonding arrays in the C$_{70}$ capsule and the nanotube, the latter being particularly stable when derived from **1**. In the cysteine case, **1**, 1 equiv. of base (per NDI) is sufficient to destroy the C$_{70}$ receptor but not the nanotube. Presumably, a partially deprotonated nanotube may co-exist in solution with deprotonated NDIs and free C$_{70}$. The subsequent addition of a further 3 equiv. of TEA results in the decrease and finally disappearance of the characteristic nanotube CD signal at 383 nm (Fig. 31a, violet trace). The reversibility of the processes was confirmed by stepwise addition of equimolar amounts of acid, which first regenerated the nanotube, and then the C$_{70}$ receptor (Fig. 31b, orange and blue traces, respectively). Furthermore, this proton-controlled morphological switching between supramolecular architectures strongly depends on the structure of the NDI component. Thus, addition of 1 equiv. of TEA to the C$_{70}$

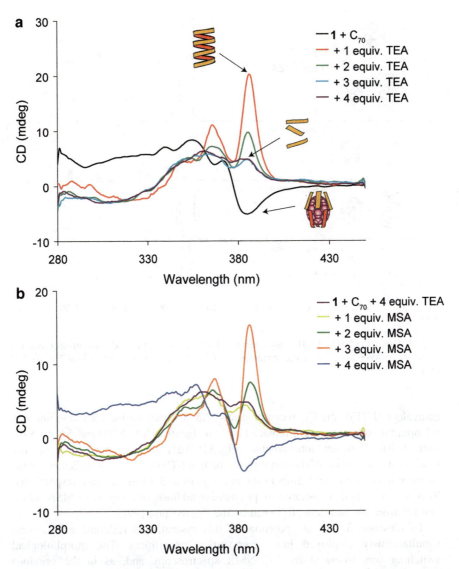

Fig. 31 Evolution in the CD spectrum of a CHCl$_3$ solution of **1** + C$_{70}$ (7 × 10^{-4} M) after addition of (**a**) 4 equiv. of TEA and (**b**) 4 equiv. of MSA [28]

receptor involving **2** resulted in complete dissociation of the supramolecular architecture giving free NDI components, completely by-passing the nanotube phase.

The nanotube, the C$_{70}$ receptor and the uncomplexed NDI molecules have distinct ^{1}H NMR spectral signatures, particularly in the aromatic region of the spectra, providing a clear window on the switching between these three architectures. Thus, starting with a solution of **3** + C$_{70}$ (ratio of NDI to C$_{70}$ was 6:1) and adding one

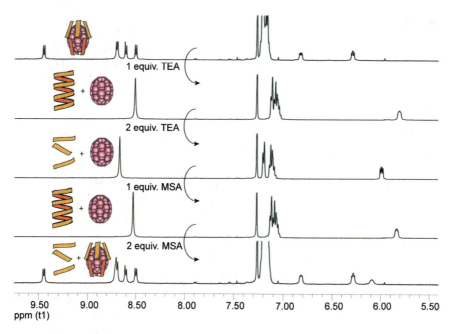

Fig. 32 Part of the 500 MHz ^{1}H NMR spectra of **3** + C_{70} showing the acid–base driven reversible switching between the C_{70} receptor, nanotube and free NDI components in CDCl$_3$ at 7×10^{-4} M [28]

equivalent of TEA, the C_{70} receptor peaks (9.5–8.4 ppm for the NDI core and 6.9, 6.1 ppm the α-protons) were replaced by two signals at 8.5 (NDI) and 5.8 ppm (α) characteristic of the nanotube structure (Fig. 32). Addition of another 1 equiv. of TEA resulted in dissociation of the nanotube to the free NDI molecules as indicated by the sharpening and downfield shifts of the two signals to 8.7 and 6.0 ppm, respectively. Reversibility was demonstrated by progressive addition of 2 equiv. of MSA which first reformed the nanotube, followed by the C_{70} receptor [29].

To illustrate further the potential of this system, two fullerene guests were simultaneously employed in a competition experiment. The morphological switching was followed by ^{13}C NMR spectroscopy and, as in the previous experiments, this showed preferential formation of the C_{70} complex over that of the C_{60}/nanotube species in the absence of base (the four signals between 143 and 148 ppm are due to the complexed C_{70}, Fig. 33a). Progressive addition of base caused, in the first instance, appearance of an additional signal at 142.9 ppm, characteristic of C_{60} within a nanotube as part of a ternary complex with triethylammonium [30] (Fig. 33b) followed by its significant amplification when more base was added (Fig. 33c). At this stage, both host–guest complexes were evident but the nanotube + C_{60} complex was dominant. Further addition of base first caused complete disappearance of C_{70} receptor signals and finally disassembly of the nanotube (Fig. 33d, e).

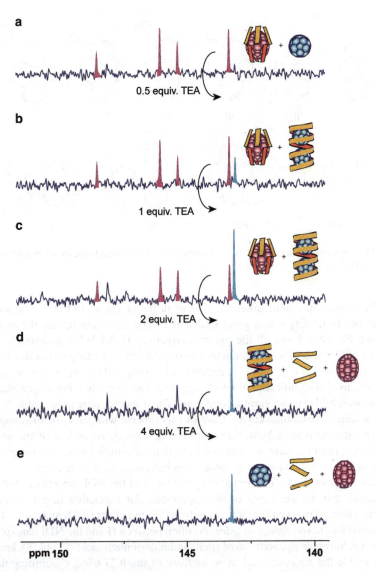

Fig. 33 Part of the 500 MHz ^{13}C NMR spectra of **3** + C_{70} + C_{60} showing the acid–base driven reversible switching between the C_{70} receptor, nanotube-C_{60} based receptor and free NDI components in CDCl$_3$ at 7×10^{-4} M

8 Complexation of Polyaromatic Hydrocarbons

Expanding our search to "fullerene fragments" we investigated the potential host–guest chemistry of the nanotubes with polyaromatic hydrocarbons shown in Fig. 34 [30].

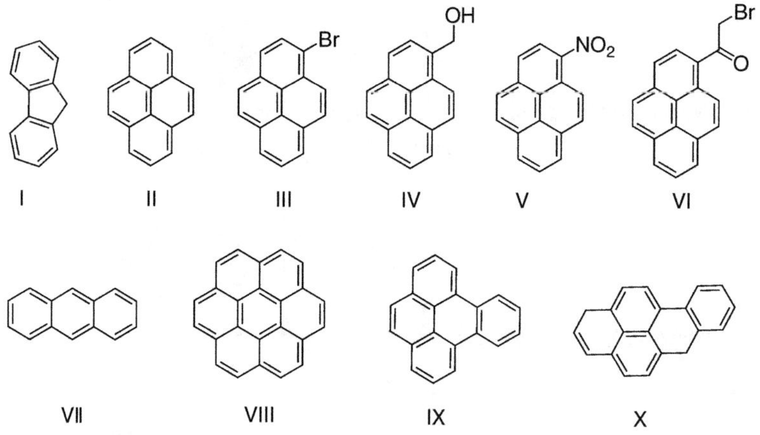

Fig. 34 Polyaromatic guests used in this investigation. (**I–VI**) were found to interact the nanotube while (**VII–X**) show no interaction

The change in colour upon addition of different guests to the preformed NDI nanotubes in CHCl₃ was a good indicator of complexation for all the examples studied. Fluorene **I** and all the pyrene derivatives **II–VI** led to a dramatic colour change of the chloroform solution of **1** from pale yellow to deep red as illustrated in Fig. 35 (inset: for pyrene), indicative of strong NDI-pyrene donor-acceptor interactions. No significant colour change was observed when the larger analogues **VII–X** were added under similar conditions. These species were presumably too large to fit within the nanotube cavity. Control experiments with the corresponding NDI-methyl ester derivative **1**-ester gave no colour change, regardless of the aromatic molecules present in solution, demonstrating that the complexation is a property of the supramolecular nanotube system rather than individual NDI monomers.

Chiro-optical studies confirmed the persistence of the NDI nanotubes in solution, indicating that the geometry of the supramolecular nanotubes is preserved upon complexation of pyrene (Fig. 36). An induced Cotton effect (ICD) at $\lambda_{max} = 440$ nm corresponding to the charge transfer transition between **II** and the NDI nanotube was observed. Surprisingly, addition of small amounts of methanol (<1 vol.%), known to disassemble the nanotube, led to an increase of the ICD band, confirming that the host–guest complexation is driven by solvophobic interactions. Increasing the MeOH concentration above 1% led to the expected destruction of the supramolecular nanotubes as manifested in a decrease in the ICD band (Fig. 36).

In a mixture containing a 2:1 molar ratio of **1** and **II** the NDI proton signal of **1** and all the protons of **II** are shifted upfield by 0.09 and 0.08 ppm, respectively, while in a similar mixture containing **1**-ester and **II** the same protons are shifted upfield by only 0.04 ppm (Fig. 37). This behaviour is consistent with the formation of an inclusion complex between **1** and **II**, where the pyrene molecules are shielded by the naphthalene aromatic cores of the NDIs, which in turn are shielded by the pyrenes complexed inside the nanotube. Spectrum *d* in Fig. 37 shows that, although there is some

Supramolecular Naphthalenediimide Nanotubes

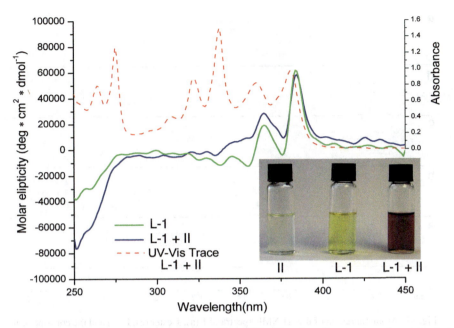

Fig. 35 CD and UV spectra of a 4:1 mixture of **1** and **II** in chloroform (*inset*: colour changes upon mixing the two solutions); [**1**] = 1.0×10^{-4} M in CHCl$_3$ at 21 °C [30]

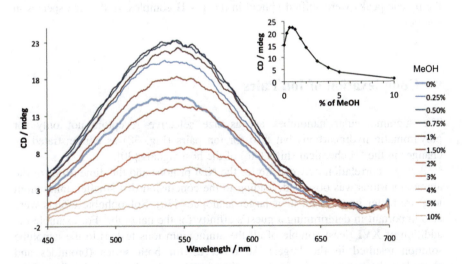

Fig. 36 CD spectra of a solution of **1** and **II** in the presence of increasing quantities of MeOH. *Inset*: change in the CD intensity at 550 nm with increasing amounts of MeOH. (**1:II** in 4:1 ratio, [**II**] = 5×10^{-3} M in CHCl$_3$ at 21 °C) [30]

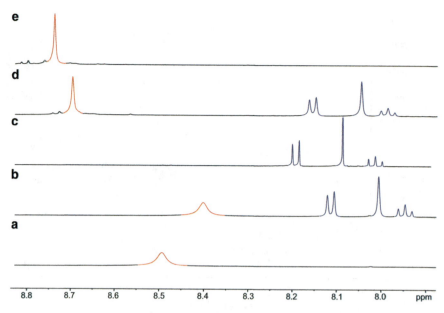

Fig. 37 Aromatic region of the ^{1}H-NMR spectra of **1** (**a**), **1**-ester (**e**), **II** (**c**) and the corresponding 2:1 mixtures: **1** + **II** (**b**) and **1**-ester + **II** (**d**). Colour coding: *red* = NDI protons of **1**, *blue* = pyrene (**II**) protons. [**1**] = [**1**-ester] = 41 × 10^{-3} M in CDCl$_3$ at 23 °C [30]

interaction between pyrene **II** and an individual NDI (**1**-ester), it is nonspecific and cannot explain the larger chemical shifts observed in the case of nanotube **1** and **II**. The ^{1}H-NMR experiments were reflected in the ^{13}C-NMR where both the NDI and the pyrene peaks were shifted upfield in the **1** + **II** complex vs the free species in solution.

9 Complexation of Ion Pairs

The supramolecular nanotubes act as size selective receptors not only for polyaromatic hydrocarbons but also for ion pairs (Fig. 38) as demonstrated by changes in the ^{1}H chemical shifts ($\Delta\delta$) of the host signals [30].

A direct correlation between $\Delta\delta$ for the NDI protons and the dimensions of the ammonium ions was observed, leading to the conclusion that smaller ammonium ions are better guests than their larger analogues [30]. Solvophobic effects were also important in determining a guest's affinity for the nanotube. For example the addition of **XVI** (least soluble of all the ammonium ions tested) to the nanotube solution resulted in the largest $\Delta\delta$ observed for both series (bromides and chlorides). Its behaviour is similar to that for C$_{60}$ whereby the solubility of the guest increases in the presence of the host. The comparison of the $\Delta\delta$ values recorded for the two series of ion pairs revealed that the size of the anion is equally

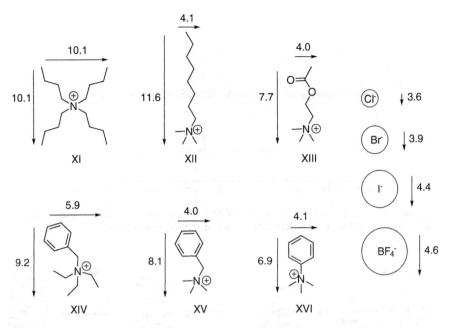

Fig. 38 Sizes (in Å) of anions and cations (ion pairs) that interact with the NDI nanotubes [25]

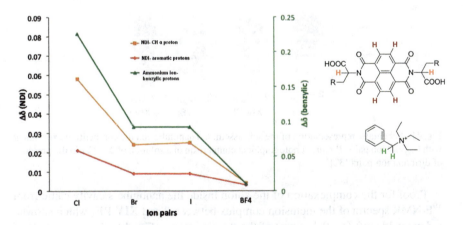

Fig. 39 Comparison of the change in ^{1}H NMR chemical shift of the NDI α proton, aromatic protons and the benzylic protons of the ammonium ion [30]

important in the complexation event (Fig. 39). Ammonium chlorides were more readily taken up by the nanotube than the corresponding larger bromides. The largest shift was observed upon addition of acetylcholine chloride (**XIII**.Cl), which is the smallest of the ion pairs tested.

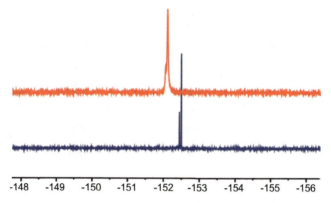

Fig. 40 ^{19}F-NMR spectra of a solution of **XIV**.BF4 in the presence (*red line*) and in the absence (*blue line*) of **2**. [**2**] = 20 mM, [**XIV**.BF4] = 5 mM at 23 °C in CDCl$_3$ [30]

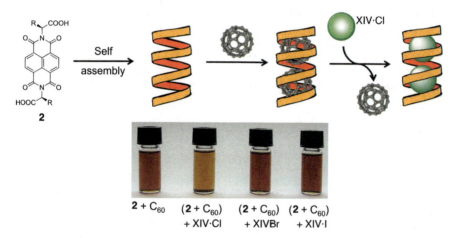

Fig. 41 Schematic representation of the self-assembled nanotube and the competition experiment with C$_{60}$ and ion pairs. *Picture*: photographical comparison of a solution of **2** + C$_{60}$ in the presence of different ion pairs [32]

Proof for the complexation of the anion inside the nanotube's cavity came from ^{19}F-NMR spectra of the inclusion complex between **2** and **XIV**·BF$_4$ which showed a downfield shift (ca. 0.41 ppm) of the fluorine atoms (Fig. 40). The magnitude of the shift observed for this system was comparable with the downfield shift observed for PF$_6$ trapped inside a tetrahedral aromatic cage (0.64 ppm) [31].

Competition experiments showed that C$_{60}$ could be displaced from the nanotube by selected ion pairs, resulting in the formation of mixed C$_{60}$-ion pair complexes [32]. The addition of 1 equiv. of **XIV**·Cl to a 10-mM solution of C$_{60}$-encapsulated **2** resulted in the precipitation of solid C$_{60}$ and a colour change of the supernatant from brown to light orange (Fig. 41). The displacement of C$_{60}$ from the nanotubular cavity was confirmed using UV–vis and CD spectroscopy; the intensity of

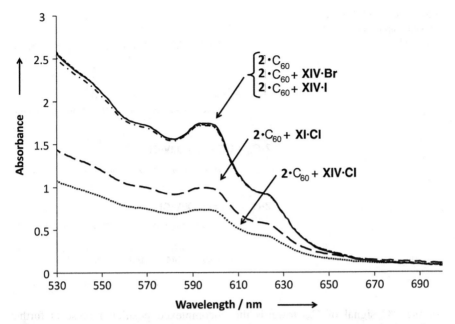

Fig. 42 UV–vis spectra of a 10 mM solution of **2** + C$_{60}$ in the presence of 1 equiv. of different ion pairs [32]

absorption and induced CD (ICD) signals at 593 and 660 nm (characteristic of C$_{60}$) decreased dramatically upon addition of 1 equiv. of **XIV·Cl** and stayed constant when **XIV·Br** and **XIV·I** were added. These results demonstrate that the interaction of **XIV·Cl** with the nanotube is stronger than that of **XIV·Br** or **XIV·I**, as the latter are incapable of displacing C$_{60}$ from the cavity of the host (Fig. 42)

Increasing the amount of **XIV·Cl** in a solution containing **2** + C$_{60}$ complex also resulted in the rapid decrease of absorption at 452 nm (attributed to fullerene–fullerene interactions in a closed-packed one-dimensional array of C$_{60}$ inside the nanotubular cavity), indicating that the encapsulation of ammonium ions led to partial disruption of the close-packed C$_{60}$ array resulting in the formation of a mixed complex ion-pair/C$_{60}$ host–guest complex, where the ion pair is intercalating between the fullerenes.

The formation of the mixed ion-pair/C$_{60}$ host–guest complex is best illustrated by ^{13}C-NMR spectroscopy. As described earlier, the ^{13}C signal of C$_{60}$ is shifted upfield by more than 1.4 ppm upon encapsulation by the nanotube due to proximal aromatic units (Fig. 43). The addition of **XIV·Cl** (0.5 and 1 equiv.) to a 10-mM solution of containing this C$_{60}$ complex in CDCl$_3$ resulted in a small but significant downfield shift and a decrease in the intensity of the ^{13}C signal, demonstrating the release of C$_{60}$ in solution. The presence of a C$_{60}$ signal even after the addition of 1 equiv. of ion pair indicated that only partial displacement of C$_{60}$ occurred; this is possible only if a mixed ion-pair/C$_{60}$ complexation takes place. The downfield shift

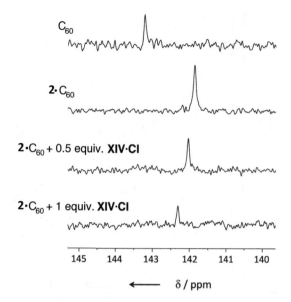

Fig. 43 C$_{60}$ ^{13}C chemical variation upon addition of **XIV·Cl** [32]

of the ^{13}C signal of C$_{60}$ towards the uncomplexed position represents further evidence for the disruption of the close packed array of C$_{60}$. The presence of the ion pair inside the tubular cavity is confirmed by the expected upfield shift observed in the ^{1}H NMR spectrum, for the cation's benzyl protons.

The reverse experiments, in which C$_{60}$ was added to solutions of **XIV·Cl** encapsulated **2**, also led to the formation of the same nanotube–fullerene–ion pair mixed complex. UV and NMR experiments confirmed that C$_{60}$ partially displaced ion pairs from the nanotube's cavity. This is indicative of the dynamic nature of the systems and the propensity of these nanotubes to form mixed complexes.

10 Conclusions

The serendipitous discovery that amino acid derived NDIs form supramolecular nanotubes in the solid state led to the unravelling of a very rich vein of supramolecular assemblies in organic solvents. This chemistry highlights how a very simple and versatile building block can form hydrogen-bonded dynamic combinatorial libraries that allow the co-existence of small oligomers, nanotubes and hexameric capsules. Achiral and mono-chiral derivatives allowed us to discover how very small changes in their structure influenced incorporation into chiral supramolecular structures. The formation of a dynamic nanoreceptor whose morphology and recognition properties can be tuned by a simple acid–base equilibrium highlighted the importance of the amino acid side chains in the formation of these supramolecular structures. The remarkable ability of NDIs to self-assemble in a receptor for C$_{60}$, C$_{70}$, polyaromatic molecules and ion pairs is unprecedented and should inspire chemists to investigate other "simple" systems whose abundant supramolecular chemistry is yet to be discovered.

References

1. Reek JNH, Otto S (2010) Dynamic combinatorial chemistry. Wiley-VCH, Weinheim
2. Lehn JM (2007) From supramolecular chemistry towards constitutional dynamic chemistry and adaptive chemistry. Chem Soc Rev 36:151–160
3. Lehn JM (1999) Dynamic combinatorial chemistry and virtual combinatorial libraries. Chem Eur J 5:2455–2463
4. Corbett PT, Leclaire L, Vial L, West KR, Weitor J-L, Sanders JKM, Otto S (2006) Dynamic combinatorial chemistry. Chem Rev 106:3652–3711
5. Herrmann A (2009) Dynamic mixtures and combinatorial libraries: imines as probes for molecular evolution at the interface between chemistry and biology. Org Biomol Chem 7:3195–3204
6. Ladame S (2008) Dynamic combinatorial chemistry: on the road to fulfilling the promise. Org Biomol Chem 6:219–226
7. Hunter CA (2004) Quantifying intermolecular interactions: guidelines for the molecular recognition toolbox. Angew Chem Int Ed 43:5310–5324
8. Crego Calama M, Hulst R, Fokkens R, Nibbering NMM, Timmerman P, Reinhoudt DN (1998) Libraries of non-covalent hydrogen-bonded assemblies; combinatorial synthesis of supramolecular systems. Chem Commun 1021–1022
9. Timmerman P, Vreekamp RH, Hulst R, Verboom W, Reinhoudt DN, Rissanen K, Udachin KA, Ripmeester J (1997) Noncovalent assembly of functional groups on calix[4]arene molecular boxes. Chem Eur J 3:1823–1832
10. Hof F, Nuckolls C, Rebek J Jr (2000) Diversity and selection in self-assembled tetrameric capsules. J Am Chem Soc 122:4251–4252
11. Wyler R, de Mendoza J, Rebek J Jr (1993) A synthetic cavity assembles through self-complementary hydrogen bonds. Angew Chem Int Ed 32:1699–1701
12. Pengo P, Pantoş GD, Otto S, Sanders JKM (2006) Efficient and mild microwave-assisted stepwise functionalization of naphthalenediimide with α-amino acids. J Org Chem 71:7063–7066
13. Tambara K, Ponnuswamy N, Hennrich G, Pantoş GD (2011) Microwave-assisted synthesis of naphthalenemonoimide and N-desymmetrized naphthalenediimides. J Org Chem 76:3338–3347
14. Anderson TW, Pantoş GD, Sanders JKM (2011) Supramolecular chemistry of monochiral naphthalenediimides. Org Biomol Chem 9:7547–7553
15. Pantoş GD, Pengo P, Sanders JKM (2007) Hydrogen-bonded helical organic nanotubes. Angew Chem Int Ed 46:194–197
16. Ponnuswamy N, Pantoş GD, Smulders MMJ, Sanders JKM (2012) Thermodynamics of Supramolecular Naphthalenediimide Nanotube Formation: The Influence of solvents, side-chains and guest templates. J Am Chem Soc doi:10.1021/ja2088647
17. Anderson TW, Pantoş GD, Sanders JKM (2010) The sergeants-and-soldiers effect: chiral amplification in naphthalenediimide nanotubes. Org Biomol Chem 8:4274–4280
18. Palmans ARA, Meijer EW (2007) Amplification of chirality in dynamic supramolecular aggregates. Angew Chem Int Ed 46:8948–8968
19. Smulders MMJ, Filot IAW, Leenders JMA, van der Schoot P, Palmans ARA, Schenning APHJ, Meijer EW (2010) Tuning the extent of chiral amplification by temperature in a dynamic supramolecular polymer. J Am Chem Soc 132:611–619

20. Smulders MMJ, Stals PJM, Mes T, Paffen TF, Schenning APHJ, Palmans ARA, Meijer EW (2010) Probing the limits of the majority-rules principle in a dynamic supramolecular polymer. J Am Chem Soc 132:620–626
21. Van Gestel J (2004) Amplification of chirality in helical supramolecular polymers. The majority-rules principle. Macromolecules 37:3894–3898
22. Van Gestel J, Palmans ARA, Titulaer B, Vekemans JAJM, Meijer EW (2005) "Majority-rules" operative in chiral columnar stacks of C_3-symmetrical molecules. J Am Chem Soc 127:5490–5494
23. Bulheller BM, Pantoş GD, Sanders JKM, Hirst JD (2009) Electronic structure and circular dichroism spectroscopy of naphthalenediimide nanotubes. Phys Chem Chem Phys 11:6060–6065
24. Green MM, Reidy MP, Johnson RJ, Darling G, Oleary DJ, Willson G (1989) Macromolecular stereochemistry: the out-of-proportion influence of optically active comonomers on the conformational characteristics of polyisocyanates. The sergeants and soldiers experiment. J Am Chem Soc 111:6452–6454
25. Cambridge Structural Database, v.5.30, 2009
26. Pantoş GD, Wietor J-L, Sanders JKM (2007) Filling helical nanotubes with C_{60}. Angew Chem Int Ed 46:2238–2240
27. Wietor J-L, Pantoş GD, Sanders JKM (2008) Templated amplification of an unexpected receptor for C_{70}. Angew Chem Int Ed 47:2689–2692
28. Harris TK, Turner GJ (2002) Structural basis of perturbed pKa values of catalytic groups in enzyme active sites. IUBMB Life 53:85–98
29. Stefankiewicz AR, Tamanini E, Pantoş GD, Sanders JKM (2011) Proton-driven switching between receptors for C_{60} and C_{70}. Angew Chem Int Ed 50:5725–5728
30. Tamanini E, Ponnuswamy N, Pantoş GD, Sanders JKM (2009) New host–guest chemistry of supramolecular nanotubes. Faraday Discuss 145:205–218
31. Glasson CRK, Meehan GV, Clegg JK, Lindoy LF, Turner P, Duriska MB, Willis R (2008) A new Fe^{II}quaterpyridyl M_4L_6 tetrahedron exhibiting selective anion binding. Chem Commun 1190–1192
32. Tamanini E, Pantoş GD, Sanders JKM (2010) Ion pairs and C60: simultaneous guests in supramolecular nanotubes. Chem Eur J 16:81–84

Top Curr Chem (2012) 322: 261–290
DOI: 10.1007/128_2011_278
© Springer-Verlag Berlin Heidelberg 2011
Published online: 15 December 2011

Synthetic Molecular Machines and Polymer/ Monomer Size Switches that Operate Through Dynamic and Non-Dynamic Covalent Changes

Adrian-Mihail Stadler and Juan Ramírez

Abstract The present chapter is focused on how synthetic molecular machines (e.g. shuttles, switches and molecular motors) and size switches (conversions between polymers and their units, i.e., conversions between relatively large and small molecules) can function through covalent changes. Amongst the interesting examples of devices herein presented are molecular motors and size switches based on dynamic covalent chemistry which is an area of constitutional dynamic chemistry.

Keywords Constitutional changes · Covalent changes · Dynamic covalent chemistry · Molecular machines · Molecular motors · Molecular switches · Polymer/monomer switches · Reversible polymers · Size switches

Contents

1 Introduction 262
2 Molecular Machines that Function Through Covalent Changes 264
   2.1 Covalent Changes as Intermediate Steps in Motional Dynamics Based on Rotation
        about Single Bonds 264
   2.2 Control of Motional Features of Molecular Shuttles based on Reversible Covalent
        Changes. Rotaxanes 273

A.-M. Stadler (✉)
Institut de Science et d'Ingénierie Supramoléculaires, Université de Strasbourg, 8 Allée Gaspard Monge, Strasbourg 67083, France

Institut für Nanotechnologie (INT), Karlsruher Institut für Technologie (KIT), 76344 Eggenstein-Leopoldshafen, Germany
e-mail: mstadler@unistra.fr

J. Ramírez
Institut de Recherche de l'Ecole de Biotechnologie de Strasbourg, UMR 7242 Biotechnologie et Signalisation Cellulaire, Groupe Oncoprotéines, Bvd Sébastien Brandt, BP 10413, 67412 Illkirch, France
e-mail: rramirez@unistra.fr

2.3 Dynamic Covalent Changes that Make Linear Molecular Motors Work. Controlling the Sense of Displacement (Walking) 276

3 Polymer/Monomer Size Switches: Switching Between Small and Large Molecular Sizes Through Reversible and Controlled Covalent Changes 280

4 Concluding remarks 283

References 285

Abbreviations

CBS	Corey–Bakshi–Shibata
CBPQT	Cyclobis(paraquat-*p*-phenylene)
DBU	1,8-Diazabicyclo[5.4.0]undec-7-ene
DCC	Dicyclohexylcarbodiimide
DMAP	4-Dimethylaminopyridine
DOSY	Diffusion-ordered Spectroscopy
IUPAC	International Union of Pure and Applied Chemistry
*m*CPBA	*m*-Chloroperbenzoic acid
MrM	Michael-retro-Michael
PDI	Polydispersion index
TFA	Trifluoroacetic acid

1 Introduction

(a) The molecular machines (e.g. shuttles, switches, molecular motors) presented in the first part of the present chapter are based on usually reversible or repeatable molecular motions that can be controlled through effectors/stimuli and generally have significant amplitude or/and directional orientation. A growing interest in this field is manifest and this is reflected in the increasing number of publications produced in the field.

Molecular machines (for reviews see [1–17]) can be of very diverse design and functioning. Several of them involve covalent changes; covalent changes (i.e. breaking and/or formation of covalent bonds; for the purpose of this chapter, only "classical" covalent bonds, excluding metal-coordinate instances, are considered) often produce changes in the constitution of the device. In IUPAC (International Union of Pure and Applied Chemistry) terminology, the constitution is "the description of the identity and connectivity (and corresponding bond multiplicities) of the atoms in a molecular entity (omitting any distinction arising from their spatial arrangement)" [18].

In the second part of this chapter are presented systems involving reversible and controlled conversions – through covalent changes – between relatively large (polymers) and small (monomers) molecules, which can be considered as size switches.

Both molecular machines and polymer/monomer switches are chemical devices that can perform a specific function. Molecular machines achieve a

Synthetic Molecular Machines and Polymer/Monomer Size Switches 263

motional function, while polymer/monomer switches achieve molecular weight and size changes (alternate increase and decrease) in response to external stimuli.

The present chapter is focused on how synthetic molecular machines and size switches can function through dynamic and non-dynamic covalent changes. The covalent bond is a strong one, much stronger than the hydrogen bond. Combining the motional dynamics of molecular machines with covalent changes requires selection of the kind of covalent bonds that may form and break (ideally they should form and break in good yield and under relatively mild conditions). Using formation and breaking of covalent bonds as a means to produce molecular machines should bring high selectivity to their functioning given that these processes would only happen under certain particular conditions.

This chapter deals with devices that do not contain units or involve species of biomolecular nature.

(b) The functioning of the motional devices herein discussed can include covalent changes conducted irreversibly and successively, but can also be based on dynamic covalent chemistry ([19]; see also [20–23]) which is a part of constitutional dynamic chemistry [24]. Moreover, the dynamic covalent chemistry can also be involved in the assembly (synthesis) of (supra)molecular machines without playing a role in their functioning. In this respect, examples of synthesis through dynamic covalent chemistry come from the field of interlocked compounds, such as catenanes and rotaxanes. Thus, in the assembly of rotaxanes (see for example [25, 26]), the dynamic features (reversible formation: hydrolysis, exchange, metathesis) of imine bonds have been particularly useful. Although very interesting, such examples [19] of the design and assembly of catenanes and rotaxanes through dynamic covalent chemistry do not really fall within the range of the current review, given that the dynamic covalent changes that occur there are not associated with the functioning of the machine.

(c) Molecular rearrangements are examples – related to switches – of motions of atoms and molecular fragments associated with covalent changes. These include prototropic tautomerization (ketone/enol, amide/imidic acid, lactam/lactim (Fig. 1a), enamine/imine), the Cope rearrangement, the Claisen rearrangement of allyl-phenyl-ethers, base-induced tautomerization of bullvalones [28], displacement of a Pt center on a stilbazole [30] (Fig. 1b), acetal migration in several sugar derivatives [31], etc.

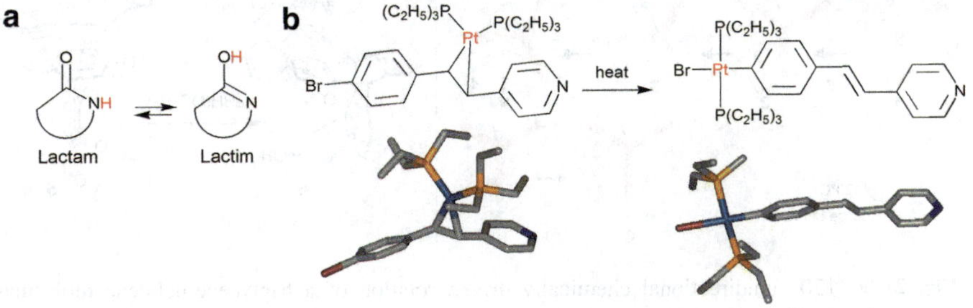

Fig. 1 (**a**) Lactam/lactim prototropic tautomerization; (**b**) displacement of a Pt center on a stilbazole

2 Molecular Machines that Function Through Covalent Changes

2.1 Covalent Changes as Intermediate Steps in Motional Dynamics Based on Rotation about Single Bonds

The covalent changes discussed in this section arise through successive chemical reactions. These sequences of reactions produce (or are designed to produce) controlled rotation-based molecular motions from a given initial station A^{1i} to a final station A^{1f}. This may be written as $A^{1i} \rightarrow A^x \rightarrow A^y \rightarrow A^z \ldots \rightarrow A^{1f}$. In a cyclic process, A^{1i} and A^{1f} are identical; A^x, A^y and A^z represent intermediates.

The control of the rotational sense of axial-rotation-based machines can operate through covalent changes in the constitution of the initial station. In the cases discussed hereafter, this type of molecular machine consists of a part of molecule that performs an axial rotation around an axis that is a single covalent bond.

Constitutional covalent changes are also performed with the aim of controlling the rotational rate.

2.1.1 Controlling the Sense of Rotation

Kelly, De Silva, and Silva [32] reported a unidirectional rotary motor (1, Fig. 2) composed of a triptycene bearing an amine group on one blade, and of a helicene having an alkyl chain with a free alcohol at one end. The triptycene and helicene are

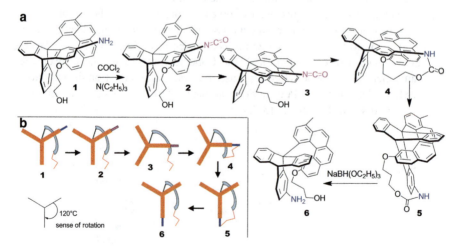

Fig. 2 (a) 120° unidirectional chemically driven rotation of a triptycene-helicene molecular motor; (b) stylized representation of the process from (a) and representation of the sense of triptycene rotation observed on the side opposite to helicene

Synthetic Molecular Machines and Polymer/Monomer Size Switches 265

connected through a single bond (Fig. 2). The motor is fuelled by chemical energy and works through successive covalent changes in its constitution. The amine is activated by reaction with phosgene. When the isocyanate **2** thus formed is close enough to the OH group connected to the helicene unit (**3**, Fig. 2), then the NCO and OH groups can react to give the urethane **4**. The isocyanate and alcohol groups from **2** and **3** react too fast and these compounds are not observed by NMR. The conformation of the resulting urethane **4** is of higher energy than that of the isocyanate **2** and at room temperature adopts, through unidirectional rotation, a form **5** where the blade bearing the NH group is behind the helicene unit. Cleavage of the urethane with sodium borohydride in ethanol gives the atropisomer **6** of the initial molecule and thus the blade bearing the NH_2 group has accomplished a unidirectional 120° clockwise rotation around a single aryl–aryl bond. A more detailed description of the motor was given later by Kelly, Silva, De Silva, Jasmin, and Zhao [33].

One may note that covalent changes that are dealt with in the functioning of this motor are, under the conditions of the reaction, irreversible (kinetically controlled reactions). Thus, the steps of this chemically driven motor do not belong to the field of dynamic covalent chemistry (that is based on covalent changes under equilibrium conditions).

The use of lactones to generate motors with unidirectional bond rotation also involves intermediate covalent changes in the constitution. Dahl and Branchaud [34] designed a biaryl lactone **7** to undergo repeatable unidirectional rotation about the aryl–aryl bond as a result of diastereoselective reactions. The functioning of the device was conceived to be as follows. The opening (90° rotation) of the lactone (Fig. 3a,b) on reaction with a nucleophile Nu^- produces two diastereomers **8** and **9**, one of them being in excess ($k_1 \neq k_3$). Subsequently, these diastereomers should interconvert thermally to reach equilibrium (270° rotation with respect to the initial position). Further they should regenerate the initial lactone (360° rotation with respect to the initial position) with different rates of lactonization ($k_2 \neq k_4$). It was established that motors able to perform very efficient directed bond rotation would require high diastereoselectivity in the opening of the lactone, fast thermal isomerization of the diastereomers relative to the other processes, and faster lactonization of the less abundant diastereomer compared to that selectively formed in the opening step (Fig. 3c). When $CH_3(CH_3O)NMgCl$ is used as a source of nucleophile for the opening of racemic **7** (Fig. 3d), two axial diastereomers D^1 and D^2 (a major one and a minor one; not shown) of the Weinreb amide **10** (Fig. 3c), in unequal amounts ($K_{eq} = 1.6$ in DMSO at room temperature) are obtained. The percentage of major diastereomer can be calculated: $K_{eq} = [D^1]/[D^2]$, then $\%D^1 = 100 \times K_{eq}/(K_{eq} + 1) = 62\%$. Trifluoroacetic acid produces lactonization of the amide **10** back to the lactone **7** (Fig. 3d). The equilibrium between the two diastereomers is too fast to allow for isolation of one of them and the direction of the rotation cannot be determined.

Lactones have been used by Flechter, Dumur, Pollard, and Feringa [35] to conceive and set up a rotary motor that uses subsequent chemical reactions to achieve a 360° unidirectional rotation (Fig. 4). The stator is a naphthalene-derived unit connected in position 1 to a phenyl ring and bearing at position 2 a group that is transformed during the motions. The rotor is a 1,3-diphenol unit connected at position 2 to the stator. In the initial station **11** of the motor, one phenol group is

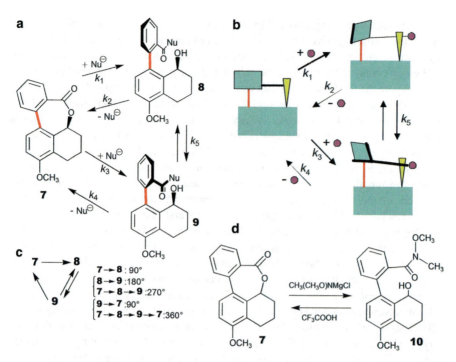

Fig. 3 (a) General scheme of a biaryl lactone opening within a molecular motor (only one enantiomer of **7** is shown); (b) stylized representation corresponding to (a); (c) functioning of the motor in a ideal case where the opening of **7** produces only **8**, the conversion between **8** and **9** is fast, and only **9** can produce the lactone **7**; (d) opening of lactone **7** (racemic) with CH$_3$(CH$_3$O)NMgCl and lactonization of the Weinreb amide **10** in the presence of TFA (trifluoroacetic acid)

protected as a *p*-methoxybenzyl ether, while the other is lactonized with the neighboring carboxyl group. The atropisomers of lactones **11** and **14** can easily racemize and each of these lactones is a mixture of interconverting atropisomers. The lactone **11** is enantioselectively (ratio 96.8:3.2) reduced (R–CO–O–R′ becomes R–CH$_2$OH and HO–R′) with (*S*)-2-methyl-CBS-oxazaborolidine, and thus a first clockwise 90° rotation is performed. The phenolic OH is allylated (**12**), then the alcohol is oxidized to an aldehyde, and further to an acid (**13**). The 180° rotation is achieved by cleavage of the *p*-methoxybenzyl ether and the subsequent lactonization (**14**). The 270° rotation is done by a new enantioselective (ratio 90.3:9.7) reduction of the lactone **14** which results in the formation of phenol and alcohol. Then, the phenol is protected as a *p*-methoxybenzyl ether (**15**). In order to perform the 360° rotation, the alcohol is oxidized to an aldehyde, then to an acid (**16**), the allylether is cleaved by using a Pd-based method, and finally the lactonization (that gives compound **11**) is achieved in the presence of DCC (dicyclohexylcarbodiimide). As can be seen, the choice of orthogonal protecting groups of the two phenolic OH groups of the rotor is crucial for the functioning of the motor.

Dahl and Branchaud [36] synthesized a biaryl lactone **17** derived from a hydroxy-diacid, that should be able to act as a molecular motor by passing through

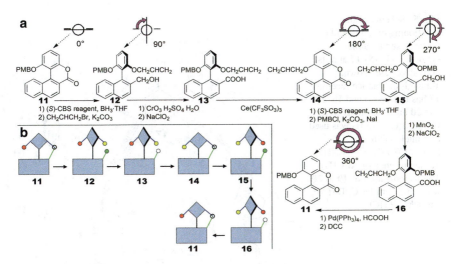

Fig. 4 A chemically driven rotary molecular motor able to achieve a 360° unidirectional rotation: (a) the sequence of chemical reactions; (b) a stylized representation of the motion

three steps – diastereoselective opening with a nucleophile, chemoselective lactonization, and chemoselective hydrolysis (Fig. 5). The racemic initial lactone **17** was diastereoselectively (at least 72%) opened with achiral reagents $C(C_6H_5)(CH_3)_2NH_2$ and $Al(CH_3)_3$. The resulting amide (consisting in principle of one minor diastereomer **18** and a major one **19**, each of them existing as a mixture of enantiomers; NMR analysis of crude amide indicated a single diastereomer) was further lactonized with DCC, DMAP (4-dimethylaminopyridine) and DMAP hydrochloride (DMAP·HCl) to give compound **20** (racemic). During these two steps, a 180° unidirectional rotation about the aryl–aryl bond took place. The chemoselective hydrolysis of the amide (unsuccessful in the presence of TFA or BF_3) as well as the synthesis of enantiopure initial lactone are under investigation.

Lin, Dahl, and Branchaud [37] designed a molecular motor (Fig. 6a) based on an achiral biaryl lactone **21** that is diastereoselectively opened with a chiral nucleophile LiNu* (Fig. 6b,c). The motor should work as follows. The opening of the lactone produces a 90° rotation (formation of compound **22** as a mixture of major and minor diastereomers, then acidification to produce **23**). Subsequent lactonization (formation of compound **24**) produces a 180° rotation with respect to the initial position. Hydrolysis of the –CONu* group of **24** (formation of the lactone **21**) followed by a new opening should produce a 270° rotation (compound **25**). The lactonization of **25** should generate a 360° rotation (formation of compound **26**). Hydrolysis of the –CONu* group of **26** will regenerate the starting compound **21**.

The initial lactone **21** was diastereoselectively opened with an excess of LiNu* nucleophiles, namely lithio(*S*)-1-phenylethanamine (Fig. 6b; the diastereomeric ratio of **22** was 3:1) and lithio(1*R*,2*S*,5*R*)-2-isopropyl-5-methylcyclohexanol (Fig. 6c; the diastereomeric ratio of **22** was 3:2). The acidification (with 2 N HCl) of the mixture of diastereomeric lithium salts **22** thus obtained led directly to the

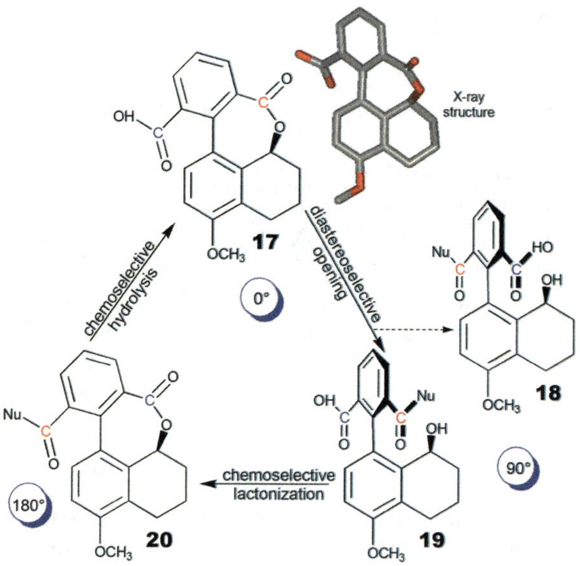

Fig. 5 Scheme of the functioning of a molecular motor based on a biaryl lactone. Only (*S*)-**17** and its derivatives are shown, although in practice racemic **17** has been used and **18, 19,** and **20** are racemic as well. $C(C_6H_5)(CH_3)_2NH_2$ was used as a nucleophile (Nu = $C(C_6H_5)(CH_3)_2NH$) in the presence of $Al(CH_3)_3$. The lactonization of **19** to **20** was done with DCC, DMAP, and DMAP·HCl

lactone of type **24** that contains the chiral group Nu* (in some cases and under certain conditions, the intermediate **23** could be observed). Within the sequence of reactions consisting of the diastereoselective opening of the lactone by the chiral nucleophile and the further lactonization in presence of acid, a 180° rotation occurred. The further functioning of the motor up to a 360° rotation requires identification of reagents and conditions for selective cleavage of the chiral nucleophile Nu* from the lactone **24**.

2.1.2 An oscillatory system

Mock and Ochwat set up a "simple oscillatory motor that spontaneously operates in solution" [38] and involves a rotary motion. It consists of the conversion of an enantiomer **27a/27b** into its mirror image **27b/27a** (Fig. 7), when acting as a catalyst for the hydration of the alkylketenimine **28a** to the corresponding carboxamide **28e**. This catalyst is the anhydride of the acid $HOOC–CH_2–O–C(CH_3)(COOH)_2$. The catalyzed reaction is $CH_3COCH=C=NC(CH_3)_3 + H_2O \rightarrow CH_3COCH_2CONHC(CH_3)_3$, the fuel of the motor being the alkylketenimine. The operation of the device requires $k_2, k_4 > k_3[CH_3COCH=C=NC(CH_3)_3] > k_1[H_2O]$. The process can be monitored by spectrophotometric measurements. This configurational change occurs through intermediate covalent changes (Fig. 7a, compounds **28b, 28c,** and **28d**) in the constitution of the initial molecule, including ring-opening and cyclization. Secondary reactions may occur along with the reactions involved in the functioning of the device, but they are kinetically negligible in the present case.

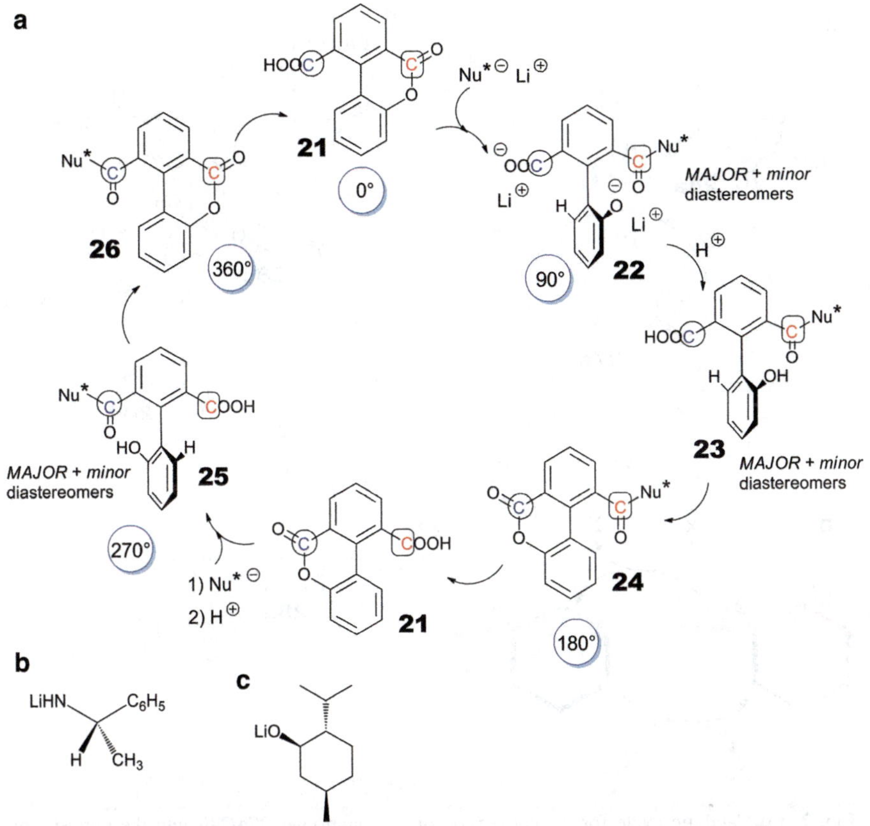

Synthetic Molecular Machines and Polymer/Monomer Size Switches

Fig. 6 (**a**) A cycle proposed for a molecular motor based on the opening of an achiral lactone **21** with a chiral nucleophile LiNu*; compound **22** is a mixture of diastereomers (same should hold for **23** and **25**) of which only one diastereomer is shown; here, **23** is the same as **25**, and **26** is the same as **24**; (**b**) structural formula of lithio(*S*)-1-phenylethanamine (a LiNu* nucleophile); (**c**) structural formula of lithio(1*R*,2*S*,5*R*)-2-isopropyl-5-methylcyclohexanol (a LiNu* nucleophile)

2.1.3 Controlling the Rate of Rotation

A point of interest in molecular machines is the control of the rotational speed. The modulation of the rotation rate around an N–Ar bond was achieved through a redox-mediated molecular brake by Jog, Brown, and Bates [39]. This consists of the oxidation of sulfide group close to the rotational axis (compound **29**) into a sulfoxide **30** and further into a sulfone **31** (Fig. 8a), so engendering successive covalent constitutional changes in strategic places of the molecule, changes that decrease the rotation. The N–Ar rotational barriers of the sulfone **31** (about 13.6 kcal mol^{-1}) and sulfoxide **30** (13.6 kcal mol^{-1}) are about 5 kcal mol^{-1} higher than in the sulfide **29** (8.6 kcal mol^{-1}). The rate of rotation is reduced from the sulfide **29** (1.6×10^6 s^{-1}) to the sulfoxide **30** (1.3×10^2 s^{-1}), or from the sulfide **29** (1.6×10^6 s^{-1}) to the sulfone **31** (1.4×10^2 s^{-1}) by a factor of about 10^4 s^{-1}.

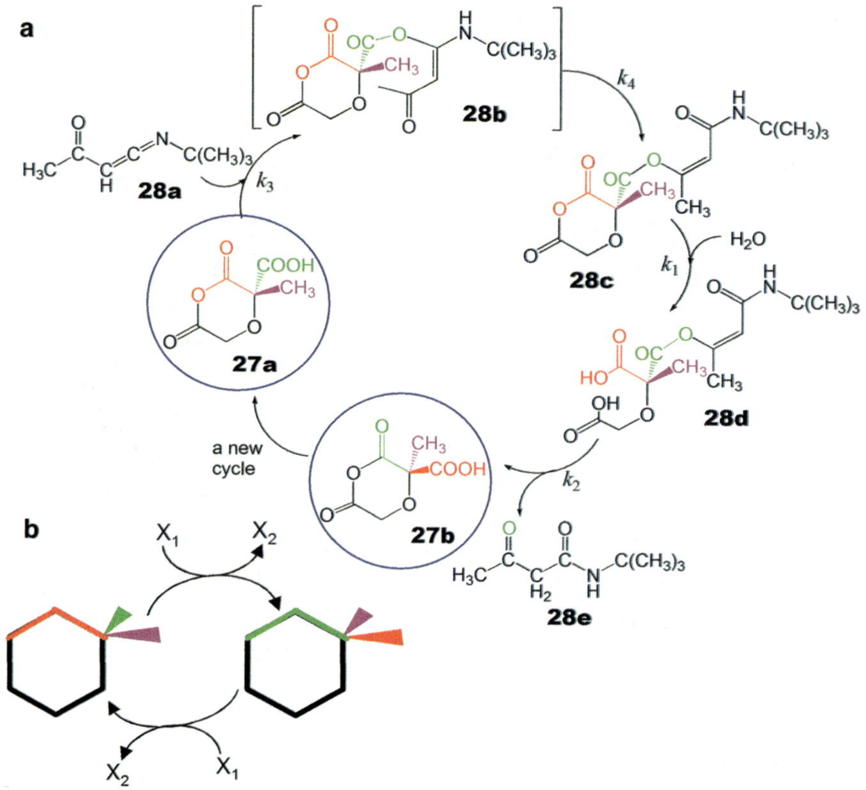

Fig. 7 (**a**) Catalytic cycle for the conversion of an enantiomer **27a/27b** into the opposite one **27b/27a** within an oscillatory device; (**b**) stylized representation of the principle of the device. $X_1 = CH_3COCH=C=NC(CH_3)_3 + H_2O$ and $X_2 = CH_3COCH_2CONHC(CH_3)_3$

The oxidation is done with m-chloroperbenzoic acid (mCPBA). The sulfoxide **30** is reduced to the corresponding sulfide **29** with Lawesson's reagent [40].

2.1.4 Molecular Switches

Zehm, Fudickar, and Linker [41] set up molecular rotary switches of bisarylanthracenes flipped by covalent binding [42], then release of oxygen (Fig. 8b,c). The stator is an anthracene unit and the rotor is an o-substituted benzene. The presence of the substituents on benzene rings prevents free rotation around aryl–aryl bonds. An example of how the switch works is shown in Fig. 8b,c for bis(o-methoxytolyl) anthracene. The *trans* isomer **32** reacts with singlet oxygen $^1O_2$ (generated by irradiation of molecular oxygen in the presence of a sensitizer, or from sodium molybdate and hydrogen peroxide) and produces the corresponding peroxide exclusively as *cis* isomer **33**, through a 180° rotation around the aryl–aryl bond.

Synthetic Molecular Machines and Polymer/Monomer Size Switches 271

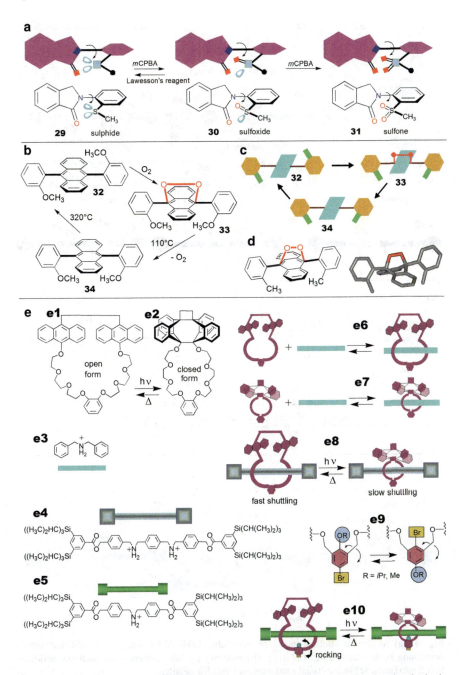

Fig. 8 (a) Chemical and stylized representation of the strategy of redox-mediated molecular brake passing from sulfide to sulfoxide and sulfone; (b) an example of oxygen-flipped rotary switch; (c) its stylized representation; (d) X-ray structure of a bisarylanthracene peroxide (H atoms were omitted for clarity); (e) control of the frequency of molecular motions in rotaxanes of which annulus (macrocycle) contains a photoisomerizable dianthrylethane group (see text for details)

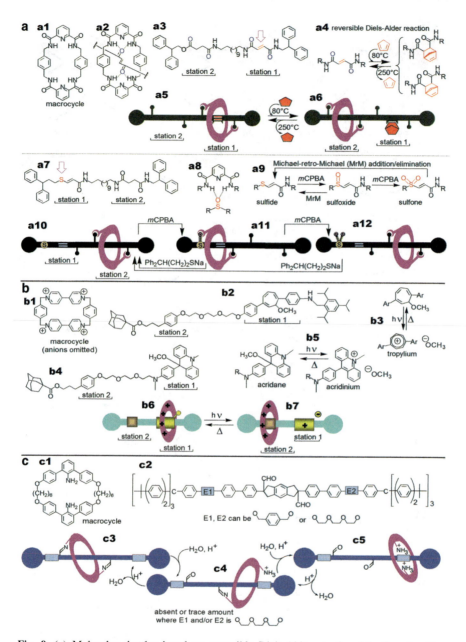

Fig. 9 (**a**) Molecular shuttles based on reversible Diels-Alder reaction (a1–a6) and sulfur-containing molecular shuttles (a7–a12); (**b**) methoxy-cycloheptatriene- and methoxy-acridane-based rotaxanes; (**c**) imine-based rotaxanes (see text for details)

The X-ray structure of such a peroxide (Fig. 8d) has been published by Zehm, Fudickar, Hans, Schilde, Kelling, and Linker [43]. Thermolysis (110 °C) of the peroxide produces the *cis* isomer of the initial molecule **34**. Its heating at 320 °C

Synthetic Molecular Machines and Polymer/Monomer Size Switches

produces thermal conversion to the initial *trans* isomer **32** through a new 180° rotation. The sense of the rotation is not controlled during these processes.

2.2 Control of Motional Features of Molecular Shuttles based on Reversible Covalent Changes. Rotaxanes

2.2.1 Controlling the Frequency of Molecular Motions in [2]Rotaxanes and Pseudorotaxanes

Hirose, Shiba, Ishibashi, Doi, and Tobe [87–89] demonstrated that reversible covalent changes induced by external physical stimuli (namely, by light) can result in changes of the size of the annulus (the macrocycle) of a pseudorotaxane or rotaxane. This size change (contraction of the macrocycle when exposed to light) allows modulation of features (in particular, the frequency) of the motional dynamics of the device. The open form of the macrocycle (Fig. 8e1) contains a dianthrylethane moiety that can photoisomerize (irradiation with a 500 W high-presure mercury lamp) through an efficient cycloaddition. It is this which reduces the size of the macrocycle. The photoisomerization (Fig. 8e1 and 8e2) is thermally reversible. The consequences of the contraction of the macrocycle are the following:

- in pseudorotaxanes with dibenzylammonium (Fig. 8e3), the frequency of threading and dethreading motions (Fig. 8e6 and 8e7) is higher with the macrocycle before photoisomerization (the open form) [87];
- the shuttling rate is slower (e.g. at least 45 times in d_8-toluene) for the rotaxane (Fig. 8e8; for the axle of the rotaxane, see Fig. 8e4) where the macrocycle is in its open form (Fig. 8e1) than for the rotaxane where it is in its closed form (Fig. 8e2) [88]. This molecular brake function is thermally reversible;
- where the *ortho*-phenylene unit of the macrocycle (Fig. 8e1) from the above studies is replaced by a *meta*-phenylene unit substituted with an alkoxy group OR at position 2 and with a Br atom at position 5 (Fig. 8e9), then the rocking motion of the *meta*-phenylene unit (Fig. 8e9, 8e10; for the axle of the rotaxane, see Fig. 8e5) is at least four times faster in the open form than in the closed form of the macrocycle where R is CH_3 (Fig. 8e9) and at least one thousand times faster in the open form than in the closed form when R is $CH(CH_3)_2$ [89].

Other photochromic devices (namely, switches that do not involve rotaxanes) where covalent changes occur are known [90].

2.2.2 Controlling the Position of Macrocycles in [2]Rotaxanes

There are examples of rotaxanes acting as molecular shuttles that work through covalent changes occurring reversibly and in response to external stimuli.

Leigh and Pérez [91] reported the first example of molecular shuttle that functions through reversible C-C bond formation due to a reversible Diels-Alder reaction

(Fig. 9a4) between cyclopentadiene and a C=C bond of the dumbbell-shaped part of the rotaxane. The dumbbell-shaped part contains two dicarbonyl stations (Fig. 9a3), one derived from fumaric acid (*trans* -CO-CH=CH-CO-, station 1), the other derived from succinic acid ($-CO-CH_2-CH_2-CO-$, station 2). The two diamide sites of the macrocycle can form four H-bonds with the two carbonyl groups of a given station (Fig. 9a1; for the interaction of the two carbonyl groups of fumaric-acid-derived station 1 with the four NH groups of the macrocycle through four H-bonds, see Fig. 9a2). Station 1 (derived from fumaric acid) has a *trans* C=C double bond; due to its preorganization, this station interacts with the macrocycle better than the station 2. Consequently, the macrocycle is initially located at station 1 (Fig. 9a5). The Diels-Alder cycloaddition (80°C, 90% yield) of cyclopentadiene to the double bond of station 1 results in a mixture of diastereomers (Fig. 9a4) and causes displacement of the macrocycle from station 1 to station 2 (Fig. 9a6). The cycloaddition is reversible and the retro-Diels-Alder reaction occurs quantitatively (250°C, reduced pressure) when cyclopentadiene dissociates from the axle of the rotaxane; this produces a displacement of the macrocycle from station 2 back to station 1.

In the above example, the rotaxane axle has two stations (sites) that can interact with the macrocycle, the interaction being stronger for one of the two stations; the best station was alternately occupied with a diene, then unoccupied and this produces the displacement of the macrocycle from one station to another. It is also possible to alternately create and disable the best binding site. Altieri, Aucagne, Carrillo, Clarkson, D'Souza, Dunnett, Leigh, and Mullen [92] invented a molecular shuttle based on this strategy (Fig. 9a7-9a12). The two stations of the axle (Fig. 9a7) in the initial step (Fig. 9a10) are *trans* -S-CH=CH-CO- (station 1) and -COCH$_2$CH$_2$CO- (station 2), the better for interacting through four H-bonds with the macrocycle (Fig. 9a1) being station 2. Oxidation of the sulfide from station 1 with *m*CPBA (Fig. 9a9) produces a sulfoxide; thus, station 1 becomes *trans* -SO-CH=CH-CO- which interacts with the macrocycle stronger than station 2, thus producing the displacement of the macrocycle from station 2 to station 1 (Fig. 9a11). Further oxidation of the sulfoxide yields the corresponding sulfone at station 2 which is now less effective for the interaction with the macrocycle than station 1. Consequently, the macrocycle goes predominantly to station 1 (Fig. 9a12). In the presence of an excess of Ph$_2$CHCH$_2$CH$_2$SNa, both sulfoxide and sulfone are reconverted into the initial sulphide via a Michael-retro-Michael (MrM) addition/elimination (equivalent to a substitution) process. For each reaction the yield is at least 90%.

The position of the macrocycle on the thread of a rotaxane can be controlled through repulsive electrostatic interactions generated in a reversible way. Photoinduced heterolysis of rotaxanes with diaryl-methoxy-cycloheptatriene- and aryl-alkoxy-acridane-based molecular threads was used in the group of Abraham to demonstrate this concept. Abraham, Grubert, Grummt, and Buck [93] synthesized a rotaxane composed of the tetracationic macrocycle cyclobis(paraquat-*p*-phenylene) CBPQT (Fig. 9b1) and a thread that incorporated a diaryl-methoxy-cycloheptatriene unit (Fig. 9b2); the thread was folded. In the initial situation (Fig. 9b6), the macrocycle is located at the diaryl-methoxy-cycloheptatriene station (station 1), but due to folding, it interacts also with the aromatic station 2 (for the sake of convenience, the thread in

Fig. 9b2 is represented in an unfolded form). Heterolytic photolysis (360 nm) of the diaryl-methoxy-cycloheptatriene unit (Fig. 9b3) produces a tropylium ion, the lifetime of the ionic state being 15 s. Repulsive interactions with the tetracationic macrocycle (Fig. 9b1) produce the displacement of the latter from station 1 to station 2 (Figs. 9b2 and 9b7). The process is thermally reversible, and it showed no fade after ten cycles.

Structural modifications (including introduction of an acetylene unit) of the thread of this type of rotaxane by Schmidt-Schäffer, Grubert, Grummt, Buck, and Abraham [94], allowed preparation of photoswitchable rotaxanes with an unfolded molecular thread. As an alternative to photolysis, treatment of the rotaxane containing the diaryl-methoxy-cycloheptatriene unit with TFA produces the diaryl-tropylium cation, and addition of solid $NaHCO_3$ in methanol can then be used to regenerate the diaryl-methoxy-cycloheptatriene station of the rotaxane.

Abraham, Wlosnewski, Buck, and Jacob [95] invented a new type of photoswitchable rotaxane based on the above principles where the diaryl-methoxy-cycloheptatriene unit is replaced by 9-aryl-9-methoxy-acridanes (Fig. 9b4) that undergo photoheterolysis (313 nm light) with formation of acridinium ions (Fig. 9b5). Duo, Jacob, and Abraham [96] demonstrated that such devices can be deposited and can operate on gold nanoparticles.

Vetter and Abraham [97] designed and synthesized rotaxanes where the axle has a 9-aryl-9-methoxy-acridane at each end, and where, consequently, the macrocycle CBPQT shuttles between one end and another. Where the axle has one aromatic ring in the central position, addition of acid produces acridinium cations at the ends of the axle, and thus forces the macrocycle to occupy the central position due to interactions with the uncharged central aromatic ring. Where the axle has two aromatic rings in the central position, addition of acid forces the macrocycle to shuttle between these two. The process can be reversed by addition of base.

Lability of dynamic covalent imine bonds towards acid-catalysed hydrolysis has been used to control the submolecular mobility of a rotaxane annulus by immobilization of the macrocycle at a given position on the axle. Kawai, Umehara, Fujiwara, Tsuji, and Suzuki [98] showed that a bis-imine rotaxane (Fig. 9c3) composed of a macrocycle with two amino groups (Fig. 9c1) and a dialdehyde axle (Fig. 9c2; E1 = E2 = p-$OCH_2C_6H_4CH_2O$) undergoes acid-catalyzed hydrolysis to the corresponding aldehyde-ammonium-monoimine (Fig. 9c4) and dialdehyde-diammonium (Fig. 9c5; in this case, the annulus shuttles between the two p-$OCH_2C_6H_4CH_2O$ units) rotaxanes. The hydrolysis results in a mixture of the three rotaxanes (Fig. 9c3-9c5) where the ratio of the dialdehyde-diammonium increases on decreasing the temperature from 40°C to −40°C. This suggests that the hydrolysis of imines is in this case enthalpy-driven, while their formation is entropy-driven. Thus, within this rotaxane the position and the mobility of the macrocycle can be controlled through imine formation and hydrolysis, as well as through temperature.

Umehara, Kawai, Fujiwara and Suzuki [99] showed that under hydrolytic conditions at 22-25°C the intermediate rotaxane of type aldehyde-ammonium-monoimine (Fig. 9c4) does not form or form in trace amounts when the equilibrium of hydrolysis is displaced towards the dialdehyde-diammonium rotaxane (Fig. 9c5) due to hydrogen bonds between ammonium and the triethylenglicol ether units (E1 and/or

E2 is $OCH_2CH_2OCH_2CH_2OCH_2CH_2O$). In this case, under dehydrating conditions, the dialdehyde-diammonium rotaxane (Fig. 9c5) regenerate the bis-imine rotaxane (Fig. 9c3). It is thus possible to control the position of the macrocycle by alternating hydrolytic and dehydrating conditions. Moreover, a control through temperature was demonstrated as well. Indeed, under hydrolytic conditions, at $100°C$ there is more than 95% bis-imine (Fig. 9c3), while at $0°C$ there is more than 95% dialdehyde-diammonium (Fig. 9c5).

2.3 Dynamic Covalent Changes that Make Linear Molecular Motors Work. Controlling the Sense of Displacement (Walking)

Concerning the chemical systems that undergo dynamic structural changes, Barboiu and Lehn [44] identified and defined three types of behavior, the third one being a combination of the first two. The first concerns *morphological dynamics*, where "molecular or supramolecular entities undergo reversible shape changes triggered by external stimuli, resulting in motional ('mechanical') processes." The second concerns *constitutional dynamics*, where "molecular or supramolecular entities undergo reversible changes in constitution by exchange, incorporation, or decorporation of components, generating pools of interconverting species such as occur in dynamic combinatorial chemistry." A type of system that has features of both the preceding types involves *combined constitutional and motional dynamics*, where "constitutional modifications by exchange, addition, or subtraction of components cause structural changes that result in motional effects by an interplay of self-assembly and disassembly, growth and decay." Lehn [24] showed that dynamic chemical processes (dynamic chemistry) encompass the areas of reactional, motional and constitutional dynamics; constitutionally dynamic chemistry encompasses both dynamic covalent and non-covalent (supramolecular) chemistry.

The small-molecule-based machine conceived by von Delius, Geertsema, and Leigh [45] is a linear (for reviews, see [46], [100]) motor based on dynamic covalent chemistry [19–24] (forming, breaking, and reforming of dynamic covalent bonds with relatively fast equilibration in response to stimuli), namely on acyl-hydrazone and disulfide exchanges. The motor consists of a "track" that has four functional groups disposed alternately aldehyde–thiol–aldehyde–thiol which are the positions 1, 2, 3, and 4 of the track, a "walker" $NH_2–NH–CO–(CH_2)_5–SH$ which has the feet A (hydrazide or acyl-hydrazine) and B (thiol), and a "placeholder" with a foot C of type thiol (Fig. 10).

In the initial station **35**, the walker is connected to the track through a hydrazone 1-A and a disulfide (for a hydrazone-disulfide macrocycle and its dynamic covalent ring-opening under acidic and basic conditions, see [47]) 2-B bond, while the placeholder $HSCH_2CH_2COOCH_3$ (the source is the corresponding disulfide) is connected through a disulfide bond 4-C to the other thiol group of the track

Synthetic Molecular Machines and Polymer/Monomer Size Switches

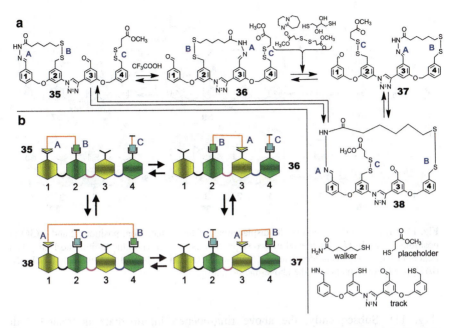

Fig. 10 (**a**) Operation of a linear motor based on hydrazone and disulfide exchanges; the conversion between **37** and **38** occurs in the same conditions as that between **35** and **36**, and that between **35** and **38** in the same conditions as that between **36** and **37**; (**b**) stylized representation of the operation of the linear motor

(Fig. 10). Under acid catalysis (TFA), there is intramolecular exchange between the initial hydrazone 1-A and the aldehyde at position 3, resulting in the formation of a new hydrazone 3-A, and in the displacement of the walker on the track (**36**; Fig. 10); the reaction is not complete and some of the initial compound is still present (**35**:**36** = 51:49). Further addition of base (1,8-diazabicyclo[5.4.0]undec-7-ene = DBU), D, L-dithiothreitol, and of the disulfide of the placeholder (SCH$_2$CH$_2$COOCH$_3$)$_2$ results in new exchanges, and the new disulfide bonds 2-C and 4-B form, which results in a new displacement of the walker (**37**; Fig. 10). In the same step, the compound **38** (Fig. 10) also forms and all four isomers are present in the reaction mixture (**35**:**36**:**37**:**38** = 45:36:11:8). All four constitutional and positional isomers are present as well at the steady state (**35**:**36**:**37**:**38** = 39:36:19:6) reached after several alternate cycles of hydrazone and disulfide exchanges. The conversion of one isomer into the subsequent isomer resulting from the displacement must occur through a ring-opened intermediate (such an intermediate **39** is shown in brackets in Fig. 11).

In order to confer a directional bias to the motion of the walker, the step that induced the migration of the thiol end of the walker (conversion of **36** into **37**) is replaced by a two-step sequence. First, the positional isomer **36** (Fig. 11) is treated with DBU and D,L-dithiothreitol to generate the intermediate **39** where the thiol groups of the track are free. Now, the thiol group of the walker is also free and the walker is connected to the track solely through a hydrazone bond 3-A

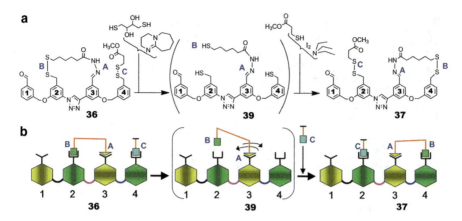

Fig. 11 (a) A directional bias of the displacement of the walker from positions 2 and 3 (**36**) to positions 3 and 4 (**37**) is achieved in two steps: treatment with D,L-dithiothreitol and DBU, then treatment with the placeholder disulfide (SCH$_2$CH$_2$COOCH$_3$)$_2$, I$_2$, and triethylamine (C$_2$H$_5$)$_3$N; (b) stylized representation of the above-mentioned sequence

(Fig. 11). Subsequently, the above ring-opened intermediate is treated with HSCH$_2$CH$_2$COOCH$_3$, I$_2$, and (C$_2$H$_5$)$_3$N, which produces, under kinetic control, an oxidation with formation of the positional isomer **37**. Thus, hydrazone exchange under acid catalysis, followed by the new conditions for disulfide exchange, then by a new hydrazone exchange, results in a mixture where **37** is in a higher amount (**35**:**36**:**37**:**38** = 24:24:43:9) than that obtained at the steady state.

This device is highly processive (as shown by labeling experiments; this means that during the motional process the walker does not detach from its track or, as defined in molecular biology, it is able "to bind to a filament and take successive steps" [48]), and operates directionally, repeatedly, and progressively.

The walker from the above-mentioned linear motor **35** (Fig. 10) synthesized and studied in the group of Leigh is of the type NH$_2$–NH–CO–(CH$_2$)$_n$–SH, where $n = 5$. Von Delius, Geertsema, Leigh, and Tang [49] investigated the role of the number of methylene groups of the walker (i.e., the spacer between the thiol and acyl-hydrazone) in the walking process. Where $n = 2$ and 3, the walker is too short and cannot really walk along the track. Where $n = 4$, 5, or 8, the walker can move from a foothold of the track to another repeatedly, directional bias being achieved for $n = 4$ and 5 (by alternating acid catalysis of hydrazone exchange with reduction of disulfide bonds, followed by oxidation to disulfide of SH groups formed in the reduction).

Barrell, Campaña, von Delius, Geertsema, and Leigh [50] combined the features of the above type of linear molecular motor with light induced $E \rightarrow Z$ and $Z \rightarrow E$ alkene isomerization. They synthesized and studied a walker-on-track-like light-driven molecular motor, where positions 2 and 3 of the track are connected through a CH = CH unit (Fig. 12). It can operate in either direction, depending on the order in which the stimuli are applied to it.

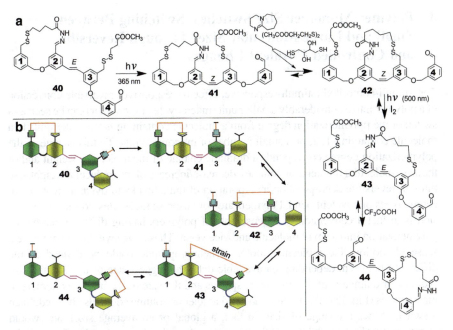

Fig. 12 (a) Operation of a linear motor driven by light and chemical stimuli; (b) stylized representation of the reactions shown in (a)

In the sequence where the walker NH$_2$–NH–CO–(CH$_2$)$_4$–SH moves from left to right (Fig. 12, i.e., from positions 1, 2 to positions 3, 4), the molecule **40** that corresponds to the initial station of the motor is the *E* isomer. After photochemical *E* → *Z* isomerization (*Z:E* = 88:12), treatment of the *Z* isomer **41** with DBU, placeholder disulfide and D,L-dithiothreitol produces the migration of the walker to positions 2, 3 (1,2 : 2,3 = 40:60). The *Z* isomer **42** with the walker at positions 2, 3 is converted with light (500 nm) and iodine to the *E* isomer **43** (*Z:E* = 75:25). This conversion produces a strain within the macrocycle that contains the walker. Treatment of this isomer with TFA (to induce acyl-hydrazone exchange) causes migration of the walker to positions 3 and 4 (compound **44**). This isomer **44** is greatly favored over the preceding one **43** where the strain of the macrocycle containing the walker is present (Fig. 12). In this way, the direction of walker migration was biased from left to right (**44**:**43** > 95:5). The displacement of the walker from right to left was achieved through the *E* → *Z* photoisomerization of **44** (365 nm), followed by treatment with TFA, then by a *Z* → *E* photoisomerization (500 nm) in the presence of I$_2$, and finally by a disulfide exchange under basic conditions, the final product that results from this sequence being **40**.

This molecular motor is an example that illustrates how the changes in configuration can have an influence on the exchanges of dynamic covalent bonds like –C=N– from acyl hydrazones and –S–S– (disulfides).

3 Polymer/Monomer Size Switches: Switching Between Small and Large Molecular Sizes Through Reversible and Controlled Covalent Changes

The stimuli-controlled (stimuli-responsive, effector-responsive) reversible conversion of a polymer into a considerably smaller unit (macrocycle or monomer) can be seen as a switch of the polymerization degree from a molecule containing n monomer units to a molecule containing 1, 2, or a small number of monomer units. To this change of the polymerization degree corresponds a dramatic change of the molecular weight and of the size (molecular diameter or volume): decrease/increase of the molecular weight/size through reversible decorporation/incorporation of units thanks to dynamic features of certain types of covalent bond. Polymers are polydisperse molecules, so such a conversion is in fact a conversion of an ensemble of polymers having different degrees of polymerization into a small molecule and vice versa. These size switches do not have a sense of motion like molecular motors. Their amplitude could be defined as the difference between, depending on available data, the biggest size or the average size (volume or diameter) of the polymeric species and the size of the initial monomers or macrocycles (Fig. 13). The hydrodynamic diameter or volume could also be used. Such a switch allows a change of size (in fact, a global or an average size), and would consequently allow modulation of the processes where the size can be a key factor. If the monomeric units bear groups with a given function, the reversible and controlled conversion between the polymer and the monomer also results in the reversible modulation of the multivalency (for reviews on multivalent molecules see, for example [101–103]).

Once such reversible systems have been identified, it is worth the effort to find conditions for their modulation through external stimuli, ideally in a repetitive way. Examples of reversible systems come from the field of dynamic covalent chemistry [19–24] which involves covalent changes with relatively fast equilibration.

Marsella, Maynard, and Grubbs [51, 52] showed (Fig. 14a) that the macrocycle **47** (Fig. 14a) undergoes (>95%) ring opening metathesis polymerization ROMP (for reviews, see [53–56]) in the presence of a Ru catalyst (Fig. 14a) and produces

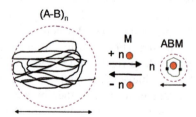

Fig. 13 The reversible conversion (here, mediated by metal ions M) of a copolymer $(A–B)_n$ into a much smaller molecule (here, a macrocycle ABM) can be seen as a size-switch which consists of a dramatic change of molecular size (molecular diameter) modulated through external stimuli, and operates through dynamic covalent chemistry. See also Fig. 15

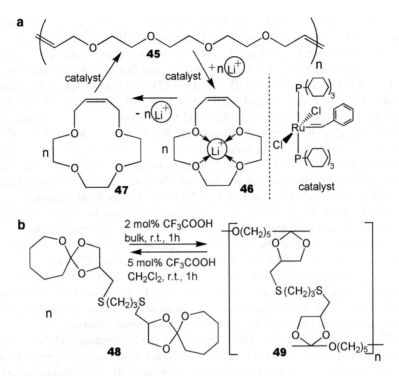

Fig. 14 (a) Metathesis-mediated reversible conversion between a polymer and a macrocycle; (b) acid catalyzed conversion between a polymer and the corresponding monomer

the polymer **45** ($M_n = 65{,}900$, PDI = 1.96, cis:trans = 1:3.7). The polymer **45** (Fig. 14a) undergoes, in dilute solutions, in the presence of LiClO$_4$, almost quantitative Ru-catalyzed conversion into a complex **46** that affords the corresponding macrocycle **47**. The cation Li$^+$ can coordinate the four oxygen atoms and consequently induces a conformation where the C = C groups are close, a fact that is favorable to the formation of the macrocycle **47** (template effect).

Endo, Suzuki, Sanda, and Takata [57] showed (Fig. 14b) that dithiol-linked bifunctional spiro orthoesters **48** can generate cross-linked polymers **49** in the bulk under acidic conditions (2 mol% CF$_3$COOH). The increase of the temperature shifts the equilibrium in favor of the monomer. Conversely, the cross-linked polymers can be depolymerized from CH$_2$Cl$_2$ suspensions to form the initial monomers, also under acidic conditions (5 mol% CF$_3$COOH). The yield of these processes is not quantitative.

A system where the conversion between a macrocycle and a polymer is modulated by chemical effectors/stimuli was described by Ulrich and Lehn [58, 59]. They set up a reversible effector-controlled constitutional switch between a polymer and a macrocycle (Fig. 15) in a dynamic covalent system. This is a sequential one-pot size-switch or polymerization-degree-switch and it involves covalent changes in the constitution through breaking and formation of covalent bonds of the imine type.

The constitutional switch is steered by a morphological switch of one of the two types of monomeric units, namely of the dialdehyde **50** derived from a pyridine–hydrazone–pyridine terpyridine-like unit. In its unmetallated form, the trigonal N atoms of this unit have a transoid orientation, and this induces the linear W-like shape of **50**. The dialdehyde **50** ($R = CH_3$ or $CH_3(CH_2)_7$) can react with diamines **51** and **52**, thus producing polymers **53** and **55**. Addition of metal ions Zn(II) (Fig. 15a1) or Pb(II) (Fig. 15a2) that bind to the trigonal N atoms produces the morphological conversion of the terpyridine-like W-shaped unit into a U-shaped unit, and the conversion of the polymer into the [1 + 1] metallamacrocycles **54** and **56**. The switch can function (50 mM concentration of each starting material in $CDCl_3/CD_3CN$ mixtures) in a reversible way. Addition of 1 equiv. of hexacyclen to the Zn(II) macrocycle or 1 equiv. of [2.2.2] cryptand to the Pb(II) macrocycle leads to complexation of the metal ions and regenerates the polymers. Subsequent addition of 1 equiv. of appropriate metal ion regenerates the macrocycle.

To have an idea about the values of the molecular diameters, one may note that reaction of dialdehyde **50** where $R = CH_3$ with the diamine **52** produces at 5 mM a [2 + 2] macrocycle (not shown) that has, according to DOSY [60, 61] studies (that provide the hydrodynamic radius of the species) in $CDCl_3$, a diameter of 16.4 Å, while at 50 mM higher diameter species with a diameter of 80 Å are observed. After several weeks even species having a diameter of 200 Å have been observed [59].

The above-mentioned examples belong to the captivating field of covalent reversible polymers. This field includes examples like reversible polymerization of cyclic oligomers from poly(ethylene terephthalate) [62], thermo-reversible polymers based on Diels–Alder and retro-Diels–Alder reaction (see, for example [63–69]), reversible photo-polymerization in the solid state of molecules

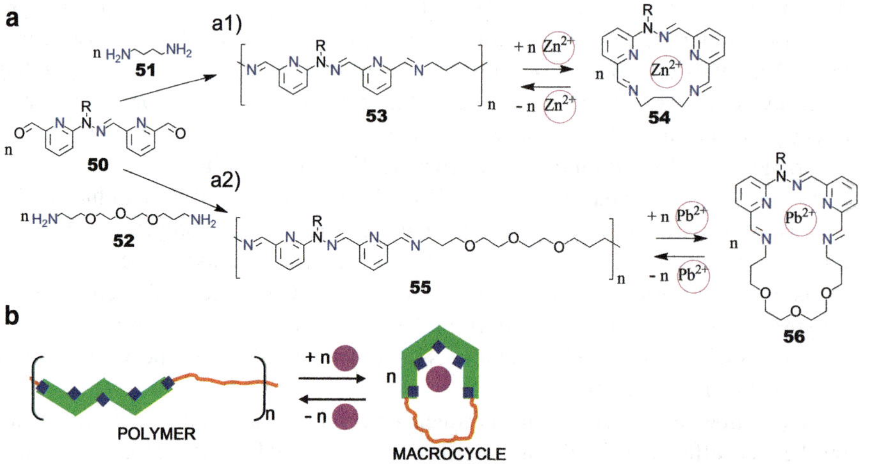

Fig. 15 (**a**) Formation of a polymer (**53** or **55**) on reaction of a pyridine–hydrazone–pyridine derived dialdehyde **50** with 1 equiv. of diamine (**51** or **52**), and its reversible conversion into a macrocycle (**54** or **56**) in the presence of the appropriate metal ions; (**b**) stylized representation of the polymer/macrocycle reversible switch

Synthetic Molecular Machines and Polymer/Monomer Size Switches

constituted of two thymine units connected through alkyl chains [70], reversible polymers of TEMPO-containing poly(alkoxyamine ester)s type [71], etc. Zhao and Moore reported on a reversible polymerization driven by folding that produces poly (*m*-phenylene ethynylene imine)s that respond (thus showing adaptability) to external stimuli (solvent and temperature) by changing their molecular weight [72, 73]. Folmer-Andersen and Lehn showed that thermoresponsive dynamers (for reviews on dynamers, see [74, 75]) consisting of a nanostructured poly(acylhydrazone) can undergo large and reversible changes in molecular weight (and so also in the degree of polymerization and size) depending on the variation of temperature, namely reversible molecular weight increase on heating [76].

A polymer/monomer (polymer/repeat-unit or polymer/macrocycle) switch may become of practical importance where a polymer decorated with certain groups has specific size-dependent properties that the monomeric units do not have. The modulation of the conversion between polymeric and monomeric (or macrocyclic) states would also result in the modulation of these properties. Moreover, such size switches, represented by polymerization/depolymerization processes that operate under the control of external events, are examples of environmentally-friendly recyclable polymers (reduction of waste treatment). As well, if the polymer has low solubility and the polymer/monomer switch can work in spite of this, then it becomes possible to reversibly generate a precipitating (solid) polymeric material from a liquid solution of monomer.

Ideally, the conversion between the monomers (subunits) and the polymer in response to stimuli should be fast, the polymers should be monodisperse, and their molecular weight should be as high as possible.

4 Concluding remarks

(a) The examples of molecular machines and polymer/monomer size switches discussed herein work thanks to various types of chemical reactions producing covalent changes.

To control the direction or the sense of molecular motors the choice of orthogonal chemical groups that play a role in the process of control, as well as of chemoselective reactions, is of particular importance.

Some of these machines and devices are based on dynamic covalent chemistry, and consequently belong to the area of constitutional dynamic chemistry.

(b) Covalent changes often produce changes in the constitution of the molecules. An attempt to classify the molecular motional devices in general can be proposed by considering constitutional changes with respect to the initial state. Consider an initial molecule which converts into a final or intermediate state in the functioning of a molecular motional device. Concerning the constitutional changes that can induce or control motions in molecular machines and devices, the following situations can be considered:

1. Processes where the constitution of intermediate (if any) and final states (molecules) of the device does not differ from that of the initial molecule. The initial molecule is converted into a new conformation or configuration and this results in a molecular motion. There is no constitutional change and no addition or elimination of any chemical species. There are included, for example, configurational changes induced by photochemical or thermal stimuli (see, for example [77]), or conformational changes induced by external fields (see, for example [27,29]);

2. Processes where the constitution of final and/or of intermediate states of the device is different from that of the initial molecule.

(2a) Within this category are certain processes (not discussed herein) that work due to supramolecular (non-covalent [104]) interactions. The initial molecule produces - on addition of chemical species (effectors) that do not interact with it covalently - species which contains it under a new conformation or configuration, and this results in a molecular motion. It may be considered that, regarded individually and formally, the constitution of the initial molecule - bound non-covalently to the effectors - incorporated into the new species does not change, but only its conformation and/or configuration. It is clear that the constitution of the initial species is different from that of the new one (made of the initial species and the effectors). Such are, for example, the molecular machines that work due to reversible binding of metal ions (for example, see [78]).

(2b) To this category also belong processes where the constitution of the initial molecule changes during the functioning of the device due to reduction/oxidation of metal centers, to covalent changes, to non-covalent exchanges etc. The covalent area of this subcategory of devices is part of the present chapter; during the motional process, the initial molecule is converted into identical or different species through a pathway involving constitutional changes consisting of reversible or irreversible formation of covalent bonds.

Of course, there can be molecular motional devices that combine processes from the above categories and subcategories.

(c) The motional devices, switches and machines, reviewed in this chapter can be divided, on the basis of mechanical criteria, in three classes:

1. Motional devices based on rotation about an axis (e.g. a bond) and generally composed of a rotor and a stator;
2. Motional devices based on linear motions (e.g. linear displacement or walking);
3. Mixed motional devices, where both rotary and linear (e.g. sliding or translation) motions should be in principle possible, like in rotaxanes.

(d) One may also consider the nature of the external stimuli that produce the covalent changes necessary for the operation of the motional devices reviewed herein: there are physical stimuli (temperature, pressure, light) and chemical stimuli (various reagents).

(e) Focusing on the design and functioning of devices, switches and machines, involving covalent changes is of importance given that covalent changes can play a role in the functioning of artificial biomolecule-based machines. To this category belong, for example, several DNA-based machines, the functioning of which involves covalent changes (e.g. hydrolysis) that are catalyzed by enzymes [79–84] (e.g. endonuclease, DNAzyme), as well as a molecular valve [85] consisting of a channel protein bearing synthetic molecules that can undergo reversible light-induced charge separation (heterolysis of a covalent bond) resulting in opening and closing of a 3-nanometer pore.

Moreover, reversible covalent changes (e.g. due to dynamic imine bonds) are of importance within the area of dynamic combinatorial chemistry, where they can serve as basis for the design and operation of functional materials and devices [86].

Acknowledgment We thank Jack Harrowfield for helpful suggestions.

References

1. Feynman RP (1960) There's plenty of room at the bottom. Eng Sci 23:22–26, 30, 34, 36
2. Davis AP (1999) Nanotechnology: synthetic molecular motors. Nature 401:120–121
3. Balzani V, Credi A, Raymo FM, Stoddart JF (2000) Artificial molecular machines. Angew Chem Int Ed 39:3348–3391
4. Feringa BL (2000) Nanotechnology: in control of molecular motion. Nature 408:151–154
5. (2001) Molecular machines special issue. Acc Chem Res 34:409–522
6. Sauvage JP (ed) (2001) Molecular machines and motors. Struct Bond 99:1–281
7. Balzani V, Credi A, Venturi M (2002) The bottom-up approach to molecular-level devices and machines. Chem Eur J 8:55246–55532
8. Balzani V, Venturi M, Credi A (2003) Molecular devices and machines – a journey into the nanoworld. Wiley-VCH, Weinheim
9. Easton CJ, Lincoln SF, Barr L, Onagi H (2004) Molecular reactors and machines: applications, potential, and limitations. Chem Eur J 10:3120–3128
10. Mandl CP, König B (2004) Chemistry in motion – unidirectional rotating molecular motors. Angew Chem Int Ed 43:1622–1624
11. Kottas GS, Clarke LI, Horinek D, Michl J (2005) Artificial molecular rotors. Chem Rev 105: 1281–1376
12. Kinbara K, Aida T (2005) Toward intelligent molecular machines: directed motions of biological and artificial molecules and assemblies. Chem Rev 105:1377–1400
13. Kay ER, Leigh DA (2005) Synthetic molecular machines. In: Schrader T, Hamilton AD (eds) Functional artificial receptors. Wiley-VCH, Weinheim
14. Kelly TR (ed) (2005) Molecular machines. Top Curr Chem 262:1–228
15. (2005) Special issue containing papers on molecular motors. J Phys Condens Matter 17: S3661–S4024
16. Leigh DA, Zerbetto F, Kay ER (2007) Synthetic molecular motors and mechanical machines. Angew Chem Int Ed 46:72–191
17. Michl J, Sykes ECH (2009) Molecular rotors and motors: recent advances and future challenges. ACS Nano 3:1042–1048
18. Moss GP (1996) Basic terminology of stereochemistry. Pure Appl Chem 68:2193–2222, precisely p 2204

19. Rowan SJ, Cantrill SJ, Cousins GRL, Sanders JKM, Stoddart JF (2002) Dynamic covalent chemistry. Angew Chem Int Ed 41:898–952, especially pp 928–937
20. Maeda T, Otsuka H, Takahara A (2009) Dynamic covalent polymers: reorganizable polymers with dynamic covalent bonds. Progr Polym Sci 34:581–604
21. Lehn J-M (1999) Dynamic combinatorial chemistry and virtual combinatorial libraries. Chem Eur J 5:2455–2463
22. Corbett PT, Leclaire J, Vial L, West KR, Wietor JL, Sanders JKM, Otto S (2006) Dynamic combinatorial chemistry. Chem Rev 106:3652–3711
23. Wojtecki RJ, Meador MA, Rowan SJ (2011) Using the dynamic bond to access macroscopically responsive structurally dynamic polymers. Nat Mater 10:14–27
24. Lehn J-M (2007) From supramolecular chemistry towards constitutional dynamic chemistry and adaptive chemistry. Chem Soc Rev 36:151–160
25. Rowan SJ, Stoddart JF (1999) Thermodynamic synthesis of rotaxanes by imine exchange. Org Lett 1:1913–1916
26. Cantrill SJ, Rowan SJ, Stoddart JF (1999) Rotaxane formation under thermodynamic control. Org Lett 1:1363–1366
27. Vacek J, Michl J (2001) Molecular dynamics of a grid-mounted molecular dipolar rotor in a rotating electric field. Proc Natl Acad Sci USA 98:5481–5486
28. Lippert AR, Kaeobamrung J, Bode JW (2006) Synthesis of oligosubstituted bullvalones: shapeshifting molecules under basic conditions. J Am Chem Soc 128:14738–14739
29. Jian H, Tour JM (2003) En route to surface-bound electric field-driven molecular motors. J Org Chem 68:5091–5103
30. Strawser D, Karton A, Zenkina OV, Iron MA, Shimon LJW, Martin JML, van der Boom ME (2005) Platinum stilbazoles: ring-walking coupled with aryl–halide bond activation. J Am Chem Soc 127:9322–9323
31. Hughes NA (1968) Further observations on derivatives of 1,6-anhydro-β-D-talopyranose; an example of acetal migration accompanying hydrolysis. Carbohydr Res 7:474–479
32. Kelly TR, De Silva H, Silva RA (1999) Unidirectional rotary motion in a molecular system. Nature 401:150–152
33. Kelly TR, Silva RA, De Silva H, Jasmin S, Zhao Y (2000) A rationally designed prototype of a molecular motor. J Am Chem Soc 122:6935–6949
34. Dahl BJ, Branchaud BP (2004) Synthesis and characterization of a functionalized chiral biaryl capable of exhibiting unidirectional bond rotation. Tetrahedron Lett 45:9599–9602
35. Flechter SP, Dumur F, Pollard MM, Feringa BL (2005) A reversible, unidirectional molecular rotary motor driven by chemical energy. Science 310:80–82
36. Dahl BJ, Branchaud BP (2006) 180° unidirectional bond rotation in a biaryl lactone artificial molecular motor prototype. Org Lett 8:5841–5844
37. Lin Y, Dahl BJ, Branchaud BP (2005) Net directed 180° aryl–aryl bond rotation in a prototypical achiral biaryl lactone synthetic molecular motor. Tetrahedron Lett 46:8359–8362
38. Mock WL, Ochwat KJ (2003) Theory and example of a small-molecule motor. J Phys Org Chem 16:175–182
39. Jog PV, Brown RE, Bates DK (2003) A redox-mediated molecular brake: dynamic NMR study of 2-[2-(methylthio)phenyl]isoindolin-1-one and S-oxidized counterparts. J Org Chem 68:8240–8243
40. Bartsch H, Erker T (1992) The Lawesson reagent as selective reducing agent for sulfoxides. Tetrahedron Lett 33:199–200
41. Zehm D, Fudickar W, Linker T (2007) Molecular switches flipped by oxygen. Angew Chem Int Ed 46:7689–7692
42. Aubry J-M, Pierlot C, Rigaudy J, Schmidt R (2003) Reversible binding of oxygen to aromatic compounds. Acc Chem Res 36:668–675
43. Zehm D, Fudickar W, Hans M, Schilde U, Kelling A, Linker T (2008) 9,10-Diarylanthracenes as molecular switches: syntheses, properties, isomerisations and their reactions with singlet oxygen. Chem Eur J 14:11429–11441

44. Barboiu M, Lehn J-M (2002) Dynamic chemical devices: modulation of contraction/extension molecular motion by coupled-ion binding/pH change-induced structural switching. Proc Natl Acad Sci USA 99:5201–5206
45. von Delius M, Geertsema EM, Leigh DA (2010) A synthetic small molecule that can walk down a track. Nat Chem 2:96–101
46. von Delius M, Leigh DA (2011) Walking molecules. Chem Soc Rev doi 10.1039/c1cs15005g
47. von Delius M, Geertsema EM, Leigh DA, Slawin AMZ (2010) Synthesis and solid state structure of a hydrazone-disulfide macrocycle and its dynamic covalent ring-opening under acidic and basic conditions. Org Biomol Chem 8:4617–4624
48. Higuchi H, Endow SA (2002) Directionality and processivity of molecular motors. Curr Opin Cell Biol 14:50–57
49. von Delius M, Geertsema EM, Leigh DA, Tang D-TD (2010) Design, synthesis, and operation of small molecules that walk along tracks. J Am Chem Soc 132:16134–16145
50. Barrell MJ, Campaña AG, von Delius M, Geertsema EM, Leigh DA (2011) Light-driven transport of a molecular walker in either direction along a molecular track. Angew Chem Int Ed 50:285–290
51. Marsella MJ, Maynard HD, Grubbs RH (1997) Template-directed ring-closing metathesis: synthesis and polymerization of unsaturated crown ether analogs. Angew Chem Int Ed 36:1101–1103
52. Maynard HD, Grubbs RH (1999) Synthesis of functionalized polyethers by ring-opening metathesis polymerization of unsaturated crown ethers. Macromolecules 32:6917–6924
53. Sutthasupa S, Shiotsuki M, Sanda F (2010) Recent advances in ring-opening metathesis polymerization, and application to synthesis of functional materials. Polym J 42:905–915
54. Slugovc C (2004) The ring opening metathesis polymerisation toolbox. Macromol Rapid Commun 25:1283–1297
55. Buchmeiser MR (2009) Ring-opening metathesis polymerization. In: Dubois P, Coulembier O, Raquez J-M (eds) Handbook of ring-opening polymerization. Wiley-VCH, Weinheim
56. Monfette S, Fogg DE (2009) Equilibrium ring-closing metathesis. Chem Rev 109:3783–3816
57. Endo T, Suzuki T, Sanda F, Takata T (1996) A novel approach for the chemical recycling of polymeric materials: the network polymer ⇌ bifunctional monomer reversible system. Macromolecules 29:3315–3316
58. Ulrich S, Lehn J-M (2008) Reversible switching between macrocyclic and polymeric states by morphological control in a constitutional dynamic system. Angew Chem Int Ed 47:2240–2243
59. Ulrich S, Buhler E, Lehn J-M (2009) Reversible constitutional switching between macrocycles and polymers induced by shape change in a dynamic covalent system. New J Chem 33:271–292
60. Macchioni A, Ciancaleoni G, Zuccaccia C, Zuccaccia D (2008) Determining accurate molecular sizes in solution through NMR diffusion spectroscopy. Chem Soc Rev 37:479–489
61. Cohen Y, Avram L, Frish L (2005) Diffusion NMR spectroscopy in supramolecular and combinatorial chemistry: an old parameter – new insights. Angew Chem Int Ed 44:520–554
62. Goodman I, Nesbitt BF (1960) The structures and reversible polymerization of cyclic oligomers from poly (ethylene terephthalate). J Polym Sci 48:423–433
63. Kuramoto N, Hayashi K, Nagai K (1994) Thermoreversible reaction of Diels-Alder polymer composed of difurufuryladipate with bismaleimidodiphenylmethane. J Polym Sci A Polym Chem 32:2501–2504
64. Watanabe M, Yoshie N (2006) Synthesis and properties of readily recyclable polymers from bisfuranic terminated poly(ethylene adipate) and multi-maleimide linkers. Polymer 47:4946–4952
65. Teramoto N, Arai Y, Shibata M (2006) Thermo-reversible Diels–Alder polymerization of difurfurylidene trehalose and bismaleimides. Carbohydr Polym 64:78–84
66. Inoue K, Yamashiro M, Iji M (2009) Recyclable shape-memory polymer: poly(lactic acid) crosslinked by a thermoreversible Diels–Alder reaction. J Appl Polym Sci 112:876–885

67. Ilhan F, Rotello VM (1999) Thermoreversible polymerization. Formation of fullerene-diene oligomers and copolymers. J Org Chem 64:1455–1458
68. Chen X, Dam MA, Ono K, Mal A, Shen H, Nutt SR, Sheran K, Wudl F (2002) A thermally re-mendable cross-linked polymeric material. Science 295:1698–1702
69. Chujo Y, Sada K, Saegusa T (1990) Reversible gelation of polyoxazoline by means of Diels-Alder reaction. Macromolecules 23:2636–2641
70. Johnston P, Hearn MYW, Saito K (2010) Solid-state photoreversible polymerization of n-alkyl-linked bis-thymines using non-covalent polymer-templating. Aust J Chem 63:631–639
71. Otsuka H, Aotani K, Higaki Y, Takahara A (2002) A dynamic (reversible) covalent polymer: radical crossover behaviour of TEMPO-containing poly(alkoxyamine ester)s. Chem Commun 2838–2839
72. Zhao D, Moore JS (2002) Reversible polymerization driven by folding. J Am Chem Soc 124:9996–9997
73. Zhao D, Moore JS (2003) Folding-driven reversible polymerization of oligo(m-phenylene ethynylene) imines: solvent and starter sequence studies. Macromolecules 36:2712–2720
74. Lehn J-M (2010) Dynamers: dynamic molecular and supramolecular polymers. Aust J Chem 63:611–623
75. Lehn J-M (2005) Dynamers: dynamic molecular and supramolecular polymers. Progr Polym Sci 30:814–831
76. Folmer-Andersen JF, Lehn J-M (2011) Thermoresponsive dynamers: thermally induced, reversible chain elongation of amphiphilic poly(acylhydrazones). J Am Chem Soc 133:10966–10973
77. Koumura N, Zijlstra RWJ, van Delden RA, Harada N, Feringa BL (1999) Light-driven monodirectional molecular rotor. Nature 401:152–155
78. Stadler AM, Kyritsakas N, Graff R, Lehn J-M (2006) Formation of rack- and grid-type metallosupramolecular architectures and generation of molecular motion by reversible uncoiling of helical ligand strands. Chem Eur J 12:4503–4522
79. Yin P, Yan H, Daniell XG, Turberfield AJ, Reif JH (2004) A unidirectional DNA walker that moves autonomously along a track. Angew Chem Int Ed 43:4906–4911
80. Bath J, Green SJ, Turberfield AJ (2005) A free-running DNA motor powered by a nicking enzyme. Angew Chem Int Ed 44:4358–4361
81. Bath J, Green SJ, Allen KE, Turberfield AJ (2009) Mechanism for a directional, processive, and reversible DNA motor. Small 5:1513–1516
82. Tian Y, He Y, Chen Y, Yin P, Mao C (2005), A DNAzyme that walks processively and autonomously along a one-dimensional track. Angew Chem Int Ed 44:4355–4358
83. Lund K, Manzo AJ, Dabby N, Michelotti N, Johnson-Buck A, Nangreave J, Taylor S, Pei R, Stojanovic MN, Walter NG, Winfree E, Yan H (2010) Molecular robots guided by prescriptive landscapes. Nature 465:206–210
84. He Y, Liu DR (2010) Autonomous multistep organic synthesis in a single isothermal solution mediated by a DNA walker. Nature Nanotechnol 5:778–782
85. Kocer A, Walko M, Meijberg W, Feringa BL (2005) A light-actuated nanovalve derived from a channel protein. Science 309:755–758
86. Moulin E, Cormos G, Giuseppone N (2011) Dynamic combinatorial chemistry as a tool for the design of functional materials and devices. Chem Soc Rev. doi: 10.1039/c1cs15185a
87. Hirose K, Shiba Y, Ishibashi K, Doi Y, Tobe Y (2008) A shuttling molecular machine with reversible brake function. Chem Eur J 14:3427–3433
88. Hirose K, Shiba Y, Ishibashi K, Doi Y, Tobe Y (2008) An anthracene-based photochromic macrocycle as a key ring component to switch a frequency of threading motion. Chem Eur J 14:981–986
89. Hirose K, Ishibashi K, Shiba Y, Doi Y, Tobe Y (2008) Highly effective and reversible control of the rocking rates of rotaxanes by changes to the size of stimulus-responsive ring components. Chem Eur J 14:5803–5811

90. (2000) Special issue on Photochromism: memories and switches. Chem Rev 100:1683–1890
91. Leigh DA, Pérez EM (2004) Shuttling through reversible covalent chemistry. Chem Commun 2262–2263
92. Altieri A, Aucagne V, Carrillo R, Clarkson GJ, D'Souza DM, Dunnett JA, Leigh DA, Mullen KM (2011) Sulfur-containing amide-based [2]rotaxanes and molecular shuttles. Chem Sci 2:1922–1928
93. Abraham W, Grubert L, Grummt UW, Buck K (2004) A photoswitchable rotaxane with a folded molecular thread. Chem Eur J 10:3562–3568
94. Schmidt-Schäffer S, Grubert L, Grummt UW, Buck K, Abraham W (2006) A photoswitchable rotaxane with an unfolded molecular thread. Eur J Org Chem 378–398
95. Abraham W, Wlosnewski A, Buck K, Jacob S (2009) Photoswitchable rotaxanes using the photolysis of alkoxyacridanes. Org Biomol Chem 7:142–154
96. Duo Y, Jacob S, Abraham W (2011) Photoswitchable rotaxanes on gold nanoparticles. Org Biomol Chem 9:3549–3559
97. Vetter A, Abraham W (2010) Controlling ring translation of rotaxanes. Org Biomol Chem 8:4666–4681
98. Kawai H, Umehara T, Fujiwara K, Tsuji T, Suzuki T (2006) Dynamic covalently bonded rotaxanes cross-linked by imine bonds between the axle and ring: inverse temperature dependence of subunit mobility. Angew Chem Int Ed 45:4281–4286
99. Umehara T, Kawai H, Fujiwara K, Suzuki T (2008) Entropy- and hydrolytic-driven positional switching of macrocycle between imine- and hydrogen-bonding stations in rotaxane-based molecular shuttles. J Am Chem Soc 130:13981–13988
100. Pérez EM (2011) Synthetic molecular bipeds. Angew Chem Int Ed 50:3359–3361
101. Baldini L, Casnati A, Sansone F, Ungaro R (2007) Calixarene-based multivalent ligands. Chem Soc Rev 36:254–266
102. Martos V, Castreño P, Valero J, De Mendoza J (2008) Binding to protein surfaces by supramolecular multivalent scaffolds. Curr Opin Chem Biol 12:698–706
103. Jayaraman N (2009) Multivalent ligand presentation as a central concept to study intricate carbohydrate-protein interactions. Chem Soc Rev 38:3463–3483
104. Schalley CA (ed) (2007) Analytical Methods in Supramolecular Chemistry. Wiley-VCH, p 2

Top Curr Chem (2012) 322: 291–314
DOI: 10.1007/128_2011_277
© Springer-Verlag Berlin Heidelberg 2011
Published online: 25 October 2011

Reversible Covalent Chemistries Compatible with the Principles of Constitutional Dynamic Chemistry: New Reactions to Create More Diversity

Kamel Meguellati and Sylvain Ladame

Abstract An approach to make chemical diversity space more manageable is to search for smaller molecules, or fragments, and then combine or elaborate these fragments. Dynamic Combinatorial Chemistry (DCC) is a powerful approach whereby a number of molecular elements each with binding potential can be reversibly combined via covalent or noncovalent linkages to generate a dynamic library of products under thermodynamic equilibrium. Once a target molecule has been added, the distribution of products can be shifted to favor products that bind to the target. Thus the approach can be employed to identify products that selectively recognize the target. Although the size of the repertoire of reversible covalent reactions suitable for DCC has increased significantly over the past 5–10 years, the discovery of new reactions that satisfy all the criteria of reversibility and biocompatibility remains an exciting challenge for chemists. Increasing the number of chemical reactions will enable the engineering of larger and more diverse DCLs, which remains a key step toward a broader use of DCC. In this review, we aim to provide a nonexhaustive list of reversible covalent reactions that are compatible with the concept of DCC, focusing mainly on the most recent examples that were reported in the literature in the past 5 years.

Keywords Dynamic combinatorial library · Constitutional dynamic chemistry · Reversible reactions · Supramolecular chemistry · Thermodynamic control

---

K. Meguellati
Institut de Science et d'Ingénierie Supramoléculaires (ISIS), Université de Strasbourg, CNRS UMR 7006, 8 allée Gaspard Monge, 67083 Strasbourg Cédex, France

S. Ladame (✉)
Department of Bioengineering, Imperial College London, South Kensington Campus, London SW7 2AZ, UK

Institut de Science et d'Ingénierie Supramoléculaires (ISIS), Université de Strasbourg, CNRS UMR 7006, 8 allée Gaspard Monge, 67083 Strasbourg Cédex, France
e-mail: sladame@imperial.ac.uk

Contents

1. Introduction 292
2. Reversible Imine Formation: New Applications for an Old Reaction 294
   - 2.1 Introduction 294
   - 2.2 Catalysis of Transimination Reactions 295
   - 2.3 Pyrazolotriazinone Metathesis 296
   - 2.4 9-Amino Acridines Amine Exchange 296
   - 2.5 Tandem Imine Metathesis and Ugi Reaction 297
3. From Imines to Hydrazones and Oximes 298
   - 3.1 Introduction 298
   - 3.2 Acyl Hydrazone Chemistry 299
4. Thiol-Based Reversible Chemistries: From Disulfides to Thiazolidines 300
   - 4.1 Introduction 300
   - 4.2 New Thiol-Based Reactions 300
5. Exchange Reactions of Carbon–Carbon Bonds: From the Olefin Metathesis to the
   Diels–Alder Reaction 302
   - 5.1 Olefin Metathesis 302
   - 5.2 Diels–Alder Reaction 304
   - 5.3 The Aldol and Henry Reactions 305
6. Acetal Exchanges and Boronic Acid Chemistry 305
   - 6.1 Introduction 305
   - 6.2 Thiohemiacetalization 306
   - 6.3 Reversible Reactions of Boronic Acids 307
7. Transamidation Reactions 308
8. Dynamic Capture of Carbon Dioxide via Reversible Ammonium Carbamate Formation 310
9. Conclusions and Perspectives 311

References 312

Abbreviations

CDC   Constitutional dynamic chemistry
DA   Diels–Alder
DCC   Dynamic combinatorial chemistry
DCL   Dynamic combinatorial library
EA   Ethacrynic acid
GSH   Glutathione
HTA   Hemithioacetal

1 Introduction

When a molecule contains covalent bonds that can form and break reversibly, its building blocks can be continuously exchanged and reorganized either to reform the same molecule or to form a novel entity. The same is true at the supramolecular level whereby supramolecular assemblies held together via easily broken and reformed (covalent and/or noncovalent) linkages can continuously reorganize.

Reversible Covalent Chemistries Compatible

This dynamic process is commonly known as constitutional dynamic chemistry (CDC). While the concept of dynamic covalent chemistry defines systems in which the molecular (or supramolecular) reorganization proceeds via reversible covalent bond formation/breakage, dynamic systems based on noncovalent linkage exchanges define the concept of dynamic noncovalent chemistry. Dynamic combinatorial chemistry (DCC) can be defined as a direct application of CDC where libraries of complementary functional groups and/or complementary interactional groups interexchange via chemical (i.e., covalent) reactions or physical (i.e., noncovalent) interactions.

In 1996, Brady and Sanders described a new approach to thermodynamic templating inspired from the general mechanism of an enzyme-catalyzed chemical reaction [1]. From this "Lock (host) and Key (guest)" principle, they suggested the possibility for a host to be formed, in the presence of a template, by the selective assembly of building blocks from a library upon binding to the template. The product should be preferentially formed in a reversible and thermodynamically controlled fashion. Simultaneously, in the mid-1990s, the group of J.M. Lehn independently reported that the distribution of a dynamic mixture of metal helicates was dictated by the nature of the counterion that binds to the center of the helicate [2]. Also in 1996, Venton and coworkers reported the use of nonspecific proteases to prepare peptides reversibly [3]. Pioneer work in the field also includes that of BL Miller using DCC for the first time to select DNA-binding ligands from a dynamic library of bis(salicylaldiminato)-zinc coordination complexes [4], that of AV Eliseev using light-induced alkene isomerization to direct the evolutionary-type formation of an anionic receptor for arginine [5], and additional work from the Sanders [6] and Lehn [7, 8] groups.

The powerful concept of DCC was born, which takes a number of molecular elements and allows them to combine reversibly via covalent or noncovalent linkages to generate a dynamic combinatorial library (DCL) of interchanging products under thermodynamic equilibrium [9–11]. These DCLs represent chemical networks, the composition of which can be modified in response to changes of the surrounding medium or through specific molecular recognition events. According to Le Châtelier's rules, upon addition of a target, the system reequilibrates as the mole fractions of individual library members are perturbed as a function of their affinity for that target. This reversible templating effect is possible due to the existence of covalent/noncovalent reversible reactions between the library components. A reversible reaction can be defined as a reaction where the interconversion of molecules is still possible via dynamic recombination of building blocks. Reversible covalent reactions are very widespread in Nature and enzymes in particular commonly make use of the dynamic nature of disulfide [12] or imine [13] bond formation for catalyzing specific reactions. However, and despite the fact that DCC has recently received considerable attention because of its successful use to identify new receptors, ligands, and catalysts, the number of covalent reactions commonly used for the formation of DCLs remains rather limited. Indeed, not only must the exchange reactions used in DCLs be reversible, but they should also occur (1) on a reasonably fast timescale and (2) under conditions (e.g., of solvent, pH, …) compatible with the template-ligand noncovalent interactions involved in the selection process. It is also important to be able to freeze the exchange prior to analysis of the DCL composition. Methods for freezing equilibration must be quick in order not to alter the relative distribution of the equilibrated library and will depend on the type of reaction being considered.

In this review, we aim to provide a *nonexhaustive* list of reversible covalent reactions that are compatible with the concept of DCC, focusing mainly on the most recent examples that were reported in the literature in the past 5 years.

2 Reversible Imine Formation: New Applications for an Old Reaction

2.1 Introduction

The reversible reaction of imine bond formation, resulting from the condensation between amine and carbonyl moieties, was first discovered in 1864 by Hugo Schiff [14]. Imine-containing compounds are of general formula $RR^1C=NR^2$. Typically, R and R^1 are an alkyl group, an aryl group, or a hydrogen atom, while R^2 is either an alkyl or an aryl group. Compounds where $R^2 = NR_2$ or OH are named hydrazones and oximes, respectively, and are also generated from a reversible reaction involving either hydrazides or hydroxylamines [15, 16]. Recent examples of their application in DCC experiments will be discussed further in Sect. 3.

In the early 1960s, seminal work by Jencks and coworkers demonstrated that formation and hydrolysis of $C=N$ bonds were proceeding via a carbinolamine intermediate, thus leading to a more general mechanism of addition reactions on carbonyl groups [17–19]. The dynamic nature of the reaction of imine formation can be exploited to drive the equilibrium either forward or backwards. Since the reaction involves the loss of a molecule of water, adding or removing water from the reaction mixture proved an efficient way to shift the equilibrium in either direction. The responsive behavior of imines to external stimuli makes the reversible reaction of imine formation perfectly suited for DCC experiments [20]. Thermodynamically controlled reactions based on imine chemistry include (1) imine condensation/hydrolysis, (2) transiminations, and (3) imine-metathesis reactions (Fig. 1).

In 1992, Goodwin and Lynn reported the first example of template-directed synthesis of DNA analogs via formation of a reversible imino linkage [21]. Five years later, Huc and Lehn described the first use of the reaction of imine condensation for the selection, by DCC, of carbonic anhydrase inhibitors from a library of amines and aldehydes [8]. Upon addition of the enzyme, the formation of one library component was strongly amplified when compared to a similar reaction carried out in the absence of template. Since then, imine exchange reaction has been applied

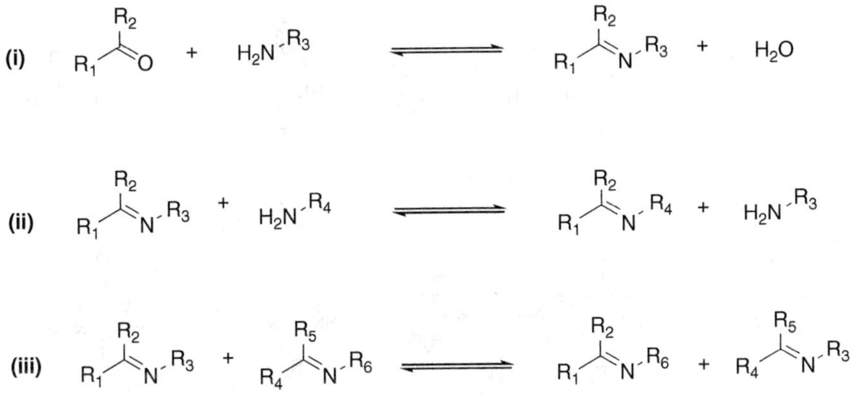

Reversible Covalent Chemistries Compatible

Fig. 1 Three examples of reversible reactions involving imines: (i) imine condensation/hydrolysis; (ii) transimination, and (iii) imine metathesis reaction

successfully to the selection by DCC of receptors and ligands of biomacromolecules [22]. Herein, we will focus on the most recent examples of dynamic systems relying, at least in part, on a reaction of imine exchange.

2.2 Catalysis of Transimination Reactions

Although many systems have taken advantage of the reversibility of the imine-type bond in water, according to the condensation/hydrolysis process, fast exchange of these key units in organic solvents remains difficult to achieve. In 2004, Giuseppone and Lehn investigated the potential of lanthanide ions as catalysts for transimination reactions. While screening a broad range of lanthanide ions, they demonstrated that exchange rates could increase linearly with a decrease in ionic radius and found that most efficient catalysis was obtained using scandium(III) ions [23]. Scandium-mediated catalysis in chloroform, for instance, proved much more efficient than attempts to catalyze the same transimination exchange reaction with protons. The following year the same group carried out a more detailed study on the scandium(III) catalysis of transimination reactions. They notably investigated structural effects on the thermodynamic distributions of products and compared the efficiency of the Lewis acid-catalyzed transimination with that of noncatalyzed or Brönsted acid-catalyzed reactions. The authors reported impressive rate accelerations up to 6×10^5 and reactions turnovers up to $3,600 \text{ h}^{-1}$ when carrying out exchange reactions in the presence of low amounts of scandium(III) [24]. Based on an analysis of the dependence of reaction rates and product distributions at equilibrium the authors proposed a mechanism involving a ternary intermediate in which ScIII is simultaneously coordinated to amine and imine (as well as solvent) molecules [24].

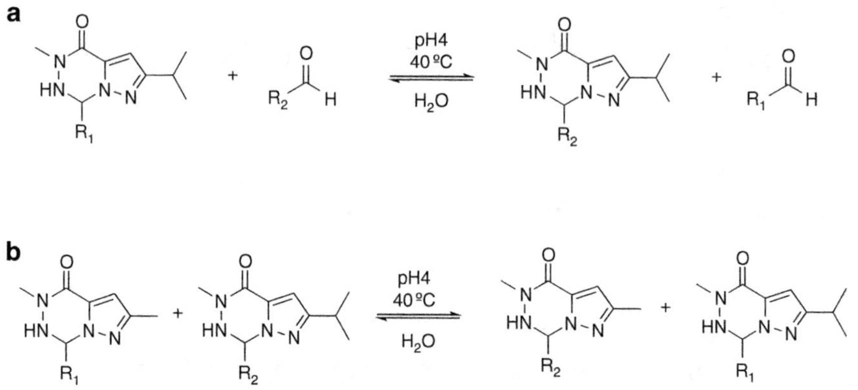

Fig. 2 (**a**) Reversible exchange reaction between a pyrazolotriazinone and an aliphatic aldehyde. (**b**) Metathesis reaction between two pyrazolotriazinones

2.3 Pyrazolotriazinone Metathesis

In 2005, Wipf and coworkers demonstrated that pyrazolotriazinones could undergo a reversible metathesis reaction in the presence of either aliphatic aldehydes (e.g., hydrocinnamaldehyde, acetaldehyde) or ketones (e.g., acetone) and under acidic (pH 4) aqueous conditions (Fig. 2a) [25]. Small libraries of five pyrazolotriazinones from reaction between one pyrazolotriazinone and four different aliphatic aldehydes were obtained while only traces (<1%) of fully hydrolyzed pyrazole-carboxylic hydrazide were formed during the exchange process. Direct side-chain metathesis of pyrazolotriazinones was also achieved under similar conditions (Fig. 2b). This reversible reaction is of particular interest because it is carried out in water and because exchange can be frozen by raising the pH of the aqueous solution up to 7, which is a much more convenient way to halt equilibrium between interconverting imines than reducing the imines into their corresponding amines with sodium borohydride, for example.

2.4 9-Amino Acridines Amine Exchange

More recently, Ladame and coworkers demonstrated that 9-amino substituted acridines could undergo an amine exchange reaction in water under near physiological conditions [26]. Proof-of-concept was first carried out by reacting 9-anilino-4-carboxyacridine with a mixture of aromatic amines (Fig. 3a). Using a large excess of free amines (50 equiv.), a small library of all possible 9-aminoacridines was obtained and thermodynamic equilibrium was reached after 6 days. This amine exchange reaction also proved to work with certain aliphatic amines and with acridines with different substitution patterns. Because of the possibility for 9-aminoacridines to coexist in solution with their

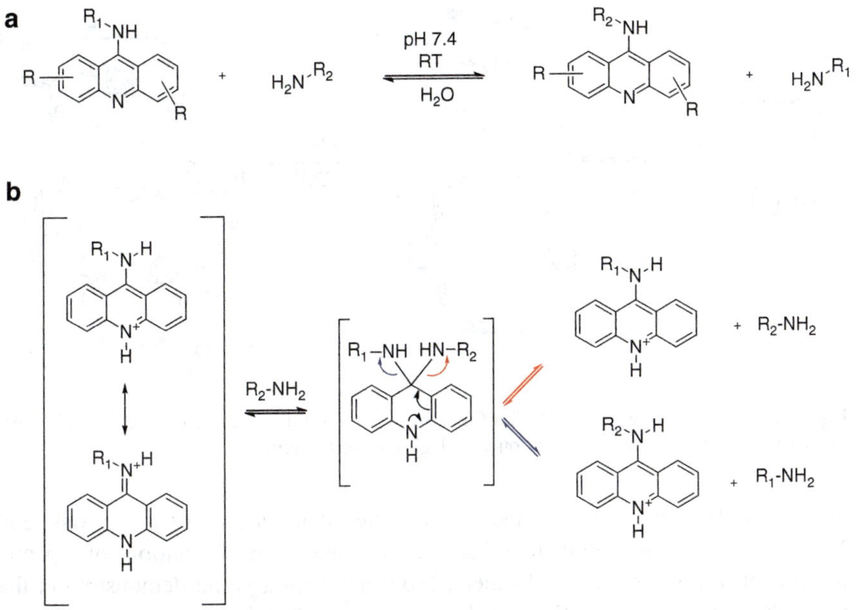

Reversible Covalent Chemistries Compatible

a

b

Fig. 3 (**a**) General reaction of reversible amine exchange between a 9-amino substituted acridine and a large excess of free amine. (**b**) Proposed mechanism of the amine exchange reaction via formation of a 9,9-diaminoacridine hemiaminal intermediate

tautomeric imino form, it was proposed that the amine exchange reaction could proceed reversibly via formation of a 9,9-diaminoacridine hemiaminal intermediate. However, formation of this unstable intermediate could proceed either via an S_NAr-like mechanism or via a transimination reaction (Fig. 3b). This original reversible reaction is of particular interest since it is carried out in water under near-physiological conditions and because 9-aminoacridines are well-known pharmacophores, commonly used as antitumor agents. Moreover, the exchange process does not involve the introduction nor the formation of free aldehyde, thus making this system highly suitable for biological assays. Also of interest, the exchange reaction can be readily quenched by lowering the pH of the solution down to 2.

2.5 Tandem Imine Metathesis and Ugi Reaction

Another recent innovation regarding the use of imine chemistry in DCC relies on an original way to freeze the equilibrating mixture by Ugi reactions [27]. In other words, this consists in conjugating a reversible reaction of imine condensation with an irreversible Ugi reaction. The latter step therefore represents an alternative to the more widespread reduction of imines with borohydrides. Wessjohann and coworkers recently prepared a library of macrocyclic oligoimines by condensation

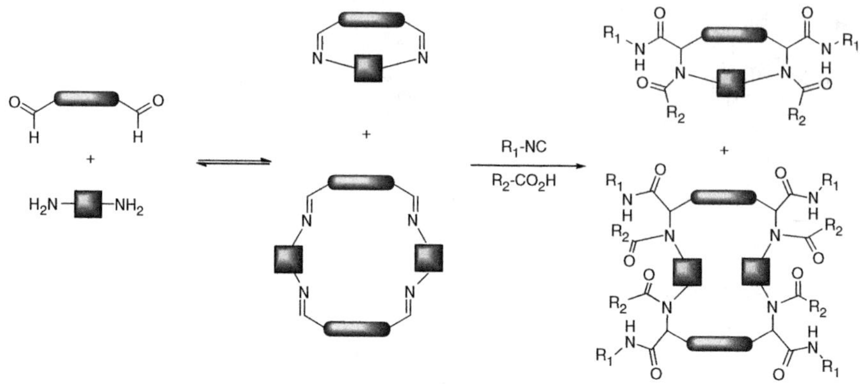

Fig. 4 Reversible formation of a DCL of macrocyclic oligoimines (*left*) followed by an irreversible freezing of the product distribution with Ugi reactions (*right*)

of dialdehydes with diamines and demonstrated that addition of metal ions (e.g., Mg^{2+}, Ba^{2+}) as potential templates could favor the formation of specific macrocycles at the expense of others. More interesting is the demonstration that those DCLs can be efficiently quenched by a series of multicomponent reactions of the Ugi type. Addition of acetic acid and terbutyl isocyanide (3 equiv.) to an equilibrated mixture of imines resulted in the polyfunctionalization of these macrocycles (Fig. 4) without modifying the relative distribution of macrocyclic library components. This new strategy represents an efficient way to create, from DCLs, static libraries of increased structural and chemical diversity.

3 From Imines to Hydrazones and Oximes

3.1 Introduction

Hydrazones and oximes ($C=N-X$) offer the advantage of significantly greater intrinsic stability than imines. This enhanced stability can be explained by multiple factors including participation of X in electron delocalization, resulting in increased negative-charge density on C, hence reducing its electrophilicity [28]. Oxime ligations typically require millimolar concentrations of each reactant or a large excess of one component to establish pseudo-first-order kinetics in water, thus limiting their applications in DCC. In 2006, Dawson and coworkers demonstrated that oxime ligations could be significantly accelerated by using aniline as a nucleophilic catalyst [29]. Rate enhancements were achieved via transient formation of a protonated aniline Schiff base intermediate which can then be rapidly converted into the desired oxime via a transimination reaction under acidic aqueous conditions. Increased ligation rates (k_{obs}) up to 400-fold in aqueous solution at pH 4.5 and up to 40-fold at pH 7 were obtained upon addition of aniline. Because aniline is soluble in a broad range of solvents, it is perfectly suited for catalyzing transimination reactions and could therefore open the way to a broader use of oxime reversible chemistry in DCC experiments.

As already described in Sect. 2.2, the use of scandium(III) ions is also an efficient way to catalyze transimination reactions. The most effective catalysis was observed with the exchange reaction between an oxime of cyclohexanone and benzylhydroxylamine. Interestingly, crossover experiments mixing hydrazones and oximes also proved successful [24].

3.2 Acyl Hydrazone Chemistry

Reaction between aldehydes and hydrazides leads to the formation of stable acylhydrazones (Fig. 5). In 1999, Sanders and coworkers demonstrated the reversibility of this covalent linkage by generating the first DCL of pseudo-peptide hydrazone macrocycles [30]. They subsequently reported the successful selection, by DCC, of ion receptors (e.g., Li^+, $R-NH_3^+$) based on hydrazone linkages [31, 32]. However, reasonably fast equilibration requires the reaction to be carried out under acidic conditions (pH < 4), which are generally not suitable for most biological applications. In order for this chemistry to be applied to biological systems, equilibrium at physiological pH must be reached in a reasonable timeframe. In 2006, Poulsen and coworkers reported that addition of an enzyme (i.e., carbonic anhydrase) to an equilibrating mixture of acylhydrazones at pH 7.2 could significantly accelerate the equilibration process [33]. Four years later (and 4 years after the work by Dawson and coworkers on aniline-catalyzed oxime formation), the groups of Greaney and Campopiano demonstrated that aniline could be used as a nucleophilic catalyst in a DCL of hydrazides, aldehydes, and hydrazones. In the presence of a significant excess of aniline (500 M equivalent with respect to the aldehyde), hydrazone-based DCLs were shown to equilibrate rapidly, even at a biocompatible pH of 6.2 [34]. After verifying that an excess (up to 20 mM) of aniline was not detrimental to the enzyme, this optimized (i.e., catalyzed) system

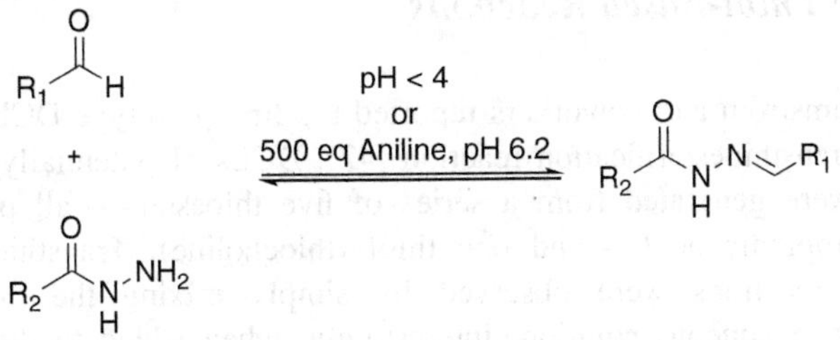

Fig. 5 Reversible formation of hydrazones under acidic conditions (for nonbiological applications) or catalyzed by aniline at pH 6.2 (for biological applications)

was then successfully applied to the selection of improved GST inhibitors. Unlike imines, the hydrazone linkage offers the advantage of a greater stability, thus facilitating the DCL characterization. As shown previously for imines, the hydrazone chemistry can also be easily switched on or off by a simple change in pH.

4 Thiol-Based Reversible Chemistries: From Disulfides to Thiazolidines

4.1 Introduction

Used for the first time for the preparation of DCLs in the late 1990s by Hioki and Still [35], the reaction of thiol–disulfide exchange (Fig. 6a) has been widely employed in the context of DCC-based experiments. The main reasons for that are because the exchange reaction is fast, can be carried out in water, and can be readily frozen by a simple change in pH (the reaction becomes extremely slow at acidic pH). Because of its biocompatibility, the disulfide exchange reaction has been used extensively in the past decade for the selection/identification of nucleic acid (DNA and RNA) binding ligands [36–38], ion receptors [39, 40], or for the construction of molecular cages for drug delivery [41]. It is also noteworthy that the largest DCLs prepared so far are based on this chemistry, with library sizes of >9,000 and >11,000 compounds prepared by the Otto group [42] and the Miller group [43, 44], respectively. Also of prime interest is the recent demonstration by the groups of Otto and Nitschke that the disulfide exchange reaction could also be used in conjunction with other reversible chemistries such as those of imine metathesis [45] or hydrazone exchange [46].

Recently, thiols have also been shown to participate in a series of new reversible reactions suitable for DCC. Such reactions include (1) the thioester exchange reaction (Fig. 6b), (2) the thiazolidine exchange reaction (Fig. 6c), and (3) the reversible Michael addition of thiols (Fig. 6d).

4.2 New Thiol-Based Reactions

In 2004, Ramström and coworkers reported the first prototype DCLs based on a reversible transthioesterification reaction [47]. DCLs of potentially ten different thioesters were generated from a series of five thioesters – all prepared from 3-sulfanylpropionic acid – and one thiol (thiocholine). Transthioesterification (exchange) reactions were observed by simply mixing the different DCL components in aqueous solution. Interestingly, when added to the equilibrated mixture of thioesters, the enzyme acetylcholinesterase was able to recognize and subsequently hydrolyze its best substrate selectively. This selective hydrolysis

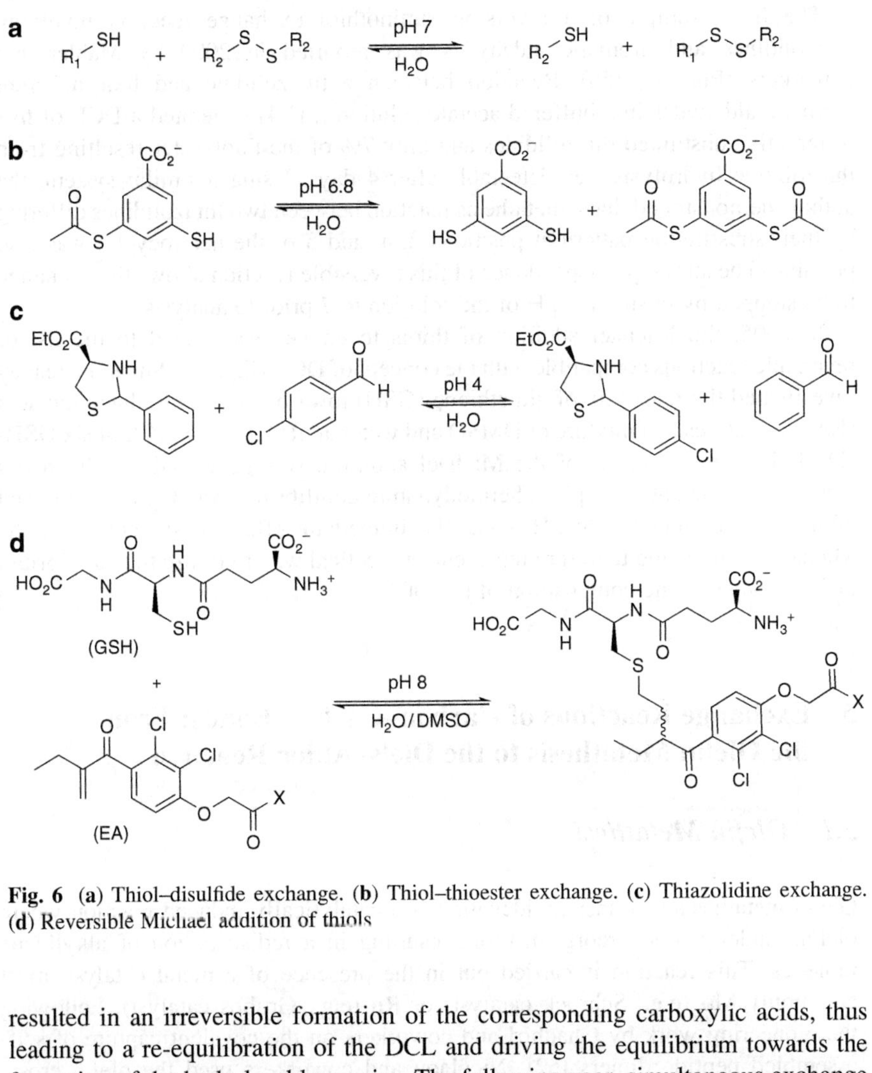

Reversible Covalent Chemistries Compatible

Fig. 6 (**a**) Thiol–disulfide exchange. (**b**) Thiol–thioester exchange. (**c**) Thiazolidine exchange. (**d**) Reversible Michael addition of thiols

resulted in an irreversible formation of the corresponding carboxylic acids, thus leading to a re-equilibration of the DCL and driving the equilibrium towards the formation of the hydrolysable species. The following year, simultaneous exchange of disulfide and thioester linkages was first reported by Sanders and coworkers [48]. Proof-of-concept was demonstrated using a single polyfunctional building block that carried, on the same aromatic scaffold, a free thiol and a thioester (Fig. 6b). Through successive thioester exchange reaction and disulfide formation, a DCL of at least eight oligomers was generated. In such a system the selection can take place on the basis of the nature of the bonds between monomeric units in addition to the units themselves. In 2010, the same thiol–thioester exchange reaction was used by Gagne and coworkers for building DCLs of cyclic thiodepsipeptides under thermodynamic control [49].

The first example of a reversible aminothiol exchange reaction involving thiazolidines and aromatic aldehydes was reported in 2009 by Mahler and coworkers (Fig. 6c) [50]. Reaction between a thiazolidine and four different aromatic aldehydes in a buffered acetate solution (pH 4) generated a DCL of five differently substituted thiazolidines and only 7% of the thiol–ester resulting from thiazolidine hydrolysis was detectable after 4 days. Using a similar system, the authors demonstrated that a metathesis reaction between two thiazolidines differing by their substitution pattern at positions 2, 4, and 5 of the heterocycles was also possible. The strong pH dependence of this reversible reaction allows the exchange to be stopped by raising the pH of the solution to 7 prior to analysis.

In 2005, the Michael addition of thiols to enones was added to the list of reversible reactions compatible with the concept of DCC (Fig. 6d). Shi and Greaney investigated the reactivity of glutathione (GSH) toward a series of ethacrynic acid (EA) derivatives in a mixture of DMSO and water at pH 8 [51]. A DCL of six GSH-EA derivatives, products of the Michael addition, was generated which proved responsive to changes in pH. Thermodynamic equilibrium was typically attained after 3 h. Acidification to pH 4 has the immediate effect of switching off the Michael addition and therefore represents a practical way to freeze the equilibrium before analyzing the composition of the DCL.

5 Exchange Reactions of Carbon–Carbon Bonds: From the Olefin Metathesis to the Diels–Alder Reaction

5.1 Olefin Metathesis

Olefin metathesis or transalkylidenation is a catalytically induced reaction where olefins undergo bond reorganization, resulting in a redistribution of alkylidene moieties. This reaction is carried out in the presence of a metal catalyst, most commonly Mo (e.g., Schrock catalyst) or Ru (e.g., Grubbs catalyst). Following the pioneering work by Ghadhiri and coworkers on the covalent capture of self-assembled peptide dimers [52], Nicolaou and coworkers used the olefin cross-metathesis reaction for the selection of vancomycin dimers with high affinity for D-Ala–D-Ala, thus demonstrating the potential of olefin metathesis in DCC experiments [53].

In 2005, Rowan, Nolte, and coworkers described an efficient and templated synthesis of porphyrin boxes using DCC and reversible metathesis reaction [54]. Cyclic tetramers were successfully prepared in good yields (62%) from an olefin-functionalized zinc porphyrin in the presence of first generation Grubb's catalyst and upon addition of a tetrapyridyl porphyrin (TPyP) serving as a template. While a mixture of linear and cyclic oligomers was obtained in the absence of template, addition of TPyP resulted in a reorganization of the DCL to favor the formation of the desired tetrameric box (Fig. 7a).

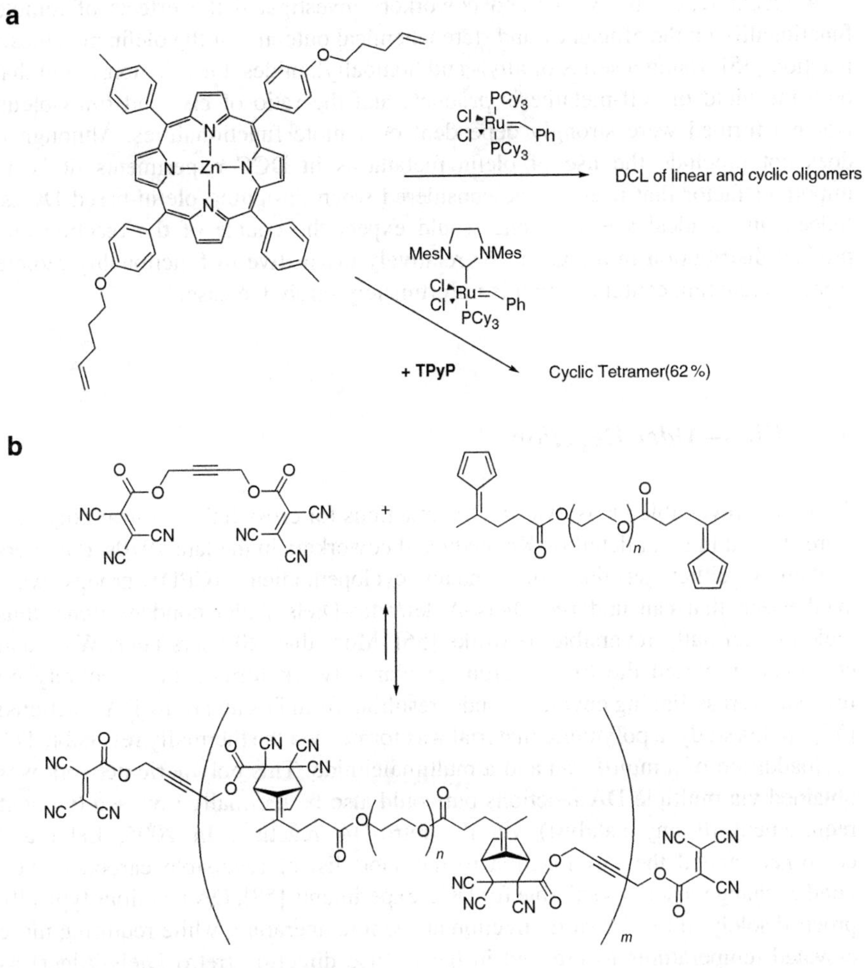

Reversible Covalent Chemistries Compatible

303

a

b

c

Fig. 7 (**a**) Selection-amplification of porphyrin tetrameric boxes using olefin metathesis. (**b**) Structure of dynamers formed via reversible Diels–Alder reaction. (**c**) Generation of a DCL of (R) and (S) nitroaldols from five differently substituted benzaldehydes

A recent report by Miller and coworkers investigated the effects of remote functionality on the efficiency and stereochemical outcome of the olefin metathesis reaction [55]. Using a series of allyl- and homoallylamides, they demonstrated that both the yield of self-metathesis products and the ratio of *cis*- and *trans*-olefin isomers formed were strongly dependent on remote functionalities. Although it does not preclude the use of olefin metathesis in DCC experiments, it is an important factor that needs to be considered when designing olefin-based DCLs. Indeed, in an ideal scenario, one would expect the course of the reaction and product distribution in a DCL to be relatively insensitive to functionality remote from the reacting centers, which is unfortunately rarely the case.

5.2 Diels–Alder Reaction

Thermally reversible Diels–Alder (DA) reactions for cross-linking linear polymers were studied in great detail by Kennedy and coworkers in the late 1970s. Polymers containing either terminal or pendant cyclopentadiene (CPD) groups were synthesized that can undergo Diels–Alder/retro-Diels–Alder condensations, thus yielding thermally reversible networks [56]. More than 20 years later, Wudl and coworkers reported the first macromolecular network formed in its entirety by reversible cross-linking covalent bonds (resulting from DA and retro-DA reactions) [57]. In this study, a polymeric material was formed via the thermally reversible DA cycloaddition of a multifuran and a multimaleimide. This polymeric network was obtained via multiple DA reactions but could also be thermally reversed (without requirement of any catalyst) via the retro-DA reaction. In 2005, Lehn and coworkers added the DA reaction to the short list of reversible carbon–carbon bond exchange reactions suitable for DCC experiments [58]. DA reactions typically proceed solely in the forward direction at room temperature, while requiring more elevated temperatures to proceed in the reverse direction (retro Diels–Alder) as shown with the polymeric materials developed by Wudl. In some particular cases, however, the Diels–Alder reaction can be reversible at room temperature. This requires the careful selection of appropriate diene and dienophile. Reversibility of the DA reaction was demonstrated by the Lehn's group using a small library of functionalized fulvenes and cyanoolefins (e.g., diethylcyanofumarate) in chloroform. Selected DA reactions proved dynamic between 25 and 50 °C and reached equilibrium in ≤ 1 min at 25 °C. More recently, Lehn and coworkers applied this reversible chemistry to the generation of dynamic polymers [59]. Diels–Alder dynamers were obtained simply by mixing bis(tricyanoethylenecarboxylate) and polyethylene-linked bis(fulvenes) (Fig. 7b). Interestingly, and due to the reversible nature (even at room temperature and in the condensed phase) of the Diels–Alder adducts linking the monomers the dynamic network proved capable to either (1) adapt in response to external stimuli such as mechanical elongation stress or (2) self-repair.

5.3 The Aldol and Henry Reactions

In the presence of acid or base catalysts the aldol reaction is reversible, and the β-hydroxy carbonyl products may revert to the initial aldehyde or ketone reactants. In the absence of such catalysts these aldol products are perfectly stable and isolable compounds.

In 2007, the group of Ramström has identified the nitroaldol (Henry) reaction as a new and efficient C–C bond-forming route to DCL formation. DCLs based on the nitroaldol reaction were generated under mild conditions from equimolar amounts of five differently substituted benzaldehydes together with 1 equiv. 2-nitropropane in dry toluene in the presence of 10 equiv. triethylamine and thermodynamic equilibrium between the initial components and all ten chiral nitroaldol adducts was reached within hours [60]. Interestingly, the stereogenic carbon center created in the reaction could also be resolved through a coupled process in the form of a kinetically controlled lipase-mediated acylation. In the same year, Coster and coworkers demonstrated through ^{1}H NMR studies that the boron-mediated ketone–ketone aldol reaction was reversible, unlike aldol reactions between boron enolates and aldehydes which proved strictly irreversible [61].

6 Acetal Exchanges and Boronic Acid Chemistry

6.1 Introduction

Reactions of acetal formation from alcohols and aldehydes and reactions of transacetalization were first applied to DCC in 2003 by Stoddart and coworkers, using a mixture of bis-acetonide D-threitol and a second diacetal in organic solvent ($CDCl_3$) and in the presence of catalytic triflic acid [62]. They described the efficient and reversible acid-catalyzed formation of a chiral polycyclic polyether which incorporates two threitol residues as bicyclic diacetals. The amplified production of the most thermodynamically favored complex(es) upon addition of a metal ion (i.e., CsPF6) serving as a template was also demonstrated. Since then, few additional examples of acetal-based DCLs have been reported in the literature. For instance, Mandolini and coworkers prepared a DCL of linear or cyclic oligomers of formaldehyde derived from the acid-catalyzed transacetalization of formaldehyde acetals [63]. The composition of these dynamic libraries proved solely dependent on the initial monomer concentration and addition of silver cations to the equilibrating mixture was shown to template selectively the formation of the cyclic dimer through stabilizing metal–ligand interactions. More recent work includes that from Berkovich-Berger and Lemcoff [64] who prepared a DCL of more than 15 cyclic and acyclic acetals from a mixture of triethylene glycol and 4-nitrobenzaldehyde (Fig. 8a). The composition of the library could also be modified upon introduction of ammonium ions. Under these new conditions, the smallest macrocycle (formed from

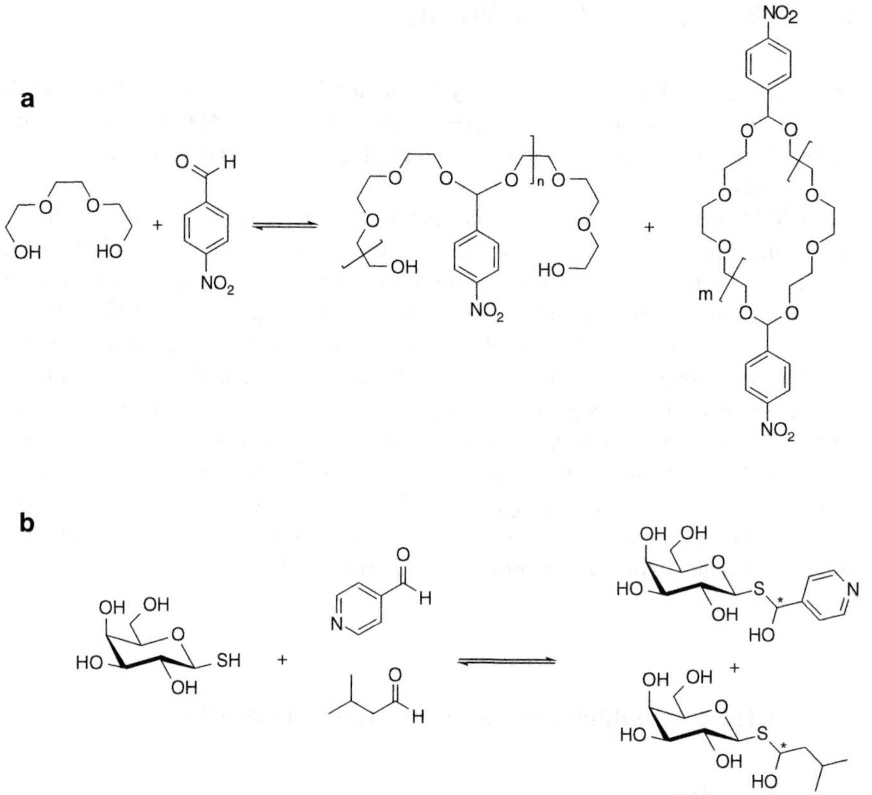

Fig. 8 (a) Reversible formation of cyclic and linear acetal oligomers from triethylene glycol and 4-nitrobenzaldehyde. (b) Reversible formation of hemithioacetal (HTA) from reaction between a thiol and an aldehyde derivative

two molecules of triethylene glycol and two molecules of 4-nitrobenzaldehyde) was significantly amplified, representing more than 50% of all the library components.

6.2 Thiohemiacetalization

In 2010, Ramstrom and coworkers showed that hemithioacetal (HTA) formation from reaction between a thiol derivative and a given aldehyde (or ketone) was a fast and reversible process in water, and therefore suitable for DCC-based systems (Fig. 8b) [65]. Because the thermodynamic equilibrium of the reaction of HTA formation is significantly displaced toward the two reactants, a virtual library of ten possible HTA was generated by mixing in water five aliphatic thiols and two aldehydes (one aromatic and one aliphatic). When carried out in the presence of an enzyme (i.e., β-galactosidase) free of any cystein residues in its active site to avoid HTA formation between the library components and the enzyme itself, the formation of a specific HTA was highly favored as determined by saturation

Reversible Covalent Chemistries Compatible

transfer difference (STD) NMR spectroscopy. Interestingly, the selected HTA proved to be the best β-galactosidase inhibitor of all ten possible/virtual derivatives, thus justifying its amplification during the dynamic selection process.

6.3 Reversible Reactions of Boronic Acids

Highly stable boronate esters can be readily obtained by reacting boronic acids with diols. This reaction is fast and can be reversible under certain conditions (e.g., under the action of external chemical stimuli). In the absence of alcohols, boronic acids can undergo a self-condensation to form six-membered boroxines. Both boronate esters and boroxines can also reversibly bind to N-donor ligands to generate reversibly coordinated adducts, thus making boronic acids potentially interesting building blocks for DCC experiments (see three examples given in Fig. 9).

In 2007, Barboiu and coworkers engineered G-quartet-type supramolecular structures using a macromonomeric guanosine dimer prepared reversibly via reaction of guanosine with a bis(iminoboronic acid) derivative [66]. K^+-templated formation of G-quartet-encoded polymeric films resulted from a double reversible covalent iminoborate connection, the imino linkage contributing to the extra stabilization of the reversible boronate–guanosine (B–O) bond via formation of a dative N–B bond. This original double "dynamer"-type system combines (1) the reversible supramolecular self-assembly properties of guanosines capable of forming G-quartet structures via Hoogsteen hydrogen bonds in the presence of monovalent cations (e. g., potassium) and (2) the reversible polycondensation of the polymer components via formation of iminoboronate bonds.

Other examples of DCLs based on reversible reactions involving boronic acids include the work from Tokunaga and coworkers on the reversible condensation of different arylboronic acids at room temperature to provide an equilibrating mixture of homo- and hetero-boroxines in solution [67]. More recently, they also achieved the dynamic formation of C_3-symmetric [4] rotaxanes containing a boroxine core obtained through trimerization of three boronic acid units [68]. In recent years, polymeric materials based on boronic acids also appeared to be very promising. For instance, boronic acid polymeric structures were prepared by Jakle and coworkers through reversible boroxine formation starting from boronic acid-end-functionalized polystyrenes [69]. Lavigne and coworkers also exploited the reversible nature of boronate esterification to create self-healing polymers through the condensation of 9,9-dihydroxyfluorene-2-diboronic acid and pentaerythritol [70, 71]. The group of Severin also developed a three-component self-assembled polymeric system where the polymeric backbone is maintained by N–B bonds. Because the N–B bonds are labile in solution, the reversible polymer can easily dissociate in solution [72]. More recent work from the same group includes the dynamic formation of cages and macrocycles via multicomponent assembly of boronic acids. In order to form such structures, the authors successfully combined the reversible formation of boronic esters with reversible imine bond formation and metal–ligand interactions [73].

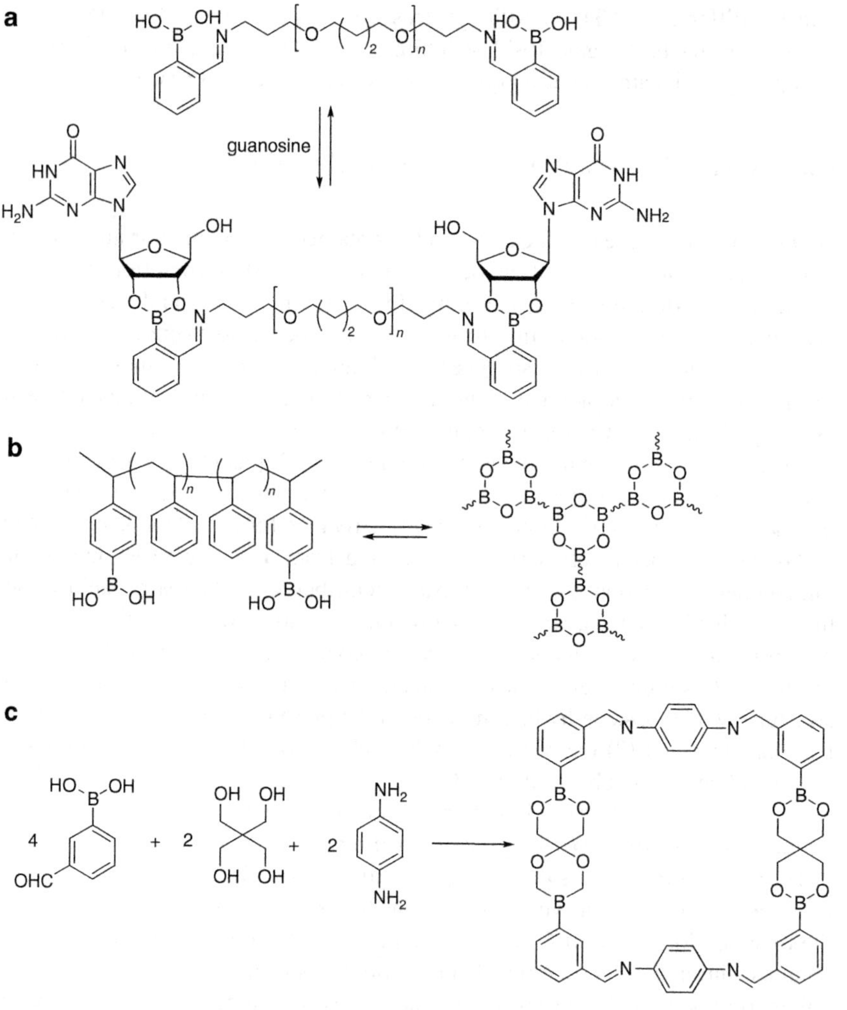

Fig. 9 (a) Reversible synthesis of a bisiminoboronate–guanosine derivative from guanosine and a bisiminoboronic acid. (b) Dynamic polymeric networks based on reversible boroxine formation. (c) Formation of a macrocycle in a [4 + 4 + 2] condensation via simultaneous reversible formation of imine bonds and boronic esters

7 Transamidation Reactions

Since the initial work in 1996 from Venton and coworkers on the dynamic screening (under conditions in which both hydrolysis and synthesis occurred in the presence of proteases) of a peptide library for the discovery of novel peptides that bind to fibrinogen [3], many efforts have been made to perform amide exchange in

Reversible Covalent Chemistries Compatible

mild conditions. Unfortunately, because thermodynamics favors amide hydrolysis rather than transamidation, making amide exchange reversible has proven very challenging. A successful approach to overcome this problem has been used by the Ulijn group and involved using a nonspecific endoprotease (thermolysin) together with self-assembling aromatic peptide amphiphiles to catalyze bond formation, hydrolysis and transamidation [74]. DCLs were composed of N-Fmoc protected dipeptides capable of assembling into nanofibers or nanotubes responsible for the formation of a gel which immobilizes water (Fig. 10). The potential of these dipeptides to assemble drives the reaction towards the peptide products, while peptide bond hydrolysis is achieved by thermolysin, thus making the reaction fully reversible [75]. In 2010, Sadownik and Ulijn increased the complexity of the enzyme-driven peptide exchange by introducing a cystein building block, thus creating a system based on two reversible covalent reactions (disulfide exchange and transamidation) [76]. Another strategy was developed by Gellman, Stahl and coworkers which requires catalysts to promote transamidation [77]. Lewis-acid catalysts such as $Sc(OTf)_3$ were able to promote thermodynamically favored transamidation but were totally ineffective for the more challenging equilibrations. Substrate-dependent variations in catalytic activities were observed during this study (e.g., alkyl vs arylamide substrates) which suggests that catalysts with functional group selectivity could be developed. However, it is noteworthy that reactions in this study were all carried out in organic solvent (toluene, xylene) and at elevated temperature.

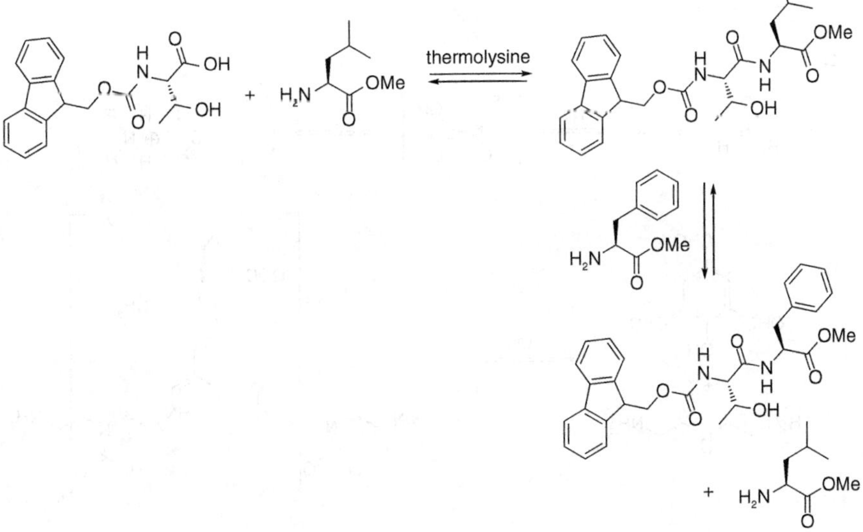

Fig. 10 Enzyme-catalyzed dipeptide synthesis with a subsequent threonine ↔ phenylalanine exchange to form a more stable assembly

8 Dynamic Capture of Carbon Dioxide via Reversible Ammonium Carbamate Formation

Within the list of reversible reactions commonly used in chemistry, only very few involve an entity which exists as a gas at ordinary temperature and pressure. The reversible formation of ammonium carbamates upon reaction of carbon dioxide with primary or secondary amines is however a very well-known mechanism responsible for CO_2 transportation during the respiratory process. Despite it being robust, the covalent carbamate linkage can also release CO_2 and the constitutive amine when under appropriate conditions of temperature and/or pressure. Thermally reversible carbamate chemistry was first employed for the preparation of reusable ionic liquids [78], organogels [79], and imprinting polymers [80]. Rudkevich and coworkers used CO_2 as a cross-linking agent for the preparation of switchable supramolecular polymers [81, 82]. By introducing and thermally releasing CO_2, they were able to convert linear, H-bonding, polymeric chains reversibly into three-dimensional networks. More recently, Leclaire and coworkers described the DCC-based selection of molecular receptors for carbon dioxide which involves for the first time a combination of covalent (amine and CO_2) and noncovalent (carbamate and ammonium) reversible interactions between the host and the guest (Fig. 11) [83]. Reversible Schiff-base chemistry using a mixture of polyamines, dialdehydes, and trialdehydes was chosen to generate a DCL of imine-based supramolecular architectures (macrocycles, capsules, branched objects). Addition of CO_2 would then be expected to template/amplify the formation of its best host(s). Although carbamate formation was confirmed by NMR spectroscopy, no shift of the transimination equilibrium upon addition of CO_2 could be detected in solution. However, addition of CO_2 to an

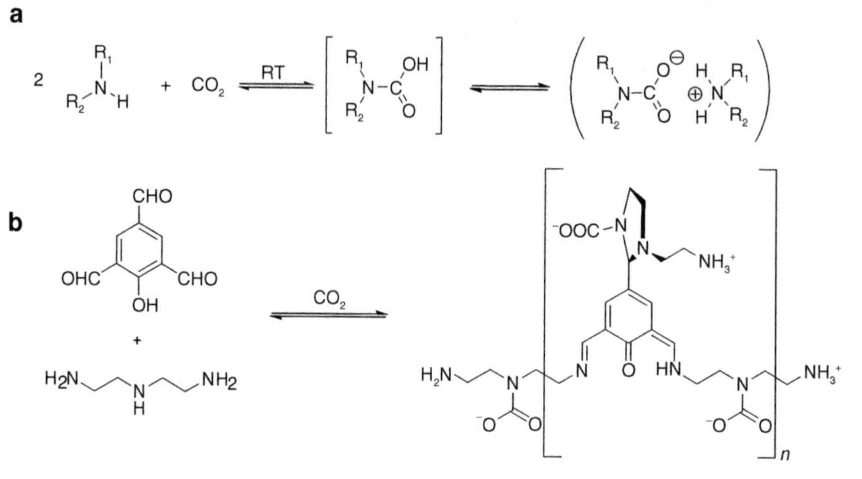

Fig. 11 (**a**) CO_2 capture by reversible ammonium carbamate formation. (**b**) Example of linear solid oligomer formed upon reaction of 2-hydroxy-1,3,5-benzenetrialdehyde and at least 3 equiv. of diethylenetriamine in the presence of CO_2

equilibrating mixture of 2-hydroxy-1,3,5-benzenetri-(n-butylimine) and diethylene-triamine quasi-quantitatively converted a DCL of soluble architectures into an abundant precipitate (S1) which was subsequently characterized by powder X-ray diffraction and ^{13}C CP-MAS solid-state NMR. n-Butylamine signals were absent from the NMR spectrum of the solid, thus indicating a CO_2-induced complete displacement of the transimination equilibrium. The appearance of characteristic carbamate signals also confirmed the immobilization of CO_2 within the solid network through carbamate linkages. Reversibility of the system was also demonstrated, for instance by converting the solid S1 back to a homogeneous solution upon gentle CO_2 desorption.

9 Conclusions and Perspectives

During the past decade, the discovery of DCC and of its multiple applications in the fields of material sciences, drug discovery, or catalysis have drawn the attention of organic chemists to a particular type of reactions they were normally trying to avoid: reversible covalent reactions. Indeed, equilibrated exchange reactions have long been ignored by synthetic chemists who preferred more complete and highly selective irreversible reactions to them. Reversibility is however a key factor in DCC and the past 5 years have witnessed a significant increase in the number of reversible covalent reactions compatible with the principles of DCC. A large, yet increasing, repertoire of single or double carbon–nitrogen, sulfur–carbon, sulfur–sulfur, carbon–oxygen, and even carbon–carbon covalent bonds can now exchange under reversible, thermodynamically controlled conditions, thus allowing the creation of DCLs of large structural and chemical diversities.

Combining different reversible covalent (or covalent and noncovalent) chemistries is of particular interest since it allows the generation of increased diversity from fewer, but polyfunctionalized, building blocks. The successful combination of thiol–disulfide exchange and either imine or hydrazone chemistry has been reported which will allow the construction of more complex molecular architectures. Even more complex DCLs have also been evolved in which three distinct kinds of linkage (disulfide, imine, and coordinative) were shown to be capable of simultaneous dynamic exchange. Macrocycle formation was also recently achieved by combining the reversible reactions of imine and boronic ester formation with metal–ligand interactions. Responsive DCLs based on a combination of reversible covalent and noncovalent (e.g., H-bonding, metallosu-pramolecular) interactions have also been reported which were not discussed here since they are beyond the scope of this chapter, which is focusing on covalent reactions only. Since one can expect that the discovery of new reversible reactions suitable for DCC will remain a slow process, creating DCLs with large chemical and structural diversity from simple but polyfunctional building blocks via simul-taneously occurring reversible covalent reactions is likely to become the most efficient way to generate complex libraries and to guarantee a broader use of DCC in the fields of drug discovery or material sciences in particular.

Recent efforts have also been made in order to increase exchange rates. Although it is essential that the covalent exchange remains slower than the actual binding of the DCL components to the target, it is also important that equilibrium can be reached within a reasonable interval that is compatible with the stability/ lifetime of the target, especially in the case of biomacromolecules (e.g., nucleic acids, proteins). Organic catalysis for instance proved an efficient way to accelerate significantly hydrazone exchange at biocompatible pH, therefore making this otherwise slow reaction more suitable for biomolecular targeting [34]. Metal catalysts (e.g., Grubb's catalyst for olefin metathesis, scandium(III) ions for transimination reactions. . .) also recently proved successful. Removing the catalyst from the equilibrating mixture can then be a simple way to halt equilibrium, allowing for characterization of the DCLs composition. However, considering that an extremely large number of turnovers may be required before equilibrium is reached, catalysts must have a sufficient lifetime to be considered, which could limit significantly the broad use of catalyzed reversible reactions.

More generally, there has been an increasing interest for reversible covalent reactions that can occur in aqueous, buffered solutions, at near physiological pH. Great progress has been made in recent years to develop biocompatible reversible chemistries. This was and remains a necessary step toward a broader use of DCC, for instance for the selection of enzyme inhibitors or nucleic acid binding ligands.

Although the size of the repertoire of reversible covalent reactions suitable for DCC has increased significantly over the past 5–10 years, the discovery of new reactions that satisfy all the criteria of reversibility and biocompatibility remains an exciting challenge for chemists. Increasing the number of chemical reactions will enable the engineering of larger and more diverse DCLs, which remains a key step toward a broader use of DCC.

References

1. Brady PA, Bonar-Law RP, Rowan SJ, Suckling CJ, Sanders JKM (1996) Chem Commun 319
2. Hasenknopf B, Lehn JM, Kneisel BO, Baum G, Fenske D (1996) Angew Chem Int Ed 35:1838
3. Swann PG, Casanova RA, Desai A, Frauenhoff MM, Urbancic M, Slomczynska U, Hopfinger AJ, LeBreton GC, Venton DL (1996) Biopolymers 40:617
4. Klekota B, Hammond MH, Miller BL (1997) Tetrahedron Lett 38:8639
5. Eliseev AV, Nelen MI (1997) J Am Chem Soc 119:1147
6. Rowan SJ, Sanders JKM (1997) Chem Commun 1407
7. Hasenknopf B, Lehn J-M, Boumediene N, Dupont-Gervais A, Van Dorsselaer A, Kneisel B, Fenske D (1997) J Am Chem Soc 119:10956
8. Huc I, Lehn JM (1997) Proc Natl Acad Sci USA 94:2106
9. Corbett PT, Leclaire J, Vial L, West KR, Wietor JL, Sanders JK, Otto S (2006) Chem Rev 106:3652
10. Lehn JM (2007) Chem Soc Rev 36:151
11. Ladame S (2008) Org Biomol Chem 6:219

Reversible Covalent Chemistries Compatible

12. Inaba K, Murakami S, Nakagawa A, Lida H, Kinjo M, Ito K, Suzuki M (2009) EMBO J 28:779
13. Larkin A, Olivier NB, Imperiali B (2010) Biochemistry 49:7227
14. Schiff H (1864) Ann Chem 131:118
15. Roberts SL, Furlan RL, Otto S, Sanders JK (2003) Org Biomol Chem 1:1625
16. Goral V, Nelen MI, Eliseev AV, Lehn JM (2001) Proc Natl Acad Sci USA 98:1347
17. Jencks WP (1959) J Am Chem Soc 81:475
18. Cordes EH, Jencks WP (1963) J Am Chem Soc 85:2843
19. Jencks WP (1964) Prog Phys Org Chem 2:63
20. Meyer CD, Joiner CS, Stoddart JF (2007) Chem Soc Rev 36:1705
21. Goodwin JT, Lynn DG (1992) J Am Chem Soc 114:9197
22. Herrmann A (2009) Org Biomol Chem 7:3195
23. Giuseppone N, Schmitt JL, Lehn JM (2004) Angew Chem Int Ed Engl 37:4902
24. Giuseppone N, Schmitt JL, Schwartz E, Lehn JM (2005) J Am Chem Soc 127:5528
25. Wipf P, Mahler SG, Okumura K (2005) Org Lett 7:4483
26. Paul A, Ladame S (2009) Org Lett 11:4894
27. Wessjohann LA, Rivera DG, Leon F (2007) Org Lett 9:4733
28. For a recent study, see Kalia J, Raines RT (2008) Angew Chem Int Ed Engl 47:7523
29. Dirksen A, Hackeng TM, Dawson PE (2006) Angew Chem Int ed Engl 45:7581
30. Cousins GRL, Poulsen SA, Sanders JKM (1999) Chem Commun 1575
31. Furlan RLE, Ng YF, Otto S, Sanders JKM (2001) J Am Chem Soc 123:8876
32. Roberts SL, Furlan RLE, Cousins GRL, Sanders JKM (2002) Chem Commun 938
33. Poulsen SA (2006) J Am Soc Mass Spectrom 17:1074
34. Bhat VT, Caniard AM, Luksch T, Brenk R, Campopiano DJ, Greaney MF (2010) Nat Chem 2:490
35. Hioki H, Still WC (1998) J Org Chem 63:904
36. Ladame S, Whitney AM, Balasubramanian S (2005) Angew Chem Int Ed 44:5736
37. McNaughton BR, Miller BL (2006) Org Lett 8:1803
38. Bugaut A, Jantos K, Wietor JL, Rodriguez R, Sanders JK, Balasubramanian S (2008) Angew Chem Int Ed 47:2677
39. West KR, Ludlow RF, Corbett PT, Besenius P, Mansfeld FM, Cormack PA, Sherrington DC, Goodman JM, Stuart MC, Otto S (2008) J Am Chem Soc 130:10834
40. Besenius P, Cormack PAG, Ludlow F, Otto S, Sherrington DC (2010) Org Biomol Chem 8:2414
41. West KR, Otto S (2005) Curr Drug Discov Technol 2:123
42. Ludlow RF, Otto S (2008) J Am Chem Soc 130:12218
43. McNaughton BR, Gareiss PC, Miller BL (2007) J Am Chem Soc 129:11306
44. Gareiss PC, Sobczak K, McNaughton BR, Thornton CA, Miller BL (2008) J Am Chem Soc 130:16524
45. Sarma RJ, Otto S, Nitschke JR (2007) Chem Eur J 13:9542
46. Rodriguez-Docampo Z, Otto S (2008) Chem Commun 5301
47. Larsson R, Pei Z, Ramström O (2004) Angew Chem Int Ed Engl 43:3716
48. Leclaire J, Vial L, Otto S, Sanders JKM (2005) Chem Commun 1959
49. Ghosh S, Ingerman LA, Frye AG, Lee SJ, Gagne MR, Waters ML (2010) Org Lett 12:1860
50. Saiz C, Wipf P, Manta E, Mahler G (2009) Org Lett 11:3170
51. Shi B, Greaney MF (2005) Chem Commun 886
52. Clark TD, Ghadiri MR (1995) J Am Chem Soc 117:12364
53. Nicolaou KC, Hughes R, Cho SY, Winssinger N, Smethusrt C, Labischinski H, Endermann R (2000) Angew Chem Int Ed 39:3823
54. Van Gerven PCM, Elemans JAAW, Gerritsen JW, Speller S, Nolte RJM, Rowan AE (2005) Chem Commun 3535
55. McNaughton BR, Bucholtz KM, Camaano-Moure A, Miller BL (2005) Org Lett 7:733
56. Kennedy JP, Castner KF (1979) J Polym Sci Polym Chem Ed 17:2039

57. Chen X, Dam MA, Ono K, Mal A, Shen H, Nutt SR, Sheran K, Wudl F (2002) Science 295:1698
58. Boul PJ, Reutenauer P, Lehn JM (2005) Org Lett 7:15
59. Reuteneauer P, Buhler E, Boul PJ, Candau SJ, Lehn JM (2009) Chem Eur J 15:1893
60. Vongvilai P, Angelin M, Larrson R, Ramstrom O (2007) Angew Chem Int Ed 46:948
61. Cergol KM, Jensen P, Turner P, Coster MJ (2007) Chem Commun 1363
62. Fuchs B, Nelson A, Star A, Stoddart JF, Vidal S (2003) Angew Chem Int Ed 42:4220
63. Cacciapaglia R, Di Stefano S, Mandolini L (2005) J Am Chem Soc 127:13666
64. Berkovich-Berger D, Lemcoff NG (2008) Chem Commun 1686
65. Caraballo R, Dong H, Ribeiro JP, Jimenez-Barbero J, Ramstrom O (2010) Angew Chem Int Ed 49:589
66. Arnal-Herault C, Pasc A, Michau M, Cot D, Petit E, Barboiu M (2007) Angew Chem Int Ed 46:8409
67. Tokunaga Y, Ueno H, Shimomura Y (2007) Heterocycles 74:219
68. Tokunaga Y, Ito T, Sugawara H, Nakata R (2008) Tetrahedron Lett 49:3449
69. Qin Y, Jakle F (2007) J Inorg Organomet Polym Mater 17:149
70. Niu W, O'Sullivan C, Rambo BM, Smith MD, Lavigne JJ (2005) Chem Commun 4342
71. Rambo BM, Lavigne JJ (2007) Chem Mater 19:3732
72. Christinat N, Croisier E, Scopelliti R, Cascella M, Rothlisberger U, Severin K (2007) Eur J Inorg Chem 2007:5177
73. Christinat N, Scopelliti R, Severin K (2008) Angew Chem Int Ed 47:1848
74. Williams RJ, Mart RJ, Ulijn RV (2009) Biopolymers 94:107
75. Williams RJ, Smith AM, Collins RF, Hodson N, Das AK, Ulijn RV (2009) Nat Nanotechnol 4:19
76. Sadownik JW, Ulijn RV (2010) Chem Commun 46:3481
77. Eldred SE, Stone DA, Gellman SH, Stahl SS (2003) J Am Chem Soc 125:3422
78. Bates ED, Mayton RD, Ntai I, Davis JH Jr (2002) J Am Chem Soc 124:926
79. George M, Weiss RG (2003) Langmuir 19:8168
80. Ki CD, Oh C, Oh SG, Chang JY (2002) J Am Chem Soc 124:14838
81. Xu H, Hampe EM, Rudkevich DM (2003) Chem Commun 2828
82. Xu H, Rudkevich DM (2004) J Am Chem Soc 69:8609
83. Leclaire J, Husson G, Devaux N, Delorme V, Charles L, Ziarelli F, Desbois P, Chaumonnot A, Jacquin M, Fotiadu F, Buono G (2010) J Am Chem Soc 132:3582

Index

A

Acetal exchanges, 305
Acetylcholine chloride, 255
Acetylcholinesterase, 300
Acridines, 9-amino substituted, 296
Acylhydrazones, 299
Adaptive chemistry, 1, 6
Adaptive materials, 11
Aldol reactions, 305
Alkylketenimine, 268
Aminoacridines, 296
Ammonium carbamate, reversible, 310
Ammonium chlorides, 255
Anions, 193
Asymmetric synthesis, 55, 77

B

Biaryl lactone, 265
Bilayers, 139
Biocatalysis, 55
Biodynamers, 13
Biologically active substances, dynamic
    search, 11
Biosensors, 139
Bisarylanthracenes, rotary switches, 270
Bis(fulvenes), polyethylene-linked, 304
Bisiminoboronic acid, 307
Bis(pyridyl) urea, 205
Bis(tricyanoethylenecarboxylate), 304
Block copolymers, dynamic assembly, 165
Boronate–guanosine, 305, 307
Boroxines, 307
Bulk/thin films, 177
Bullvalones, 263

C

Carbinolamine, 294
Carbon dioxide, dynamic capture, 310
Carbonic anhydrase, 33
Carboxamide, 268
Catenanes, 263
C–C bonds, exchange reactions, 302
C60 encapsulation, 240
Chemical evolution, 87
    prebiotic, 9
Circular dichroism, 217
Combinome, 8
Complementary functional/interactional
    groups, 7
Constitutional changes, 261
Constitutional dynamic chemistry (CDC), 1ff
Constitutional dynamic library (CDL), 8
Constitutional dynamic networks, 23
Constitutional dynamic synthesis, 21
Conversion, 8
Covalent changes, 261
C70 receptors, self-assembled, 243
Crystallization, 55
Cyanoolefins, 304
Cyclopentadiene, 304

D

Diaminoacridine, 297
Dicyclohexylcarbodiimide (DCC), 266
Diels–Alder (DA) reactions, thermally
    reversible, 304
Diethylcyanofumarate, 304
Dihydroxyfluorene-2-diboronic acid, 307
4-Dimethylaminopyridine (DMAP), 267

Index

DNA, 106
    analogs, 294
Dynamers, 11
    self-healing, 13
Dynamic assembly, block copolymers, 165
Dynamic combinatorial chemistry (DCC), 1ff, 7, 33
Dynamic combinatorial libraries (DCLs), 87, 293
Dynamic coordination chemistry, 10
Dynamic covalent chemistry (DCC), 6, 261
Dynamic crystallization, anion recognition, 205
Dynamic interactive systems, 33
Dynamic interfaces, 139
Dynamic networks, 1, 20
Dynamic non-covalent chemistry (DNCC), 6
Dynamic polymers, 1
Dynamic systemic resolution, 55
DyNAs, 14

E
Effector binding, 19
Enzymes, 55
μ-EOF, 176
Expression, 8

F
Fibrinogen, 308
Fluorenes, 252
Fullerenes, 217
    fragments, 251
Fulvenes, 304
Functional recognition, 7

G
Genotypes, 8
Glutathione (GSH), 302
Glycodynamers, 14
G-quadruplex, 106
Graphoepitaxy, 184
Grubb's catalyst, 302
Guanosines, 307

H
Hamburger micelles, 176
Helicenes, 264
Hemithioacetal (HTA), 306
Henry reactions, 305
Homoallylamides, 304
Host–guest, 217
Hybrid materials, 33
Hydrazides, 22, 128, 294

Hydrazones, 21, 126, 202, 279, 298
    disulfide exchanges, 134, 277
2-Hydroxy-1,3,5-benzenetrialdehyde, 310
Hydroxylamines, 294

I
Imines, 21, 294
Information storage, 22
Ion pairs, 217
    complexation, 254

K
Kinetic selection, 11

L
Lactam/lactim prototropic tautomerization, 263
Lactones, 265
Lithio(1R,2S,5R)-2-isopropyl-5-methylcyclohexanol, 269

M
Membrane asymmetry, 176
Metallosupramolecular architecture, self-assembly, 10
Micelles, cylindrical, 174
    preparation, 169
    spherical, 171
Molecular machines, 261
Molecular motors, 261, 264
    triptycene-helicene, 264
Molecular recognition, 1, 3
    anions, 193
Molecular shuttles, reversible Diels–Alder reaction, 272
Molecular switches, 261, 270
Multiple dynamics, 1, 21

N
Nanoparticles, 139
Nanotubes, naphthalenediimide, 217
    self-sorting, 242
Naphthalenediimides (NDIs), 217
Naphthalenemonoimide (NMI), 219
Naphthalenetetracarboxylic dianhydride (NDA), 219
Networks, dynamic, 20
Nitroaldols, 303
    Henry reaction, 305
4-Nitrobenzaldehyde, 305
Nucleic acids, recognition, 106

Index

O
Olefin metathesis, 302
Oligoimines, macrocyclic, 298
Oscillatory motor, 268
Oximes, 298

P
PAA-*b*-PMA-*b*-PS, 176
PAI-PS-PAI, 173
PB-*b*-PEO, 174
Pentaerythritol, 307
PEO-*b*-PEP, 174
Peptoids, dynamic, 14
Phase change selection, 33
Phenotypes, 8
PLA-*b*-PDMA-*b*-PS, 180
Poly(acrylic acid) (PAA), 176
Poly(acylhydrazone), 283
Polyaromatic hydrocarbons, complexation, 251
Polybutadiene (PB), 180
Poly(ethylene-co-butylene), 180
Poly(ethylene oxide)-*b*-poly(ethylene), 175
Poly(ethylene terephthalate), 282
Polyethylethylene-*b*-poly(ethylene oxide), 176
Polymer/monomer switches, 261, 280
Polymersomes, 175
Polymers, reversible, 261
  supramolecular, 217
Polystyrene-*b*-(4-vinylpyridinium decyl
  iodide), 175
PPO, 175
Prebiotic chemical evolution, 9
Prebiotic Darwinism, 10
Programmed chemical systems, 4
PS-*b*-PAA, 171
PS-*b*-PDMS, 183
PS-*b*-PEB-b-PMMA, 179
PS-*b*-PEO, 171
PS-*b*-PI, asymmetric, 171
Pseudorotaxanes, 273
P2VP-*b*-PAA-*b*-PnBMA, 174
Pyrazolotriazinones, metathesis, 296
Pyridine-2-carboxaldehyde, 22
Pyridyl-acyl hydrazone, 22

Q
Quartz crystal microgravimetry, 139

R
Receptors, anions, 193
  dynamic generation, 8
Resin-bound DCC, 106

Resolution, dynamic systemic, 55
Reversible reactions, 291
RNA, 106
Rotation, bonds, 264
  rate, 269
Rotaxanes, 263, 273

S
Scandium(III), 295
Self-assembly, anion hosts, 193
Self-optimization, 9
Self-organization, 1, 4
Self-replication, 87
Sergeants-and-soldiers, 217, 236
Shape switching, 19
Size switches, 261
  polymer/monomer, 280
Stilbazole, 263
Substrates, 8
3-Sulfanylpropionic acid, 300
Supramolecular chemistry, 1
Surface plasmon resonance (SPR), 139
Switches, 261, 270
Systems chemistry, 87

T
Tandem imine metathesis, 297
Tetrapyridyl porphyrin (TPyP), 302
Thermodynamic control, 291
Thiazolidines, 300
Thin films, self-assembly, 180
Thiocholine, 300
Thiodepsipeptides, cyclic, 301
Thiohemiacetalization, 306
Thiol–disulfide exchange, 300
Thiol–thioester exchange, 301
Threitol, 305
Transamidation, 308
Transimination, catalysis, 295
Transition metal complexes, 106
Transthioesterification, reversible, 300
Triethylene glycol, 305
Triptycene, 264
Tris(pyridylurea) ligand, 208

U
Ugi reaction, 297

V
Virtual combinatorial libraries (VCLs), 6, 7,
  195, 205